JSP从零开始学

（视频教学版）

刘 鑫 编著

清华大学出版社

北 京

内 容 简 介

JSP 依靠强大的 Java 基础，成为世界上最流行的 Web 开发利器。本书通过大量的实例，循序渐进地为读者介绍了有关 JSP 开发所涉及的各类知识，所有版本一律采用最新版本，是目前市场上学习 JSP 技术的首选。

本书分为 14 章，首先介绍网页开发的基础原理，然后搭建 JSP 开发环境，再介绍 JSP 基础、基本语法、内置对象、Servlet、JavaBean、MySQL、XML 文件、资源国际化等，最后通过一个完整的在线购物网站案例，回顾前面所学的 JSP 技术。

本书利用实例贯穿所有的语法，具有很强的操作性，适合 JSP 初学者、Web 开发者和所有前端人员学习。

图书在版编目（CIP）数据

JSP 从零开始学：视频教学版 / 刘鑫编著. - 北京：清华大学出版社，2016
ISBN 978-7-302-42188-7

I. ①J… II. ①刘… III. ①JAVA 语言－网页制作工具 IV. ①TP312②TP393.092

中国版本图书馆 CIP 数据核字(2015)第 272718 号

责任编辑：夏非彼
封面设计：王　翔
责任校对：闫秀华
责任印制：李红英

出版发行：清华大学出版社
网　　址：http://www.tup.com.cn，http://www.wqbook.com
地　　址：北京清华大学学研大厦 A 座　　邮　　编：100084
社 总 机：010-62770175　　邮　　购：010-62786544
投稿与读者服务：010-62776969，c-service@tup.tsinghua.edu.cn
质 量 反 馈：010-62772015，zhiliang@tup.tsinghua.edu.cn
印 刷 者：三河市君旺印务有限公司
装 订 者：三河市新茂装订有限公司
经　　销：全国新华书店
开　　本：190mm×260mm　　印　张：27.5　　字　　数：704 千字
版　　次：2016 年 1 月第 1 版　　印　　次：2016 年 1 月第 1 次印刷
印　　数：1～3000
定　　价：69.00 元

产品编号：065613-01

前言

互联网时代，Web 无所不在，开发更好的网页和 Web 应用成为很多企业的追求。鉴于使用 Java 开发的用户最多，JSP 也称为最热门的 Web 开发技术。JSP 实现了动态页面与静态页面的分离，脱离了硬件平台的束缚，而且还可以编译后运行，这大大提高了它的执行效率。在网络信息量巨大的今天，使用 JSP 技术提高效率是网站的首选。鉴于学习 JSP 的人数众多，我们特地编写这样一本 JSP 入门书籍，内容安排由浅入深，由理论到实践，适合 JSP 初学者逐步学习和完善自己的知识结构。

为什么选择 JSP

（1）JSP 语言的可扩充性。JSP 技术依靠众所周知的 Java 技术为载体，可以很轻松的将应用平台进行扩展，类似于插板上的插头，只要插孔够用，可以很容易的扩展出很多的电器。

（2）JSP 服务器的高稳定性。JSP 服务器的运行环境主要是 Linux 服务器平台，这种平台在大型应用服务中采用甚多，国内著名的电商平台：淘宝、凡客、京东以及各种 OA 办公自动化平台都采用 Linux 服务器作为应用环境。

（3）并发性能。JSP 在用户访问量的负载程度上具有很大的优势，在多数用户同时访问的情况下不会出现崩溃、服务器瘫痪的危险。

（4）安全性。由于 JSP 语言最初便定位于 OA 办公自动化和电子商务平台等大型应用平台，因此对程序本身的安全性做出了极高的要求。

（5）JSP 数据量的吞吐及负载能力。基于 JSP 技术操作的数据库连接在信息访问量上要更加快速、庞大，效率更高、数据的稳定性也更强。

（6）JSP 的高安全性。JSP 技术主要依赖于 Java 语言，所以在信息的安全性上要更加优越，程序的开发更加缜密，大大降低了网站黑客被入侵的概率。

（7）JSP 网站访问的高速性。JSP 技术制作的网站因其代码量更加优化、冗余度低，所以访问时会更加高效，速度会更快。

本书的内容安排

本书共分为 14 章，主要章节规划如下。

第 1 章介绍网页最基本的知识，包括什么是动态、静态网页，网页浏览的原理，网络传输的协议，流行的网页开发语言，JSP 网页的执行顺序等。

第 2 章详细介绍要使用 JSP 必须配置的开发环境，包括 JDK 的安装与配置、Tomcat 的安装与配置，MyEclipse 的安装与配置。最后还介绍如何用 MyEclipse 来开发第一个 JSP 程序。

第 3 章介绍 JSP 最最基础的语法知识，包括 JSP 的注释、声明、表达式、指令、动作等。

第 4 章介绍 JSP 的内置对象 request、response、session、application、out、page、config，这些对象实现了网页请求的一些最基本的应用。

第 5 章是 Servlet 技术的应用，这是 JSP 实现更多功能的关键，本章着重介绍了 Servlet 的生命周期、编写和部署、应用程序事件、监听器、过滤器和异步处理等等。

第 6 章是 EL 标签的应用，标签可以让 JSP 更高效的工作，本章重点介绍了如何使用 EL 标签和如何禁用 EL 标签。

第 7 章是网页的请求、响应与会话管理，这在整个网页浏览周期中是最最关键的地方，一个网页从请求到响应的过程中，我们如何保存数据、传输数据都在本章体现。

第 8 章介绍了如何使用 Java Bean 读取数据库，网页信息量这么大，数据的存储和读取非常关键，本章重点是数据库的操作。

第 9 章介绍了 JSTL 标签库，这是 JSP 常用的一个标签库，包括了 5 大标签种类，本章通过实例的方式详细介绍了这些标签的使用。

第 10 章教会我们实现自定义标签，如果官方提供的标签不够用，我们还可以使用自己制作的标签。

第 11 章是使用 JDBC 连接数据库，JDBC 使用非常广泛，也是 JSP 操作数据库最常用的方式。

第 12 章介绍了 XML 文件格式，如今遍地都是 XML 文件，这类文件如何编写，如何使用，都是本章介绍的重点。

第 13 章介绍了资源国际化，这是保证页面中文和英文都能更清晰表达内容的关键所在，使用资源国际化，我们可以开发面向全球的网页。

第 14 章通过一个完整的网上购物系统，详细介绍了使用 JSP 开发 Web 网站的整个流程，包括后台类的编写、页面的设计和公共页面的处理。

本书读者与作者

- Java Web 开发初学者
- JSP 初学者
- 从事 JSP 编程的 Web 开发人员
- 高校、大中专院校的师生

本书第 1~10 章由湖南铁道职业技术学院的刘鑫编写，其他参与编写的还有李阳、张学军、陈士领、陈丽、殷龙、张鑫、赵海波、张兴瑜、毛聪、王琳、陈宇、生晖、张喆、王健，排名不分先后。

代码、课件与教学视频下载

本书配套代码、课件与教学视频下载地址（注意数字和字母大小写）如下：

http://pan.baidu.com/s/1dDJksGp 密码：3y17

如果下载有问题，请电子邮件联系 booksaga@163.com，邮件主题为“JSP 从零开始学”。

编　者

2015 年 12 月

目 录

第 1 章

网页制作与浏览原理

为人们打开外部世界的是互联网，而人们能看得见的是存在于互联网中的无数个网页，这些网页虽然变幻多端，但无论怎么变化，它必须遵循一定的 Web 浏览规则，如本章介绍的 HTTP 协议。本章的目的是让读者了解什么是网页，网页的发展，网页的浏览原理，以及网页开发中使用的脚本语言。

1.1 我们所理解的网页

网页是读者上网浏览时看到的页面，其通过浏览器，呈现在电脑上。用户在浏览器的地址栏，输入一个网站地址，如 www.baidu.com，则打开一个页面，这个页面就是我们常说的“网页”。网站是一个具有多个网页的站点，如 www.baidu.com 就是一个网站的地址，用户通过这个地址来访问网站，网站包含多个相关的网页。图 1.1 是打开的百度网站的一个网页，也是百度网站的主页。

图 1.1　百度网站的主页

网页分为静态网页和动态网页，本节将具体介绍这两种网页。

1.1.1 静态网页

静态页面是指网络上内容和外观总是保持不变的页面。这些页面的文件名后缀通常为.htm 或者.html。这些网页的制作最为简单，由 HTML（超文本标记语言）实现，适合表现相对固定的内容，如网站的联系方式、公司简介等等。下面通过编写一个名为 hello.html 的静态页面，帮助读者理解静态页面的工作方式。

新建一个文本文件，手动输入如下代码，然后保存为 hello.html。

```
<html>
<!--标题-->
<head><title>Welcome!</title></head>
<!--页面主体-->
<body>
    Hello，静态网页!
</body>
<html>
```

由于某些系统设置为不显示文件扩展名，所以有时文件的名字显示为“hello.html”，但其实并不是.html 类型的文件，而是“hello.html.txt”文件，所以需要设置系统显示扩展名。

对于该页面，用户可以双击打开，不管用户何时何地以怎样的方式访问，页面的内容都不会再改变，这就是静态页面，也就说，页面的内容是固定的。

1.1.2 动态网页

静态网页有很多问题，最明显的就是无法与服务器进行交互，用户无法从服务器获取信息并自动更新，也无法将用户的信息提交到服务器。这就出现了动态网页，动态网页一般由两部分组成：静态页面和动态操作。通过 HTML 来显示页面，然后通过动态操作完成信息的更新。

动态网页的实现依靠浏览器端和服务器端的互动。服务器端可以实时处理浏览器端的请求，然后将响应的内容传给浏览器（这些内容可能来自数据库，每次请求的内容都不同）。这样，动态页面就显示在浏览器中了。

由于本书主要就是讲解动态网页，所以这里不再举例，读者可以看本章最后的 JSP 示例。

1.1.3 浏览器和服务器对应的 B/S 模式

B/S 模式（Browser/Server）也就是通过浏览器来访问服务器。用户可以在互联网的任何一个角落，甚至可以是个无线终端（PDA 等）。B/S 模式所用的业务逻辑及数据支持都是在服务器上，当用户通过页面提出请求时，服务器及时响应，并把运行后的数据及时送回。B/S 模型图如图 1-2 所示。

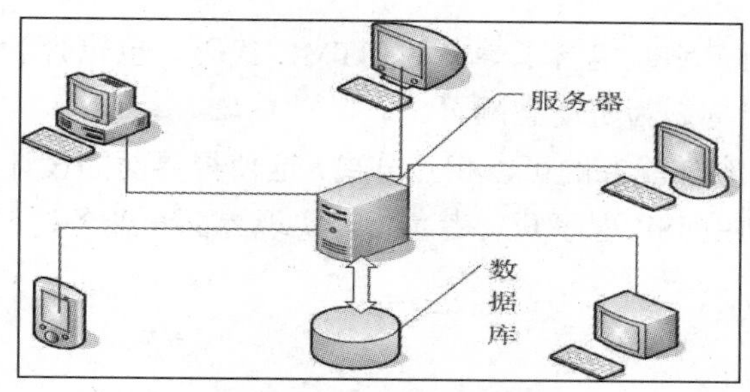

图 1.2　B/S 模型图

B/S 模型的优势：

（1）易于维护。基于 B/S 模型的系统当需要升级或维护时，只需修改服务器程序即可。

（2）易于实现。B/S 模型的表现层可以用制作网页的 HTML 来实现，浏览器和网页设计技术已经相当成熟。而且用 Java 技术开发的 Web 系统可以安装在任意一种服务器系统平台上，也就是常说的具有跨平台性。

（3）使用方便。无论客户在什么地方，只要服务器正在运行，客户就能通过网络进行连接，实现对数据的访问和操作。

B/S 模型的不利因素：增加了服务器的压力。把业务实现都放在服务器上，当有大量的用户访问时，势必会给服务器带来很大的负担，更有可能造成系统崩溃，所以一定要做好系统数据的备份。

1.2 常见的动态网页开发语言

目前最常见的动态网页开发语言有 JSP 和 PHP，关于谁是最好的语言一直争论不休，笔者建议还是根据项目的安排或性能来做决定，每种语言都有自己独特的优势。

1.2.1 JSP

JSP 技术是由 SUN 公司（现被 Oracle 收购）提出，多家公司参与的，于 1999 年推出的一款建设动态网页的方法。它基于 Java Servlet 技术来开发动态的、高性能的 Web 应用程序。JSP 的网页实际上是在 HTML 文件中加入 Java 代码片段和 JSP 特殊的标记构成的。

因为 JSP 是 Java 的成员，所以 JSP 具有平台无关性即实现跨平台功能，实现了用户界面和程序代码的解耦合，使得业务逻辑和代码的耦合度更低，开发人员可以在不更改 JSP 程序下修改用户的界面。

JSP 页面实质也是个 HTML 页面，只不过它包含了用于产生动态网页内容的 Java 代码，这些 Java 代码可以是 Java Bean、SQL 语句、RMI（远程方法调用）对象等。例如：一个 JSP 页面包含了用于产生静态网页的 HTML 代码，同时也包含了连接数据库的 JDBC 代码，那么当网页在浏览器中显示时，它既包含了静态的 HTML 代码，也包含了从数据库中取得的动态内容，也正因为这样才能称之为是动态网页。

JSP 页面中动态的内容与静态的可以相互分离，这使得界面的设计者可以完全专注于界面的美化，而动态的部分则由 JSP 程序开发者负责，实现界面与业务逻辑的分离，可以实现 JSP 代码的高度复用。

1.2.2 PHP

PHP（Hypertext Preprocessor）是一种开源的脚本语言，它具备简单而独特的语法，这些语法混合了 C、Java、Perl 以及 PHP 自创的语法，这样的混合既吸收了其他语言的优点，又便于普通开发人员学习，所以 PHP 的应用也越来越广泛。

用 PHP 开发的动态网页与其他语言开发的页面相比，PHP 是将程序嵌入到 HTML（标准通用标记语言下的一个应用）文档中去执行，执行效率比完全生成 HTML 标记的 CGI 要高得多。

PHP 语言的优势包括：

- 编辑简单，实用性强，更适合初学者。
- 本身免费且是开源代码。
- 由于是运行在服务器端的脚本，所以可在 UNIX、LINUX、WINDOWS、Mac OS、Android 等平台上运行。
- 消耗相当少的系统资源。

1.3 网页的浏览原理

前面介绍了 B/S 模型的优越性，这种模式就是在 Web 开发中，通过浏览器和服务器的交互来实现系统的业务逻辑。这种模式组成的应用程序一般称为 Web 应用程序。本节就对 Web 的一些基础知识进行描述。

1.3.1 Web 是什么

Web 出现于 1989 年 3 月，由欧洲粒子物理研究所（CERN）的科学家 Tim Berners-Lee 发明。1990 年 11 月，第一个 Web 服务器正式运行，这时通过 Web 浏览器看到了最早的 Web 页面。1991 年 Web 技术标准正式发布。1993 年，第一个图形界面的浏览器 Mosaic 开发成功，1995 年著名的 Netscape Navigator 浏览器问世。随后，微软公司推出了著名的 IE 浏览器（Windows 操作系统默认安装 IE 浏览器）。目前，与 Web 相关的各种技术标注都由著名的 W3C 组织（World Wide Web Consortium）管理和维护。

Web 是一个分布式的超媒体信息系统，它将大量的信息分布在网上。目的就是为人们提供更多的多媒体网络信息服务。从技术层面上看，Web 技术的核心有 3 点：

- 超文本传输协议（HTTP）协议，实现网络的信息传输。
- 统一资源定位符（URL），实现互联网信息的定位的统一标识（如 http://www.sohu.com 中的“www.sohu.com”）。
- 超文本标记语言（HTML），实现信息的表示与存储。

1.3.2　HTTP 超文本传输协议是什么

超文本传输协议 HTTP（HyperText Transfer Protocol）是专门为 Web 设计的一种应用层协议。常用的 Web 服务器软件有 Apache、IIS 等，Web 浏览器软件包括 IE、Netscape、Mozilla Firefox 等。

当用户在浏览器地址栏中输入网址或通过超链接访问目的网站时，都向目标主机（Web 服务器）发送一个 HTTP 请求。HTTP 定义的信息交互处理由以下 4 步组成：

- 浏览器与 Web 服务器建立连接。
- 浏览器向服务器提出请求。
- 如果请求被接受，则服务器送回响应，在响应中包括状态码和所需要的文件。
- 浏览器和 Web 服务器断开连接。

HTTP 向服务器发出一段请求也就是一段报文，是由以下 4 个部分组成的文本。

请求行	请求头标	空行	请求数据

（1）请求行。请求行有 3 个标记组成，即请求方法、请求 URL 和 HTTP 版本，它们用空格分隔。例如，GET /Index.html HTTP 1.0 的意思就是检索当前文件夹下的 Index.html 文件，使用的 HTTP 版本是 HTTP1.1。其中 HTTP1.1 规范定义了 8 种请求的方法。

- Get：检索 URL 中标识资源的一个简单请求。
- Head：与 Get 方法相同，服务器只返回状态行和头标，并不返回请求文档。
- Post：服务器接收被写入客户端输出流中的数据请求。
- Put：服务器保存请求数据作为指定 URL 新内容的请求。
- Delete：服务器删除 URL 中命名的资源的请求。
- Options：关于服务器支持的请求方法信息的请求。
- Trace：Web 服务器反馈 HTTP 请求和头标的请求。
- Connect：已文档化但当前未实现的一个方法，预留做隧道处理。

在 Web 应用中，通常只使用 Get 和 Post 方法。

（2）请求头标。由关键字和值对组成，每行一对，关键字和值用冒号（:）分隔。请求头标通知服务器关于客户端的功能和识别，典型的请求头标有：

- User-Agent：客户端厂家和版本。
- Accept：客户端可识别的内容类型列表。

- Content-Length：附加到请求的数据字节。

（3）空行。最后一个请求头标之后是一个空行，发送回车符和退行，通知服务器不再有头标。

（4）请求数据。使用 Post 发送数据。

服务器接到请求后，解析请求，如果请求的是静态资源，如文档、图片等，则将请求的资源返回给浏览器；如果请求的动态的服务器程序，如：Servlet、JSP、ASP、CGI 等，则在服务器端运行程序后返回运行的结果，通常的运行结果是生成的一个 HTML 文档。一个响应由 4 个部分组成，这些部分与请求报文的部分基本相同，本节只介绍状态行部分。

状态行	响应头标	空行	响应数据

状态行由 3 个部分组成：HTTP 版本、响应代码和响应描述。

- HTTP 版本：向客户端指明其可以理解的最高版本。
- 响应代码：三位的数字代码，指出请求的成功或失败，如果失败则指明原因。
- 响应描述：为相应代码的可读性解释。例如：

```
HTTP/1.1 200 OK
```

HTTP 响应代码的说明如下：

- 1xx：信息请求收到，继续处理。
- 2xx：成功，行为被成功的接收、理解和采纳。
- 3xx：重定向，为了完成请求，必须进一步执行动作。
- 4xx：客户端错误。

1.4 用 JSP 进行网页开发

上一节简单介绍了 Web 开发的一些背景知识，读者已经了解了 Web 访问的基本原理，HTTP 超文本传输协议，静态网页与动态网页的区别，以及主流的浏览器和 Web 服务器，本节将介绍 JSP 的基本概念、执行过程等，让读者了解 JSP 是什么，能做些什么。

1.4.1 JSP 与其他语言相比的优势

JSP 可以看作是 Java Servlet 的一种扩展，JSP 在使用前必须被编译为 Servlet，也就是 Java 类，然后被调用执行，Servlet 所产生的 Web 页面是不能包含在 HTML 标签中的，因为它离不开 Java 类文件的支持。随着学习的深入，使用 JSP 将带给用户很多明显的优点：

（1）开发简单方便

在 JSP 中的编辑跟编写 HTML 文件基本一样，在处理表单方面极为方便。设置 HTTP 报头，JSP 同样提供了丰富的方法。使得 JSP 开发者在编写通用功能时很便捷，就能有更多的时间花费在业务逻辑上。

（2）跨平台

Java 本身就有跨平台的特性，因此 JSP 程序可以在支持 Java 的平台上开发运行。显然这对平台移植极其有利。当 JSP 在更换服务平台时，如若不涉及数据库等相关操作，几乎可以不做任何变动就能完成服务平台的迁移。当需要更换 Web 服务器时，JSP 同样可以做到不修改或者少量修改就能在新的 Web 服务器中编译、运行。

（3）高效率和高性能

上文提到过，JSP 可以是 Servlet 的扩展，因此 Java 虚拟机为每一个请求创建一个单独的线程，而不是进程，如此系统能很快地处理请求。同时 JSP 只会被编译一次，只是在首次的加载时需要编译，这样加快了系统的响应速率。当一个请求处理结束之后，相关的 JSP 映射的 Java 类并不会从内存中删除，会被保留在内存中，当下次同样的请求发生时，系统会提供更快的响应速度。

（4）低成本

众所周知 Java 是开源的开发语言，JSP 也是基于 Java 的开源环境开发的动态网页技术，所以这就省去了商业的付费项目。再有，开发者可以从众多的 Java IDE 中选择一款适合自己的开发工具来进行项目研发，当然了，也可以直接用文本编辑器直接编写，只是这样比较耗时而且易出错。还有许多的商业软件可以使用，但是通常来说使用 JSP 开发总成本比采用其他技术要低廉些。

综上所述，采用 JSP 动态网页技术是目前 Web 开发者的最佳选择。

1.4.2　JSP 网页的执行顺序

在编写 JSP 程序时，要了解它的执行顺序，这样对于后续的学习会有很大的帮助。JSP 程序的执行过程大致如下。

首先，客户端向 Web 服务器提出请求，然后 JSP 引擎负责将页面转化为 Servlet，此 Servlet 经过虚拟机编译生成类文件，然后再把类文件加载到内存中执行。最后，由服务器将处理结果返回给客户端。整个流程如图 1.3 所示。

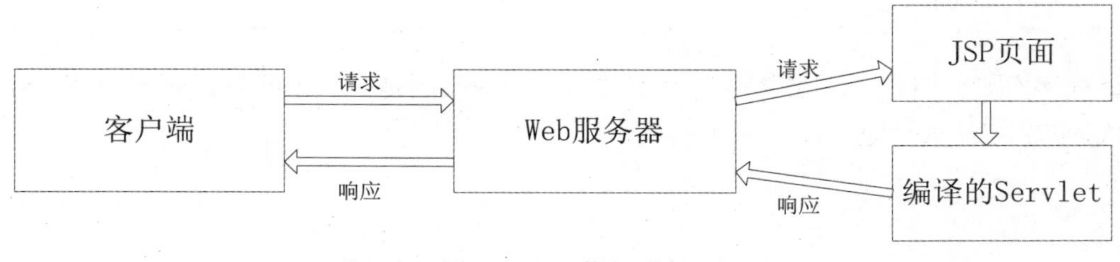

图 1.3　JSP 执行顺序

JSP 页面代码会被编译成 Servlet 代码，所以从执行效率上说肯定是没有 Servlet 快的，但并不是每一次都需要编译 JSP 页面。当 JSP 第一次被编译成类文件后，重复调用该 JSP 页面时，JSP 引擎发现该 JSP 页面没有被改动过，那么就会直接使用编译后的类文件而不会再次编译成新的 Servlet。当然，如果页面被修改过，则需要重新加载编译。

1.4.3 实例：第一个 Hello JSP 网页

本章因为只是熟悉网页的基本机构和浏览器浏览网页的原理，还没有搭建 JSP 开发的基础环境，所以这里只给一个简单的例子，读者先仔细看看，等搭建完环境后，再来分析具体每段代码的功能。

下面是 JSP 网页的一个简单例子，功能是循环输出 1~10 的输出结果，代码如下：

```
-----------------------Index.jsp--------------------------
01  <%@ page language="java" import="java.util.*" pageEncoding="UTF-8"%>
02  <!DOCTYPE HTML PUBLIC "-//W3C//DTD HTML 4.01 Transitional//EN">
03  <html>
04    <head>
05      <title>JSP 简单例子</title>
06    </head>
07    <body>
08      <%
09          int count=0;
10          for(int i=1;i<10;i++)
11          {
12             count+=i;
13          }
14          out.print("1到10的相加结果："+count);
15       %>
16    </body>
17  </html
```

第 1 行中，pageEncoding 标签可以设定字符类型，在工作中这一行会自动生成。第 8~15 行为 Java 代码。

该程序主要作用是利用 JSP 输出 1~10 的和，其中代码是由简单的 HTML 代码和 JSP 表达式构成，JSP 表达式中是一段 Java 程序段。

1.5 上机实践

本章还没有开始搭建 JSP 环境，所以读者从练习 HTML 静态网页开始做练习。

1. 创建一个静态网页，在浏览器正中间显示“Hello，JSP”。

2. 创建一个静态网页，在浏览器中显示一首唐诗：

```
春眠不觉晓，处处闻啼鸟。
夜来风雨声，花落知多少。
```

第 2 章

搭建JSP开发环境

学习一门语言前，首先需要搭建好这门语言的开发环境。JSP 的开发环境主要包括 3 部分：Java 环境、Tomcat 服务器和 MyEclipse 开发工具。因为 JSP 后台使用的 Java 语言，所以必须安装 Java 环境。本章将按顺序来逐步安装这 3 部分。

2.1 安装并配置 Java 环境

Java 环境需要从官方网站下载，只要部署好 Java 环境，让 Windows 系统能自动匹配 Java 的各种命令，在开发 JSP 时才能得心应手，本节首先来搭建 Java 环境。

2.1.1 下载并安装 JDK

JDK（又名 SDK）是由 SUN 公司（现被 Oracle 收购）提供的 Java 开发工具和 API。首先从 Oracle 公司的官网（http://www.oracle.com/technetwork/java/index.html）下载 Java SE。需要注意的是，JDK 的版本较多，各版本之间还存在不兼容问题和操作系统的兼容性问题，本书使用的是 Windows 7 x86 位的操作系统，所以下载了 jdk-8u45-windows-i586.exe 版本，如图 2.1 所示，下载前一定要先接受官方协议。读者可以根据自身的操作系统下载不同的版本。

API 在 Java 世界中指 Java 类库和方法，也表示 Java API 参考文档.

Java SE Development Kit 8u45

You must accept the Oracle Binary Code License Agreement for Java SE to download this software.

Accept License Agreement　Decline License Agreement

Product / File Description	File Size	Download
Linux x86	146.89 MB	jdk-8u45-linux-i586.rpm
Linux x86	166.88 MB	jdk-8u45-linux-i586.tar.gz
Linux x64	145.19 MB	jdk-8u45-linux-x64.rpm
Linux x64	165.24 MB	jdk-8u45-linux-x64.tar.gz
Mac OS X x64	221.98 MB	jdk-8u45-macosx-x64.dmg
Solaris SPARC 64-bit (SVR4 package)	131.73 MB	jdk-8u45-solaris-sparcv9.tar.Z
Solaris SPARC 64-bit	92.9 MB	jdk-8u45-solaris-sparcv9.tar.gz
Solaris x64 (SVR4 package)	139.51 MB	jdk-8u45-solaris-x64.tar.Z
Solaris x64	95.88 MB	jdk-8u45-solaris-x64.tar.gz
Windows x86	175.98 MB	jdk-8u45-windows-i586.exe
Windows x64	180.44 MB	jdk-8u45-windows-x64.exe

图 2.1　下载正确的 JDK 版本

下载完成后双击执行 jdk-8u45-windows-i586.exe，根据安装向导提示安装完成，若用户选择的安装目录是 C:\Java\jdk1.8.0_45，安装结束后应该在 C:\Java\jdk1.8.0_45\bin 目录下出现图 2.2 所示的文件。

该目录文件夹下包含了很多可执行文件，例如：java.exe、javac.exe、javadoc.exe 等等。javac.exe 是 Java 的编译器，用来编译 Java 文件将它变为字节码。在 DOC 环境下，用命令“javac test.java”的形式编译一个 Java 文件。java.exe 是运行编译后的 Java Class 文件。java.exe 也可以运行 jar 文件，运行的命令如下：java –jar test.jar，这样就可以运行 test.jar 文件。

名称	修改日期	类型	大小
appletviewer.exe	2015/5/31 22:04	应用程序	16 KB
extcheck.exe	2015/5/31 22:04	应用程序	16 KB
idlj.exe	2015/5/31 22:04	应用程序	16 KB
jabswitch.exe	2015/5/31 22:04	应用程序	30 KB
jar.exe	2015/5/31 22:04	应用程序	16 KB
jarsigner.exe	2015/5/31 22:04	应用程序	16 KB
java.exe	2015/5/31 22:04	应用程序	187 KB
javac.exe	2015/5/31 22:04	应用程序	16 KB
javadoc.exe	2015/5/31 22:04	应用程序	16 KB
javafxpackager.exe	2015/5/31 22:04	应用程序	79 KB
javah.exe	2015/5/31 22:04	应用程序	16 KB
javap.exe	2015/5/31 22:04	应用程序	16 KB
javapackager.exe	2015/5/31 22:04	应用程序	79 KB
java-rmi.exe	2015/5/31 22:04	应用程序	16 KB
javaw.exe	2015/5/31 22:04	应用程序	187 KB
javaws.exe	2015/5/31 22:04	应用程序	266 KB
jcmd.exe	2015/5/31 22:04	应用程序	16 KB
jconsole.exe	2015/5/31 22:04	应用程序	16 KB
jdb.exe	2015/5/31 22:04	应用程序	16 KB
jdeps.exe	2015/5/31 22:04	应用程序	16 KB
jhat.exe	2015/5/31 22:04	应用程序	16 KB
jinfo.exe	2015/5/31 22:04	应用程序	16 KB
jjs.exe	2015/5/31 22:04	应用程序	16 KB
jli.dll	2015/5/31 22:04	应用程序扩展	156 KB
jmap.exe	2015/5/31 22:04	应用程序	16 KB
jmc.exe	2015/5/31 22:04	应用程序	312 KB
jmc.ini	2015/5/31 22:04	配置设置	1 KB
jps.exe	2015/5/31 22:04	应用程序	16 KB
jrunscript.exe	2015/5/31 22:04	应用程序	16 KB
jsadebugd.exe	2015/5/31 22:04	应用程序	16 KB

图 2.2　JDK 安装目录下的文件

2.1.2　配置环境变量

安装完 JDK 后，接着就是配置 Java 的环境变量，配置的作用如下：

- 让系统自动查找 Java 编译器路径；
- 服务器安装时需要知道 Java 路径；
- 编译和执行时指定 Java 路径。

配置环境变量的步骤如下：

以 Windows 7 为例，右键单击开始菜单中的“计算机”，在弹出的快捷菜单中选择“属性”

命令，打开“系统”窗口，单击左侧的“高级系统设计”选项打开“系统属性”对话框，选择“高级”选项卡，单击“环境变量”按钮，如图 2.3 所示。

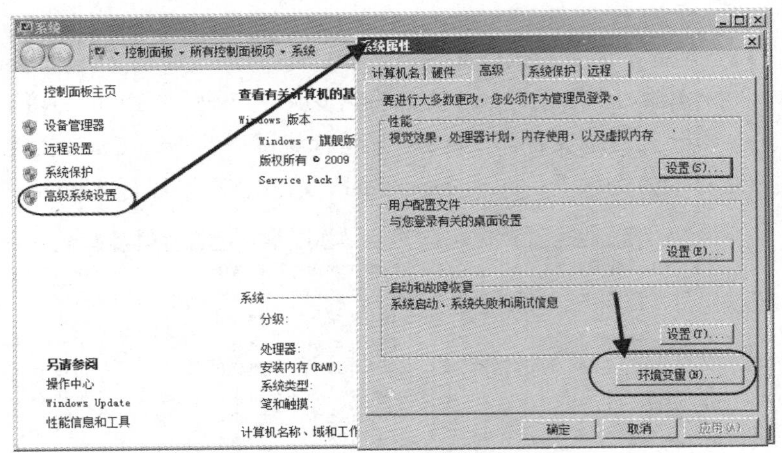

图 2.3　配置环境变量

在弹出的“系统变量”对话框中单击“新建”按钮，分别设定 JAVA_HOME、PATH 和 CLASSPATH 3 个新的环境变量。

PATH 环境变量如果已经存在，则只需要在后面添加变量内容即可，两个变量之间用英文的分号间隔。

3 个变量的值分别如下：

```
JAVA_HOME= C:\Java\jdk1.8.0_45
CLASSPATH= .;C:\Java\jdk1.8.0_45\lib;C:\Java\jdk1.8.0_45\lib\tools.jar
PATH= C:\Java\jdk1.8.0_45\bin
```

CLASSPATH 中要注意前面有一个符号“.”。

2.1.3　测试 Java 配置结果

当环境变量设置好后，就可以检验是否设置成功。打开操作系统的命令行窗口（在“开始”菜单的“运行”框中输入“cmd”可打开），输入命令“javac”并按下 Enter 键，如果出现如图 2. 4 所示的输出，则表明环境变量已经配置成功。

如果出现的是“未找到指令”的错误提示，则说明上一节中的环境变量配置没有成功，可检查 JAVA_HOME、PATH 和 CLASSPATH 这 3 个变量的值是否与安装的 JDK 目录相匹配。

```
C:\Windows\system32\cmd.exe
Microsoft Windows [版本 6.1.7601]
版权所有 (c) 2009 Microsoft Corporation。保留所有权利。

C:\Users\tina6>javac
用法: javac <options> <source files>
其中, 可能的选项包括:
  -g                         生成所有调试信息
  -g:none                    不生成任何调试信息
  -g:{lines,vars,source}     只生成某些调试信息
  -nowarn                    不生成任何警告
  -verbose                   输出有关编译器正在执行的操作的消息
  -deprecation               输出使用已过时的 API 的源位置
  -classpath <路径>            指定查找用户类文件和注释处理程序的位置
  -cp <路径>                   指定查找用户类文件和注释处理程序的位置
  -sourcepath <路径>           指定查找输入源文件的位置
  -bootclasspath <路径>        覆盖引导类文件的位置
  -extdirs <目录>              覆盖所安装扩展的位置
  -endorseddirs <目录>         覆盖签名的标准路径的位置
  -proc:{none,only}          控制是否执行注释处理和/或编译。
  -processor <class1>[,<class2>,<class3>...] 要运行的注释处理程序的名称; 绕过默
认的搜索进程
  -processorpath <路径>        指定查找注释处理程序的位置
  -parameters                生成元数据以用于方法参数的反射
  -d <目录>                    指定放置生成的类文件的位置
  -s <目录>                    指定放置生成的源文件的位置
```

图 2.4　测试环境变量

2.2 安装并配置 Tomcat 服务器

一段写出的代码，要想转换成网页，就需要一个 Web 应用服务器。与 ASP.NET 匹配的 Web 服务器一般首选是 IIS，而与 JSP 匹配的首选是 Tomcat，本节就介绍 Tomcat 服务器。

2.2.1 下载并安装 Tomcat 服务器

Tomcat 是轻量级的 Web 应用服务器，可以从官网 http://tomcat.apache.org/下载最新的 Tomcat 服务器版本，本书用的是 Tomcat 7.0 版本。下载完成后直接解压 Tomcat 文件到指定的目录下，例如：C:\Tomcat\apache-tomcat-7.0.62 中。Tomcat 目录结构如图 2.5 所示。

名称	修改日期	类型	大小
bin	2015/5/31 22:30	文件夹	
conf	2015/5/31 22:30	文件夹	
lib	2015/5/31 22:30	文件夹	
logs	2015/5/7 20:15	文件夹	
temp	2015/5/31 22:30	文件夹	
webapps	2015/5/31 22:30	文件夹	
work	2015/5/7 20:15	文件夹	
LICENSE	2015/5/7 20:17	文件	57 KB
NOTICE	2015/5/7 20:17	文件	2 KB
RELEASE-NOTES	2015/5/7 20:17	文件	9 KB
RUNNING.txt	2015/5/7 20:17	文本文档	17 KB

图 2.5　Tomcat 目录结构

下面介绍 Tomcat 的目录结构：

- bin 文件夹，包含 Tomcat 服务器启动和终止服务器的批处理文件。例如：startup.bat、startup.sh、shutdown.bat、shutdown.sh、catalina.bat、catalina.sh 等。其中 startup.bat、 shutdown.bat、catalina.bat 是 Windows 中的批处理文件；startup.sh、shutdown.sh、catalina.sh 是 Linux 中的脚本文件。

- conf 文件夹，包含 Tomcat 的配置信息。主要有 server.xml 和 web.xml 这两个配置文件。在 server.xml 中可以更改服务端口和改变 Web 默认的访问目录，后面的小节将介绍如何修改。
- lib 文件夹，存放 tomcat 运行中需要的 jar 包文件，例如：catalina.jar、servlet-api.jar、tomcat-dbcp.jar 等 jar 包，正因为有这些包的支持，Tomcat 才可以运行 Web 应用程序。
- logs 文件夹，存放执行 Tomcat 的日志文件。
- temp 文件夹，存放 Tomcat 的临时文件信息。
- webapps 文件夹，是 Tomcat 默认的 Web 文件夹。本身自带两个 admin 应用和 manager 应用。开发人员可以直接将 Web 应用存放在该文件夹下。
- work 文件夹，存放 Tomcat 执行应用后的缓存。

2.2.2 Tomcat 的文件结构

Tomcat 服务器中，要经常修改其配置信息来满足系统的需求，例如在 server.xml 中可以更改服务端口和改变 Web 默认的访问目录。

修改端口号，方法如下：

```
<Connector port="8080" protocol="HTTP/1.1" connectionTimeout="20000"
           redirectPort="8443" URIEncoding="UTF-8"/>
```

修改 port 端口为 8088。修改完毕，保存 server.xml，然后重启 Tomcat 服务器，这样服务器的端口就更改为 8088 了。

修改 Web 默认的访问目录，方法如下：

```
<Host name="localhost"  appBase="webapps"
        unpackWARs="true" autoDeploy="true"
        xmlValidation="false" xmlNamespaceAware="false">
```

修改 appBase 中的文件夹地址。例如：将 appBase 的属性值 webapps 改为 d:\test，修改后的文件如下：

```
<Host name="localhost"  appBase="d:\test"
        unpackWARs="true" autoDeploy="true"
        xmlValidation="false" xmlNamespaceAware="false">
```

这样就可以将 Web 默认的访问目录更改为 d:\test。那么以后加载 Web 应用程序就会在该目录下再创建目录。

建立自身的 Web 目录。开发人员可以将应用部署在 Tomcat 服务器的默认 webapps 目录下，也可以部署在自己创建的目录下。方法如下：

首先创建自身的目录 D:\test，其次配置 Web 目录，在 server.xml 文件的末尾</HOST>中加入如下语句：

```
<Context path="text" docBase="D:\test" debug="0" reloadable="true"></Context>
```

该语句的作用是将目录 D:\test 设置为 Tomcat 服务器的 Web 目录，将该文件的访问路径设置为“/text”。属性 docBase 的值为“D:\test”，它是指应用的物理路径。修改后将 server.xml 文件进行保存。假设现在有个 test.jsp 页面在 D:\test 目录下，那么页面的访问路径就为：http://localhost:8080/text/test.jsp。

在 bin 文件夹下，可以修改 catalina.bat 或者 catalina.sh 来更改 Tomcat 启动配置信息。例如增加 Java 运行内存：

```
set JAVA_OPTS=-XX:PermSize=512M -XX:MaxPermSize=512m -Xms512m -Xmx1024m
```

更多的修改内容请参见 Tomcat 官网说明。

2.2.3　Tomcat 的工作原理

前面我们说过 Tomcat 是一个 Web 服务器，那它的作用是什么？它又是如何帮助外部网页访问我们自己服务器上的网页的？

如果想让某台计算机上的一个目录内容，被外面的 Web 浏览器通过 HTTP 协议访问，就要在这台计算机上安装和启动一个 Web 服务器软件，还要将该目录映射成一个虚拟的 Web 站点目录，这个 Web 站点目录通常称为虚拟目录。

一个 Web 站点上只能有一个虚拟目录的根目录，其他的虚拟目录都必须是该根目录的子目录。一般根目录用“/”符号表示，根目录与该根目录下的子目录或资源文件也以“/”符号进行分割。如图 2.6 所示的映射中，假设 D:\test 文件夹映射成 Web 站点的根目录，当计算机的域名为 http://www.student.com 时，那么浏览器访问 http://www.student.com/test.html 就可以得到 D:\test\test.html 文件的内容，如果访问 http://www.student.com/test.html/student/test.html 就可以得到 D:\test\student\test.html 文件的内容。

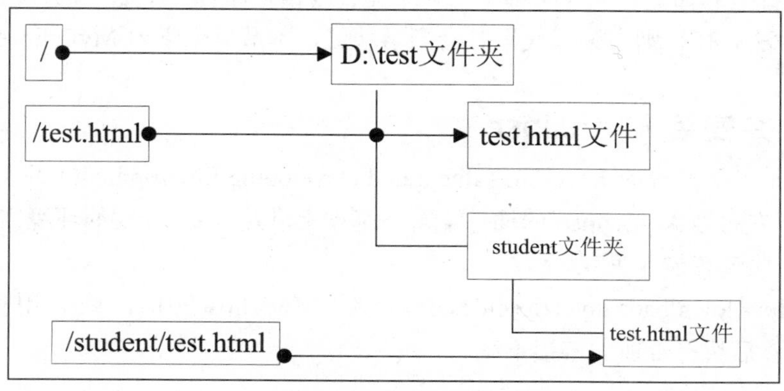

图 2.6　映射示意图

一个文件系统目录可以被映射成一个或多个虚拟目录，但是一个虚拟目录只能对应一个本地文件系统目录，如图 2.7 所示。

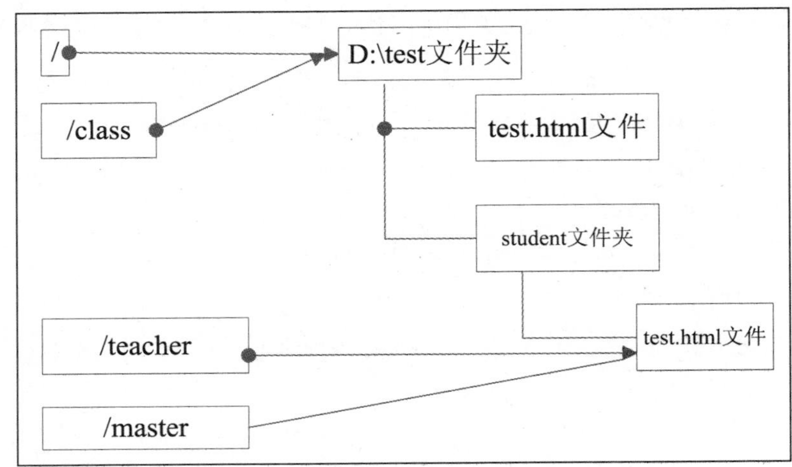

图 2.7　映射路径与本地文件系统多对一关系

当浏览器请求某个路径资源时，一般按照最长路径匹配原则进行处理。首先以请求路径中最深子目录作为一个虚拟子目录名称，查看是否存在这个虚拟子目录：如果不存在则查找是否存在上一级目录名称的虚拟子目录；如果找到匹配的虚拟目录后，则进入该虚拟目录中查找指定名称的资源。

Tomcat 中默认的根目录是 webapps，上一节已经介绍了如何修改根目录，这里不再赘述。

2.3 使用 MyEclipse 开发工具

JSP 有很多标签和命令，如果全部手写，难免会写错，而有一个顺手的开发工具，它又具备自动完成和语法高亮，则在编写代码时会事半功倍，本节就来学习 MyEclipse 的使用。

2.3.1 下载并安装 MyEclipse

MyEclipse 是当下流行的 Java IDE（Integrated Developing Environment）开发工具，深受开发者的青睐，一方面是 MyEclipse 的插件丰富，基本上满足 Web 开发的环境需求；另一方面安装简单，界面比较智能和美观。

从官网 http://downloads.myeclipseide.com/ 下载 MyEclipse2014，或者用搜索引擎搜索 MyEclipse2014 然后选择合适的链接下载。

myeclipseide 网站有中文版，对英文不太熟悉的读者可以切换语言，如图 2.8 所示。

图 2.8　下载 MyEclipse 界面

下载完后双击安装包，安装界面如图 2.9 所示。

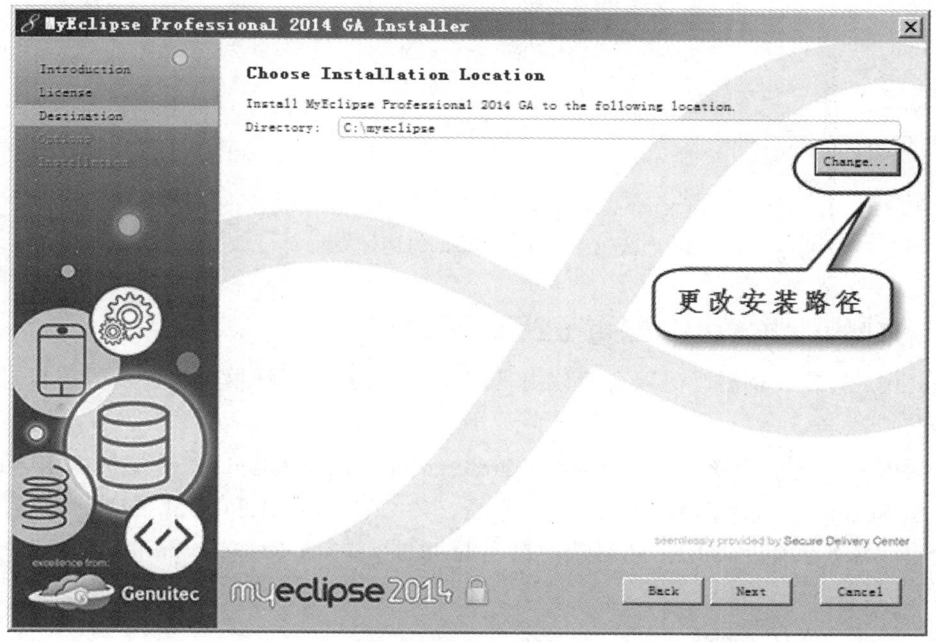

图 2.9　安装 MyEclipse 的示例界面

安装完成后首先要求设置 workspace，我们的项目都会保存在这个目录下，读者可根据实际项目情况来设置，如 C:\Workspaces\MyEclipse。然后单击 OK 按钮，如图 2-10 所示。

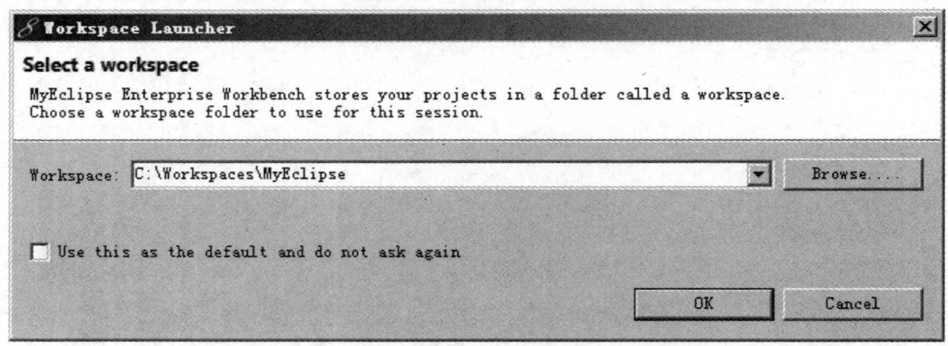

图 2.10　设置 workspace

配置完成后打开项目的界面如图 2.11 所示。

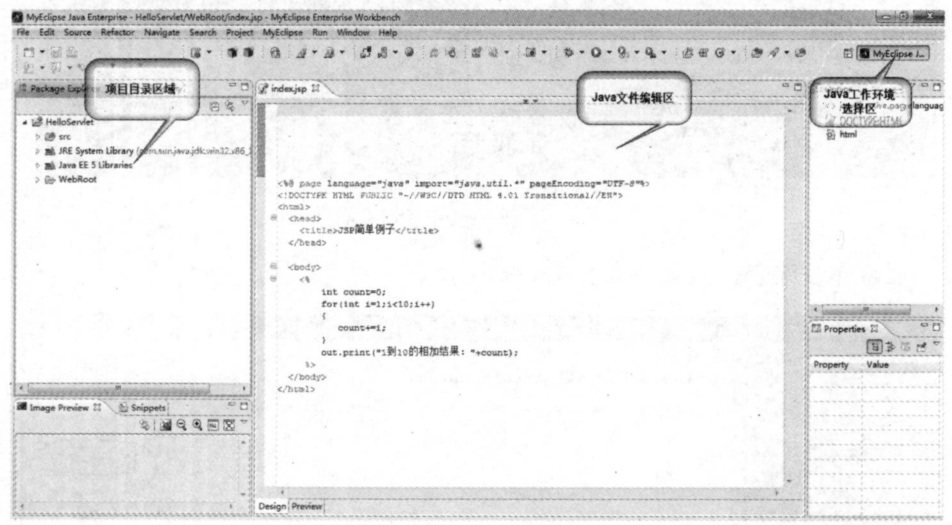

图 2.11　MyEclipse 的运行界面

2.3.2　在 MyEclipse 中配置 J2EE 环境

MyEclipse 安装成功后，可以直接使用其默认的 J2EE 运行环境，也可以配置自身的 J2EE 运行环境。

如果要配置自身的 JDK，选择 Window|Preferences|Java|Installed JRES 增加 JDK，如图 2.12 所示。单击 Search 按钮，找到 JDK 的安装路径，如 C:\Java\jdk1.8.0_45，单击“确定”按钮，列表中会自动添加我们的 JRE，然后选中自己的 JRE，单击 OK 按钮。

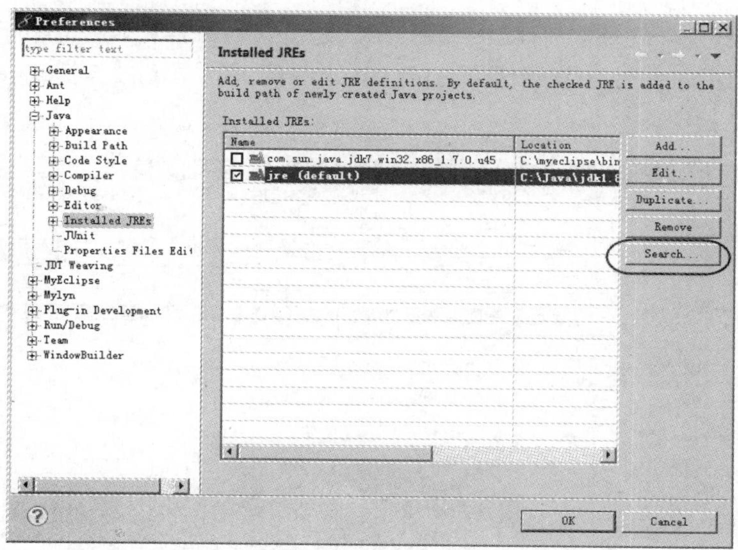

图 2.12 MyEclipse 查找 JDK 的选择界面

2.3.3 在 MyEclipse 中配置 Tomcat

要在 MyEclipse 测试 Web 程序，也需要配置自己的 Tomcat。选择 Window|Preferences|MyEclipse|Servers|Tomcat 7.x 中选择 Tomcat 的安装目录，如图 2.13 所示。首先选中 Enabled 单选框，然后单击第 1 个 Browse 按钮，找到自己的 Tomcat 安装目录，单击“确定”按钮后，第 2、3 个文本框会自动添加，然后单击 OK 按钮。

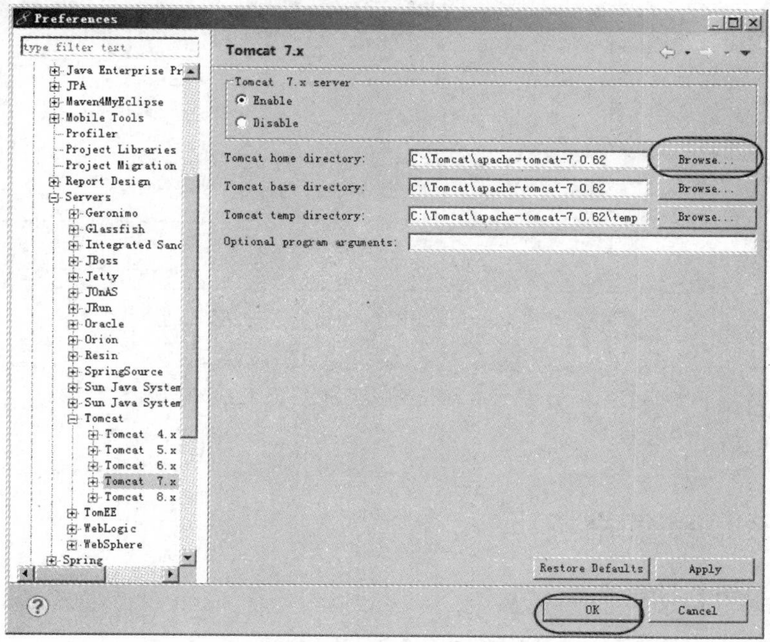

图 2.13 MyEclipse 选择 Tomcat 的界面

设置好 Tomcat 后，还需要将 Tomcat 的 JDK 设置成自己的 JDK。在 Window|Preferences|MyEclipse|Servers|Tomcat 7.x|JDK 中选择 jdk1.8.0_45（配置时默认的名字是 jre），如图 2.14 所示，然后单击 OK 按钮。

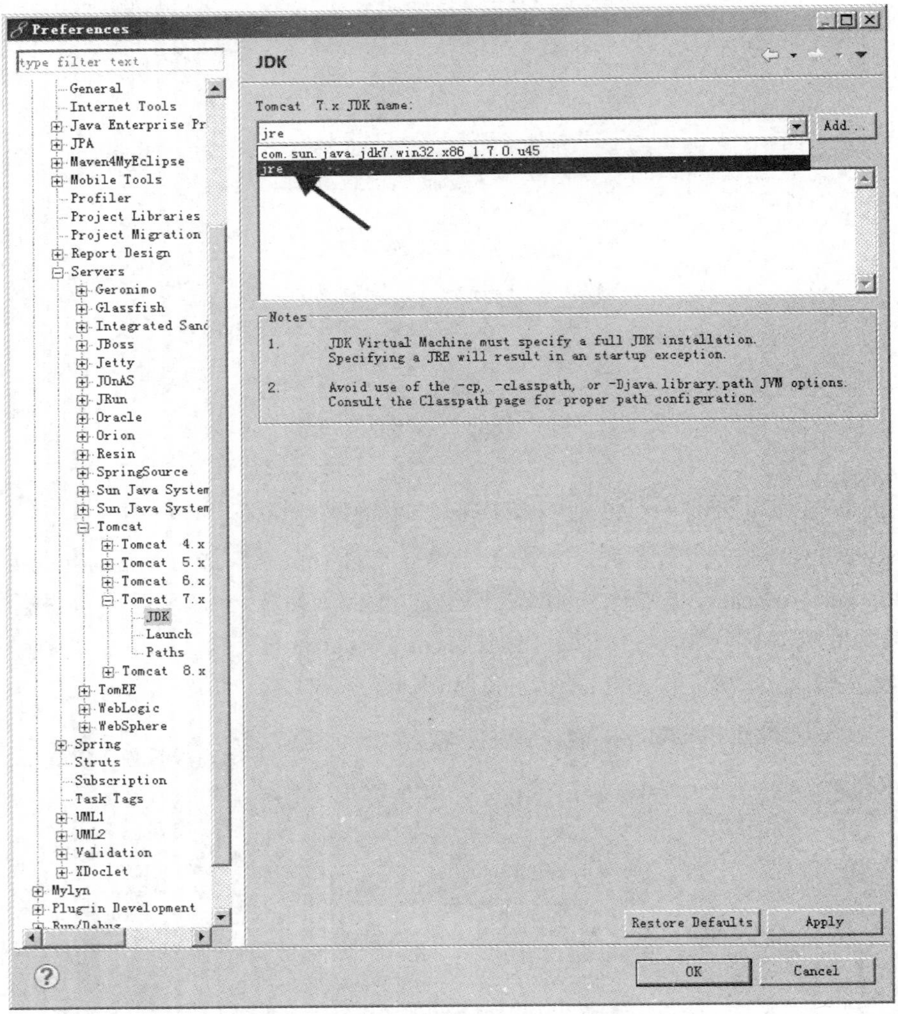

图 2.14　MyEclipse 中 Tomcat 选择 JDK 界面

如果安装了多个 JDK，为方便后期识别，可在配置 J2EE 环境的界面，为自己的 JDK 重新命名，加上版本号。

2.3.4　MyEclipse 使用技巧

1. 关闭 maven 自动更新

自动更新程序会在不知不觉中占用网络流量或系统进度，这里需要关闭 maven 的自动更

新功能。单击 Window|Perferences|MyEclipse|Maven4MyEclipse，选中 Enable Maven4MyEclipse featrures 复选框，然后取消选中 Download repository index updates on startup 复选框，如图 2.15 所示。

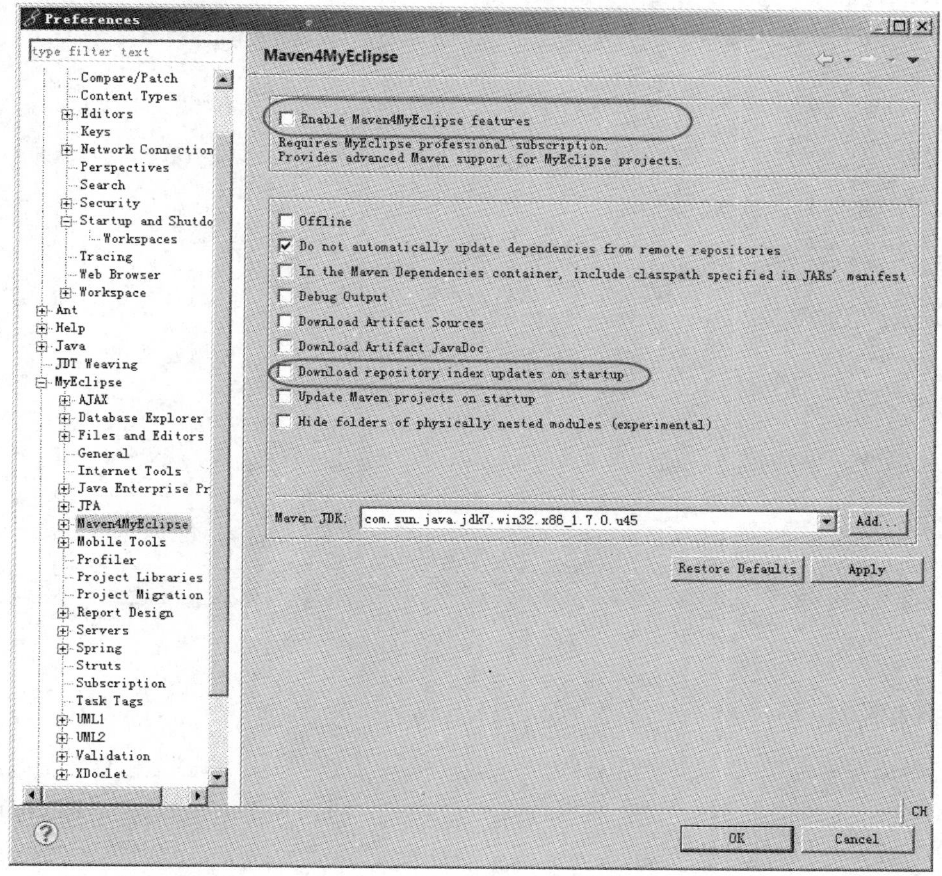

图 2.15　关闭 maven 自动更新

2. 优化启动项

MyEclipse 启动时会自带很多默认的模块，但有些根本不需要。这里单击 Window|Preferences|General|Startup and Shutdown，默认右侧会列出所有启动时加载的模块，可以关闭不需要使用的模块，提高 MyEclipse 的启动速度。图 2.16 是作者的设置，读者可根据自己的需要进行设置。

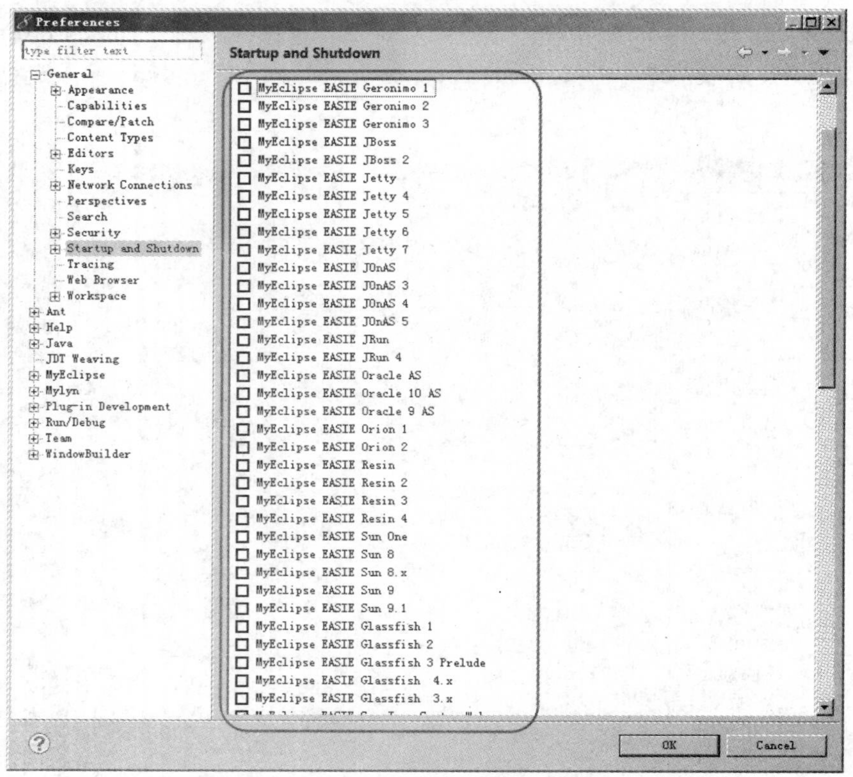

图 2.16　优化启动项

3. 添加自动联想

之所以选择开发工具，就是因为开发工具能提高开发速度，尤其是当输入一些关键字的前几个字母时，工具可以帮给出完整的关键字。单击 Window|Perferences|Java|Editor|Content Assist，在右侧下方界面看到 Auto activation delay(ms)，这个改为 10，这是提示的延迟时间，值越小的话提示越快。还有 Auto activation triggers for Java 文本框，这里的英文“.”表示输入“.”会出现自动提示功能，如图 2-17 所示。

如果觉得输入“.”出现提示很麻烦，可以设置为输入字母出现提示，把这个“.”修改为“abc.”。abc 字母可以是 a、b、c…..z 的所有 26 个字母。

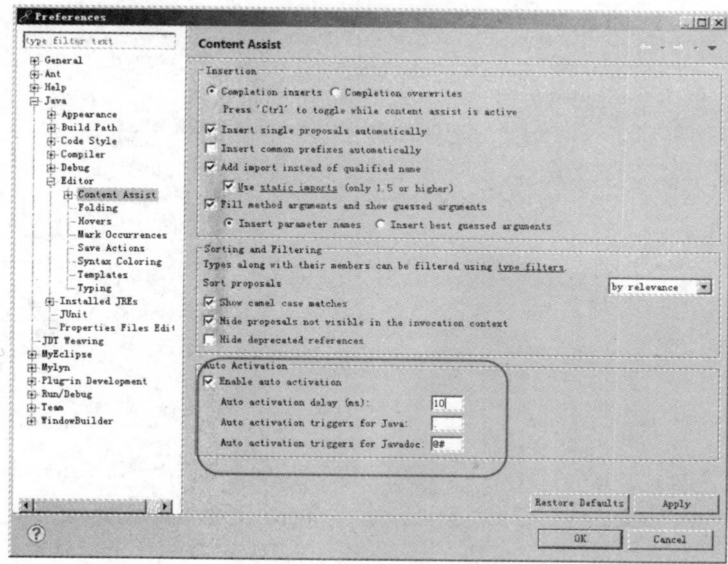

图 2.17　设置自动联想

4. 设置默认编码

经常在开发网页时碰到网页中显示乱码的情况，这一般是整个工程的编码设置问题。为了避免后期的编码出现这类问题，可以提前进行全局设置。

（1）设置全局编码为 UTF-8，单击 Window|Preferences|General|Workspace 找到 Text file encoding 修改为 UTF-8。如图 2.18 所示。

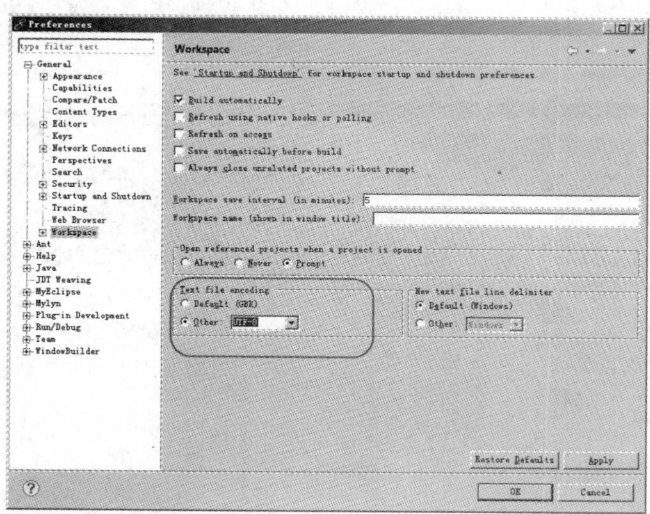

图 2.18　设置全局编码

（2）设置 JSP 页面默认编码方式为 UTF-8，单击 Window|Preferences|MyEclipse|Files and Editors|JSP，找到右侧 Encoding 列表框，设置为 ISO10646/Unicode(UTF-8)。如图 2.19 所示。

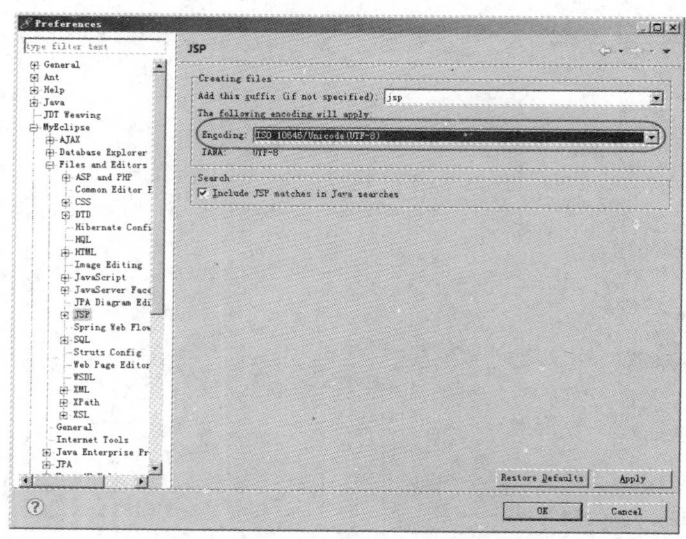

图 2.19　设置 JSP 页面默认编码

5. 取消 Validation 自动校验

MyEclipse 在启动时会自动验证项目的配置文件，这往往会耗费比较长的时间，而有些文件根本不需要每次都去自动校验，所以需要取消对一些文件的自动校验功能。单击 Window|Perferences|Myclipse|Validation，将 Build 一列下的复选框全部选中取消即可，效果如图 2.20 所示。

单独验证某个文件时，在要验证的文件上右键鼠标，然后单击 MyEclipse|Run Validation 命令。

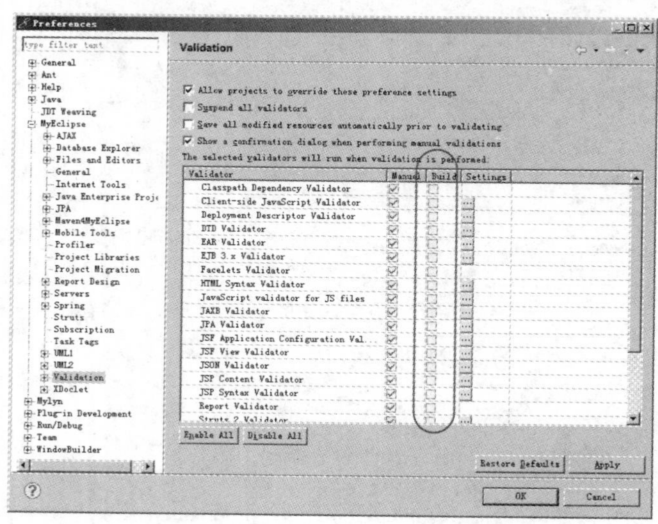

图 2.20　取消 Validation 自动校验

2.3.5 其他 IDE

当然，除了 MyEclipse 外还有很多 Java IDE 工具，例如：NetBeans、JCreator、JRun 等。编辑 JSP 页面的工具还有 Dreamveaver MX、Editplus 等。

单从学习角度来说，建议选择轻便型的 IDE 工具，如果是企业开发那么选择 MyEclipse 是最佳选择。

2.4 实例：使用 MyEclipse 开发一个完整的 Java Web 网页

现在已经安装好 MyEclipse，可以开始用它来开发自己的第一个 Java Web 项目了。因为读者刚接触 MyEclipse，这里给出创建项目的详细步骤。

步骤 01 单击 File|New|Web Project 菜单，弹出新建 Web 工程对话框，如图 2.21 所示。输入项目名称：HelloWorld，然后选择 Java 版本，再选择 Tomcat 版本。

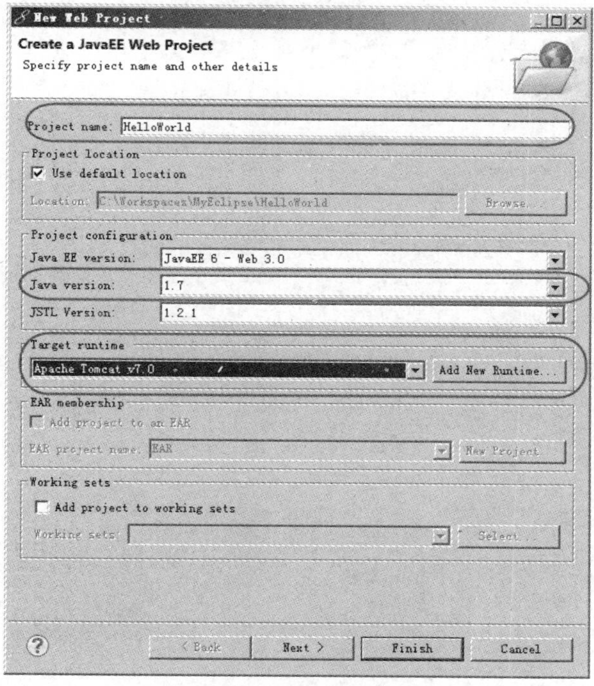

图 2.21 新建项目对话框

步骤 02 单击 Next 按钮，这里是项目的文件结构，不做任何修改，直接单击 Next 按钮，如图 2-22 所示。

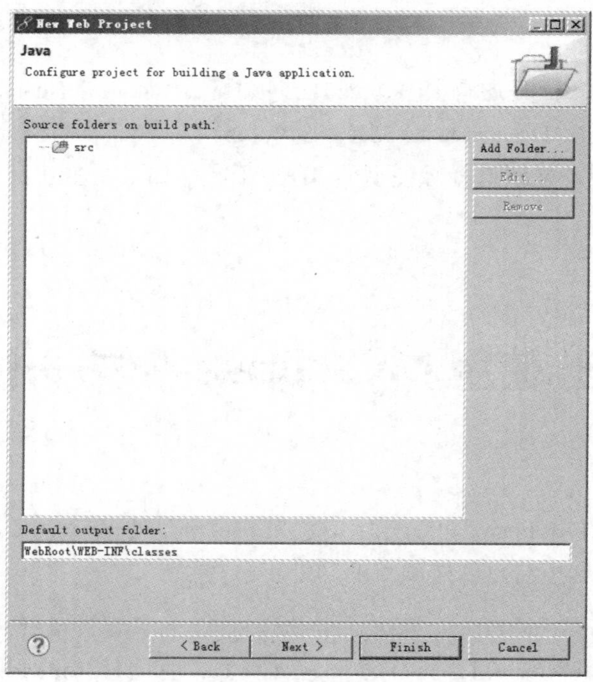

图 2.22 默认的目录结构

步骤 03 此时 Web 配置界面，如图 2.23 所示，这里要选中两个复选框，第 1 个是自动生成 index.jsp 文件，第 2 个是自动生成 web.xml 文件。

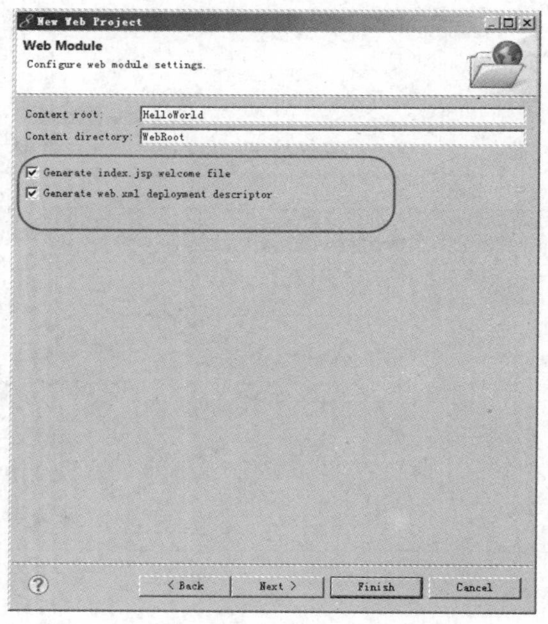

图 2.23 自动配置文件

步骤 04 继续单击 Next 按钮，这里配置加载的一些库，不做任何修改，如图 2-24 所示。

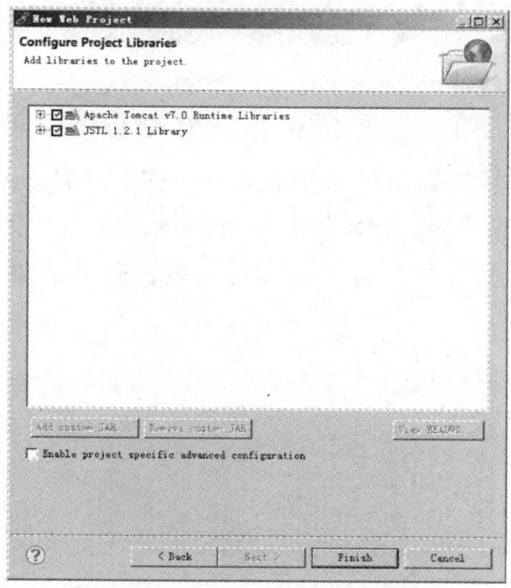

图 2.24　加载的库

步骤 05　单击 Finish 按钮，测试在 MyEclipse 左侧列表会生成刚创建的 HelloWorld 项目，结构如图 2.25 所示。

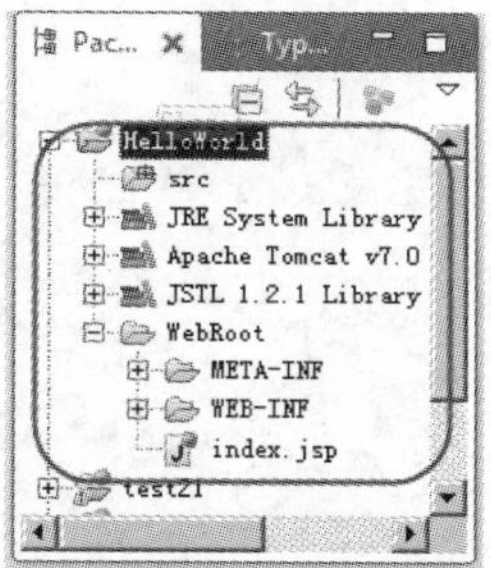

图 2.25　HelloWorld 项目结构

步骤 06　此时已经默认创建一个 index.jsp 文件，双击此文件，修改<body></body>内的文字为 Hello World，如图 2.26 所示。

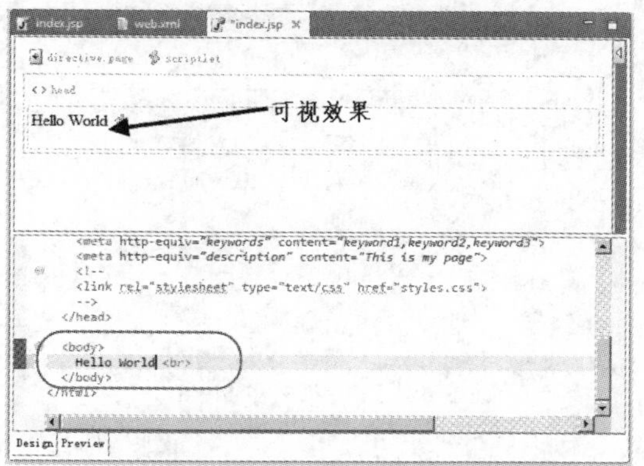

图 2.26　修改代码

步骤 07 单击工具栏的 按钮，或者按 Ctrl+S 快捷键，保存创建好的项目。下面的两个步骤很关键，一定不能省略。

步骤 08 单击工具栏的 按钮，打开如图 2.27 所示的配置对话框，我们需要配置项目到 Tomcat 中。首先选择工程下拉框中我们的项目，如 Hello World。然后单击 Add 按钮，打开如图 2.28 所示的“选择服务器”对话框，这里从下拉列表框中选择 Tomcat 7.X 服务器。单击 Finish 按钮，回到配置对话框，如果出现图 2.29 下面的提示，则表示配置成功。

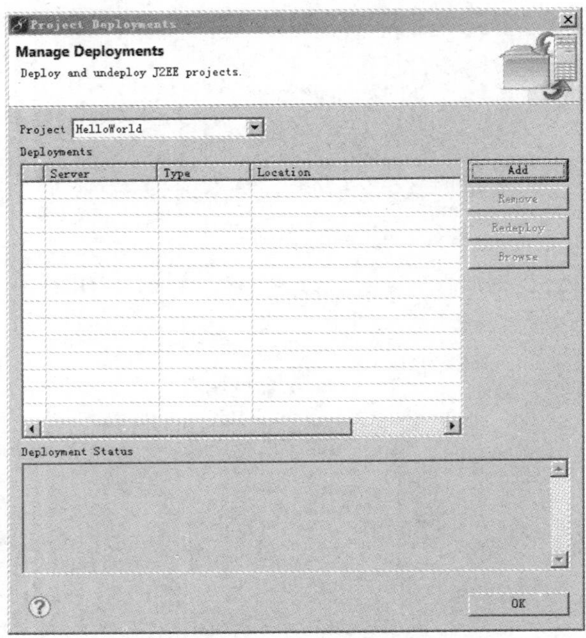

图 2.27　配置项目

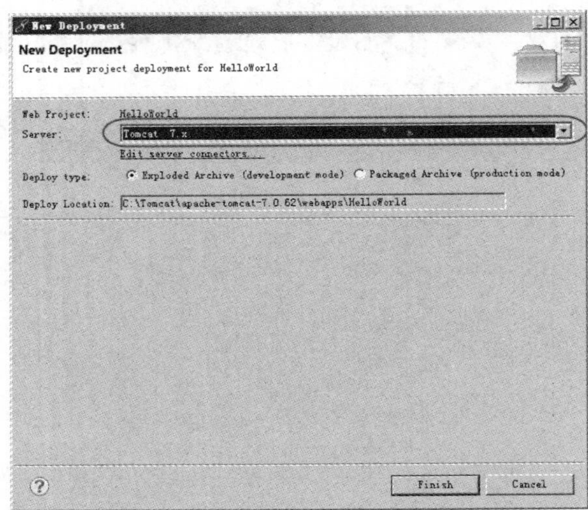

图 2.28　选择服务器

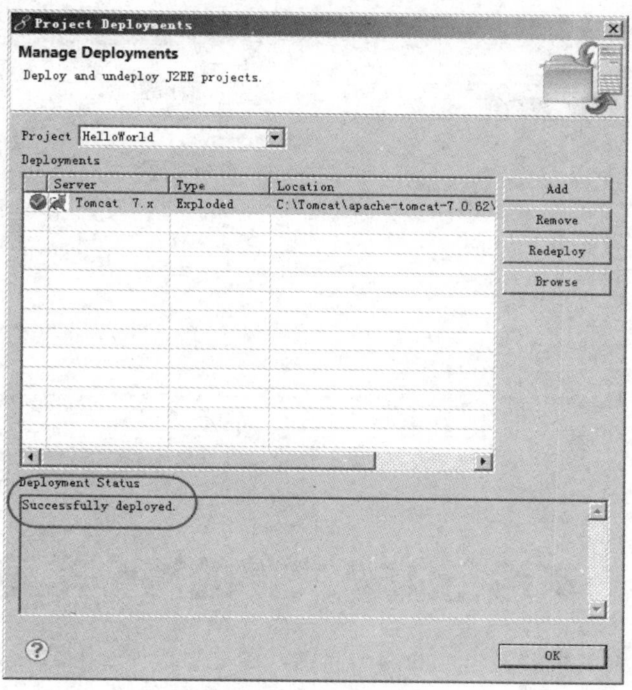

图 2.29　配置成功

步骤 09 单击工具栏的按钮，会出现一个下拉列表，选择 Tomcat，这里作者选择 Tomcat 7.x，然后单击快捷菜单的 Start，开启 Tomcat 服务器。开启成功如图 2.30 所示。

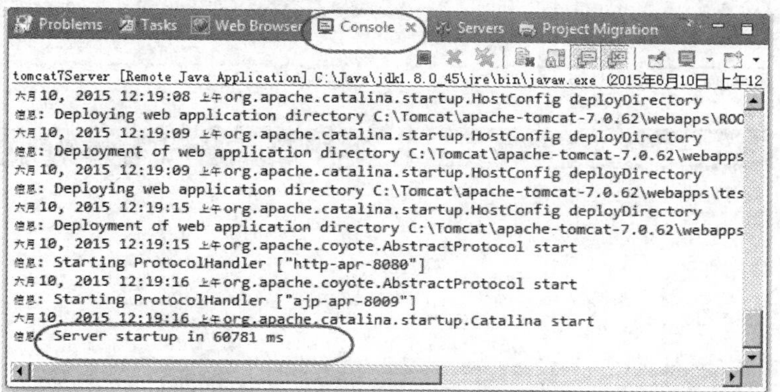

图 2.30　开启成功

步骤 10 好了，一切准备就绪，在 MyEclipse 下方的 Web Browse 浏览器中，输入 http://localhost:8080/HelloWorld，如果出现图 2.31 所示的页面，则表示第一个 JSP 页面测试成功。

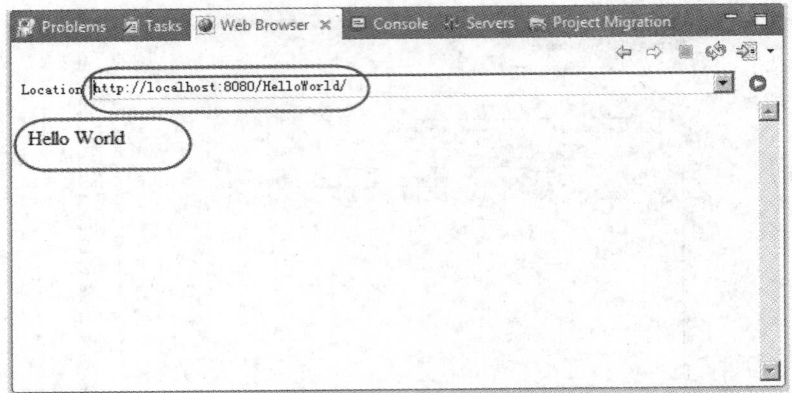

图 2.31　第一个 JSP 页面

2.5 第一次运行 JSP 文件的两个常见问题

初次运行 JSP，读者最容易出现的两个问题，这里简单说一下。

1. 未部署代码到 Tomcat 中

如果忘记 2.4 节的第 8 步骤，当测试页面时，会出现图 2.32 所示的页面。

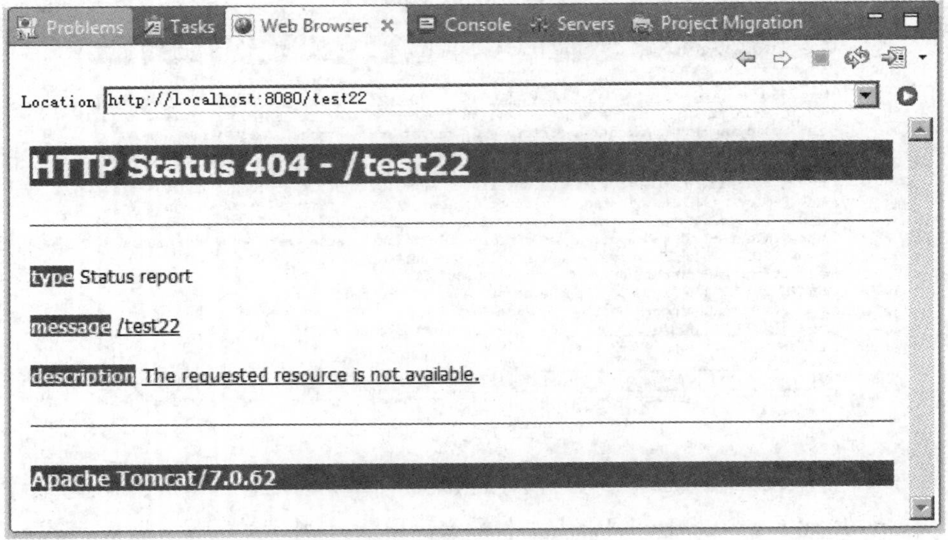

图 2.32　错误页面 1

2. 未运行 Tomcat 服务器

如果忘记 2.4 节的第 9 步，则会出现图 2.33 所示的页面。

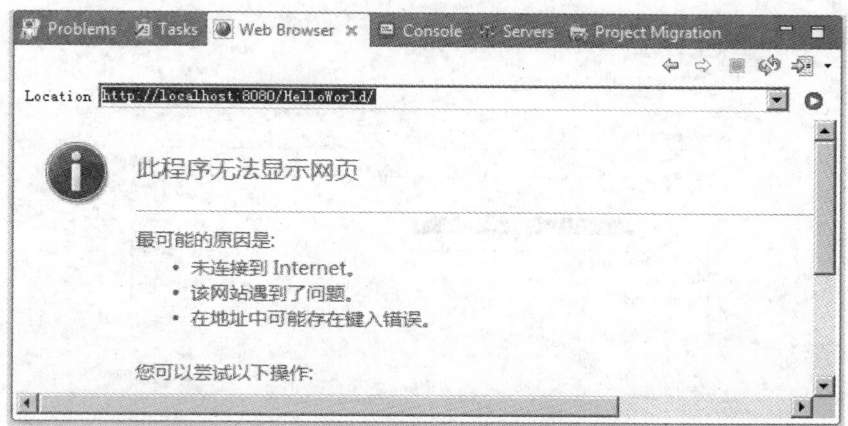

图 2.33　错误页面 2

如果要测试 Tomcat 是否正常，在浏览器中输入：http://localhost:8080，如图出现图 2.34 所示的页面，说明 Tomat 没有问题。

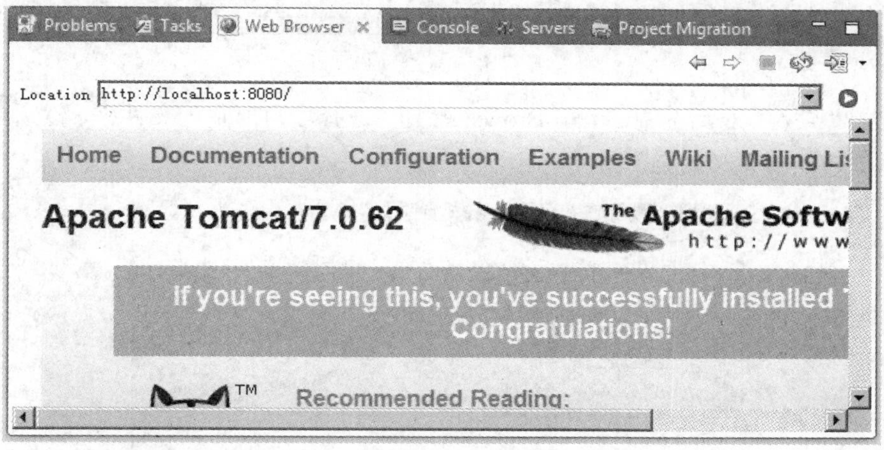

图 2.34　Tomcat 正常访问

2.6 在 MyEclipse 中导入原来的项目

如果读者已经有原来的 Web 项目，或者想导入本书源码中的项目，则可以右击左侧项目列表的空白处，在快捷菜单中选择“Import”命令，选中图 2.35 中圈住的位置，表示选择已经存在的项目，然后单击 Next 按钮。

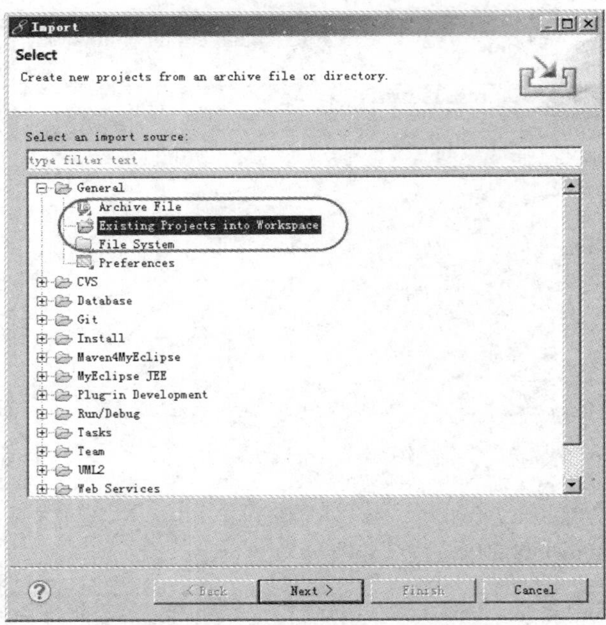

图 2.35　导入项目

在出现的“导入”对话框中，单击 Browse 按钮，选择项目所在的文件夹，然后单击 Finish 按钮就可以了，如图 2.36。

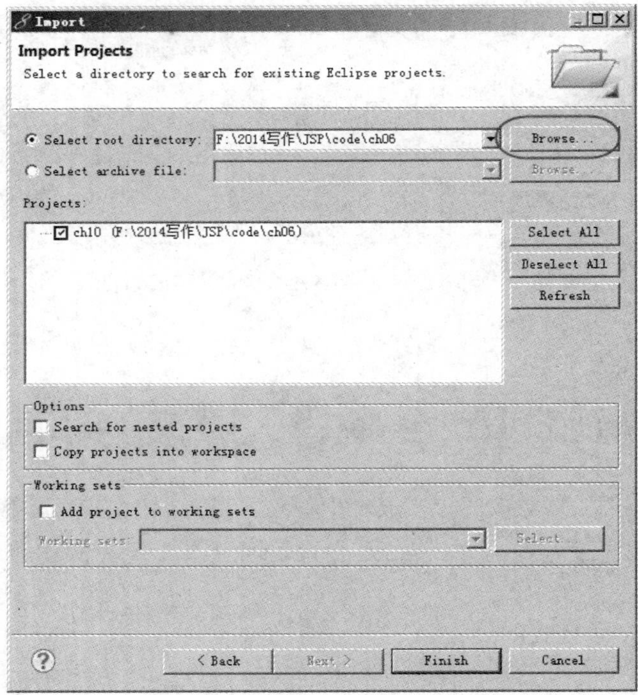

图 2.36 选择项目

2.7 上机实践

把第 1 章最后的代码案例添加到本章创建的 HelloWorld Web 项目中，并自己测试运行效果。

【提示】右击项目下的 WebRoot 文件夹，会出现一个 New 菜单，里面设计好了允许我们添加的所有文件类型。

第 3 章

◂ JSP的基础语法 ▸

前面两章介绍了网页开发的一些基本概念和实现原理，然后搭建好了 JSP 的开发环境，并介绍了 MyEclipse 开发工具的使用。作为一门动态网页开发语言，JSP 有一些基础语法必须了解。从本章开始意味着读者将正式开始学习 JSP 技术，随着学习的深入，本书也会逐步介绍如何编写高效的 JSP 界面。

3.1 JSP 的注释和声明

注释是一门语言最基本的功能，为了让别人或自己能看懂编写过的代码，注释必不可少。JSP 中的声明是表示“这是 JSP 语言”的一个关键。本节就来详细介绍 JSP 的这两部分基础。

3.1.1 JSP 中的注释

在编写规范的代码时，合理和必要的注释是至关重要的，这也是对开发人员的一项最基本要求。编程语言的注释起到了说明解释的作用，在实际执行的过程中并不执行。同样，JSP 也是一样的。在 JSP 页面中，注释可以归纳为两种：一种是原有的 HTML 注释；另外一种是 JSP 的注释。

（1） HTML 的注释如下：

```
<!--  注释的内容 -->
```

例如：

```
<!--  注释的内容会被照搬到浏览器中 -->
```

（2）JSP 的注释语法如下：

```
<%-- 注释的内容 --%>
```

例如：

```
<%-- 注释的内容不会照搬到浏览器中 --%>
```

在实际应用中，JSP 通常会引入 Java 的注释，二者混着用。因为 Java 的注释在脚本中注释，而 JSP 的注释在脚本外注释。

【例 3.1】JSP 中不同的注释用法

```
-----------------------myjsp.jsp---------------------------
<%@ page language="java" import="java.util.*" pageEncoding="UTF-8"%>
<!DOCTYPE HTML PUBLIC "-//W3C//DTD HTML 4.01 Transitional//EN">
<html>
  <head>
    <meta content="text/html;charset=utf-8" http-equiv="Content-Type">
    <title>JSP 注释例子</title>
  </head>
      <body>
        <!-- 该 JSP 注释可以在浏览器文件中看到 -->
        <br/>
        <center>
            <table>
                <tr>
                    <th>
                        <b>JSP 注释</b>
                    </th>
                </tr>
            </table>
        </center>
        <table>
            <tr>
                <th>
                    <b>JSP 注释</b>
                </th>
            </tr>
        </table>
        <table align="right">
            <tr>
                <th>
                    <b>JSP 注释</b>
                </th>
            </tr>
        </table>
       <%--该 JSP 注释无法在浏览器中看到 --%>
       <%
       //  这是脚本中的 Java 注释
       /*
           这也是脚本中的 Java 注释
       */
        %>
     </body>
```

```
42        </html>
```

上述代码介绍了 JSP 的各种注释格式，其中代码第 9 行是“<!--　-->”格式，第 34 行是“<%-- --%>”格式。因为都是注释，所以界面运行的结果看不到注释中的内容，通过浏览器“查看源文件内容”功能可以看到 HTML 注释的内容。本例运行的结果如图 3.1 所示。

图 3.1　JSP 注释运行结果界面

第 36 行的代码说明在 JSP 页面中可以嵌套 Java 注释。

3.1.2　JSP 中的声明

JSP 声明用于定义 JSP 中的变量、方法以及静态方法，实际上它跟 Java 中定义一个全局变量、共用方法一样，JSP 声明部分的变量和方法是可以在 JSP 页面中被调用和使用的。

JSP 声明的基本语法有两种，分别是：

```
(1) <%! 变量定义/方法定义/类 %>
(2) <jsp:declaration>变量定义/方法定义/类</jsp:declaration>②
```

②中的语法现在已经过时，现在开发一般采用第一种声明语法。

JSP 声明的结尾跟 Java 的结尾一样以“;”结束，可以一次定义多个变量，用“,”分割。声明部分的变量和方法只在当前的页面中有效。

【例 3.2】演示变量、方法和类的声明。

```
-----------------------shengming.jsp---------------------------
<%@ page language="java" import="java.util.*" pageEncoding="UTF-8"%>
<!DOCTYPE HTML PUBLIC "-//W3C//DTD HTML 4.01 Transitional//EN">
<html>
  <head>
    <meta content="text/html;charset=utf-8" http-equiv="Content-Type">
    <title>JSP 声明例子</title>
  </head>
  <%!
     int x,y=60,z;              //多个声明以“,”分割
     String name="John";
     Date date = new Date();
   %>
```

```
  <%!
    int add(int m,int n){        //计算两个数的和
        int result=0;
        result = m+n;
        return result;
    }
   %>
   <%!
    int chengji(int m,int n){   //计算两个数的乘积
        int result=0;
        result = m*n;
        return result;
    }
   %>
   <%!
     class Circle{             //计算圆的面积
       double r;
       Circle(double r){
       super();                //继承空的构造方法
       this.r = r;
       }
      double area(){         //取整
          return  Math.floor(Math.PI*r*r);
       }
     }
    %>
<body>
  <%
      out.println("我的名字："+name);
      out.println("<br/><br/>");
      out.println("x 的值为："+x);
      out.println("<br/><br/>");
      out.println("y 的值为："+y);
      out.println("<br/><br/>");
      out.println("z 的值为："+z);
      out.println("<br/><br/>");
      out.println("现在的时间为："+date);
      out.println("<br/><br/>");
      out.println("10与20的和："+add(10,20));
      out.println("<br/><br/>");
      out.println("10与20的积："+chengji(10,20));
   %>
  <br/>
  <br/>
  <%
```

```
            Circle c = new Circle(6);
            out.println("半径为6的圆面积为: "+c.area());
        %>
    </body>
</html>
```

页面效果如图 3.2 所示。

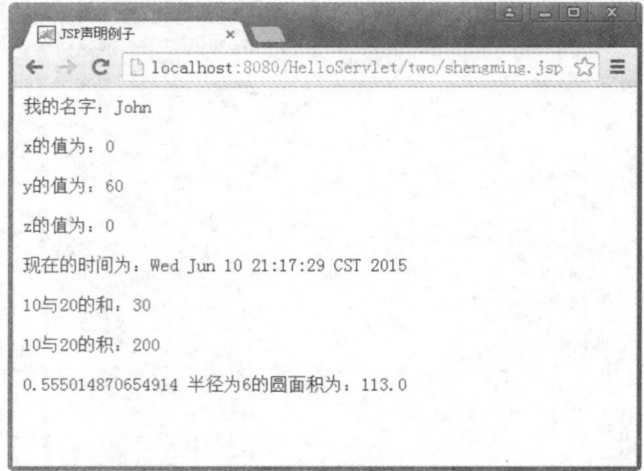

图 3.2 declar.jsp 页面的运行结果

3.2 JSP 表达式

JSP 表达式的作用是将动态信息显示在页面中，它的语法形式也有两种：

```
<%=变量或者表达式%>
```

或者

```
<jsp:expression>变量或者表达式</jsp:expression>
```

上述形式中第 2 种所示的表达式现在已经不用，一般的 IDE 工具中也不提供这种形式的表达式，第 1 种形式是目前主要的写法，本书例子也是用该形式进行书写。

表达式的值由服务器负责计算，计算结果以字符串的形式发送到客户端。

下面看一个例子，JSP 页面使用类 Date 输出当前的时间。

```
-----------------------date.jsp--------------------------
<%@ page language="java" import="java.util.*" pageEncoding="UTF-8"%>
<!DOCTYPE HTML PUBLIC "-//W3C//DTD HTML 4.01 Transitional//EN">
<html>
```

```
04     <head>
05       <meta content="text/html;charset=utf-8" http-equiv="Content-Type">
06       <title>JSP 表达式例子</title>
07     </head>
08
09     <body>
10                      当前的时间为：</br>
11         <%=new Date()%>
12     </body>
13   </html>
```

上述代码中第 1 行导入 Date 类库，第 11 行直接引用 Date 对象。页面效果如图 3.3 所示。从页面的显示结果看，文本“当前的时间为：”被正常显示，其后的“</br>”HTML 标记使得后面显示的内容换行，只有“<%=new Date()%>”一行被替换成当前的时间。

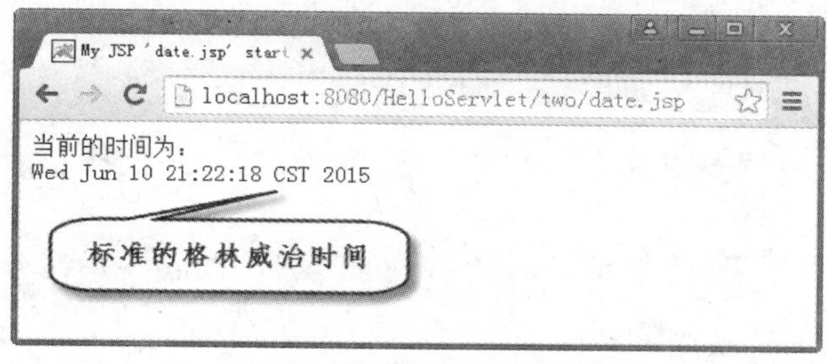

图 3.3　JSP 表达式运行结果

以下是上面 JSP 页面生成的源代码，与 JSP 页面代码对比可以发现，只有首行和第 11 行不同，其他的代码都一样。

```
<!DOCTYPE HTML PUBLIC "-//W3C//DTD HTML 4.01 Transitional//EN">
<html>
  <head>
    <meta content="text/html;charset=utf-8"
http-equiv="Content-Type">
    <title>JSP 表达式例子</title>
  </head>

  <body>
                 当前的时间为：</br>
      Wed Jun 10 21:22:18 CST 2015
  </body>
</html>
```

通过对比发现，JSP 页面中的 HTML 元素在源代码中被原样保留，只有 JSP 代码会发生改变。因此， JSP 页面中的静态代码都是用 HTML 模板来写。

查看 JSP 页面生成的源代码，一般是通过浏览器来查看。对于 IE 浏览器或者是 IE 内核的浏览器，可以在待查看的页面上右键单击，然后在弹出的菜单中选择“查看源文件”命令。

3.3 JSP 指令

上一节介绍了 JSP 表达式的用法以及示例，读者已经可以应用前面章节的知识编写简单的 JSP 动态网页了，但要编写出功能强大、复杂的 JSP 页面还需要学习更多的内容。本节介绍 JSP 的另一元素：指令。JSP 指令包括：page 指令、include 指令、taglib 指令。

3.3.1 与页面属性相关的 page 指令

page 指令用来设置 JSP 页面的属性和相关功能，基本语法形式有两种，如下：

```
<%@ page attribute1=”value1” […attribute2=”value n”]%>
```

或者

```
<jsp:directive.page attribute1=”value1” […attribute2=”value n”] />
```

第 2 种形式的表达现在很少使用，大多使用第 1 种表达形式。

page 指令有多种属性，但使用最多的是 language、import、pageEncoding 这 3 个，其中 language 是必须设置的，目前 JSP 页面使用 Java 语言，所以其默认值是 java。Import 用来声明需要导入的包，pageEncoding 属性设置页面的编码。在所有的属性中除 import 可以声明多个外，其他的属性都只能出现一次。page 其他的属性见表 3.1。

表 3.1 page 指令的常用属性

属性和属性值	说 明
session=”true|false”	限定 session 对象是否可用，默认 true
autoFlush=”true|false”	指明缓冲区域是否自动清除，默认 true
info=”text”	描述该 JSP 页面的相关信息
errorPage=”URL”	当页面产生异常时，跳转的路径
isErrorPage=”true|false”	指定该 JSP 页面是否为处理异常错误的页面，当设定为 true，才能使用 exception 对象，默认 false
isThreadSafe=”true|false”	是否允许多线程使用，默认 true
buffer=”8kb”	输出流是否有缓冲区，默认 8KB
contentType="text/html; charset=UTF-8"	设定 MIME 类型和编码属性, 编码属性一般设置为 UTF-8, MIME 类型还有很多, application/vnd.ms-excel 表示 Excel 电子表格，image/gif 表示 GIF 图像等等
Extends=”class”	指明由该 JSP 页面产生的 Servlet 所继承的父类

3.3.2 引入文件的 include 指令

include 指令是在 JSP 页面生成 Servlet 时引入需要包含的页文件，可以是 HTML 文件也可以是 JSP 文件，还可以是其他文件（例如 js 文件）。include 指令的作用是在标签插入的位置插入静态的文件内容，使其与 JSP 文件组合成新的 JSP 页面，然后由 JSP 引擎翻译成 Servlet 文件。这样做有两个优点：

页面的代码可以复用，因为被引入的文件是静态文件，所以在其他的 JSP 页面中要是需要也可以导入；

JSP 页面的代码结构显得清晰易懂，维护也比较简单。

include 指令的基本语法如下：

```
<%@include file="URL"%>
```

file 属性是指向要包含的文件，一定要注意引入的路径是否正确，一旦路径出错在编译的时候将不能通过。

include 指令经常用来包含网站中经常出现的相同页面。例如，一般网站为每个页面都设置导航栏，把它放在页面的顶端或者左边，这部分的代码在每个页面都重复。那么就可以用 include 来解决这类问题，它能够为开发者省去这些重复的动作。下面的例子将对 include 指令进行演示。

【例 3.3】演示 include 指令。

index.jsp 是首页，其源代码如下：

```
----------------- index1.jsp ----------------
<%@ page language="java" import="java.util.*" pageEncoding="UTF-8"%>
<!DOCTYPE HTML PUBLIC "-//W3C//DTD HTML 4.01 Transitional//EN">
<html>
    <head>
      <meta content="text/html;charset=UTF-8" http-equiv="Content-Type">
      <title>JSP include 指令演示</title>
    </head>
    <body>
        <%@include file="john.html" %>
        <br/>
        <div align="center">JSP include 指令演示</div>
        <%@include file="copyRight.jsp" %>
    </body>
  </html>
```

John.html 是内嵌页面，用来显示内容，其源代码如下：

```
----------------- john.html ----------------
<table align="center" border="1" width="600">
    <tr>
```

```
            <td align="center" height="18" width="100%">
                <span>This is top page.</span>
            </td>
        </tr>
    </table>
```

copyRight.jsp 是显示版权信息的页面，其源代码如下：

```
----------------- copyRight.jsp ----------------
<%@ page language="java" import="java.util.*,java.text.SimpleDateFormat"
         pageEncoding="UTF-8"%>
<%
    Date d = new Date();
    SimpleDateFormat sdf = new SimpleDateFormat("yyyy");
    String t = sdf.format(d);
    String copyRightsMess = "John 版权所有 2010-"+t;
%>
<br/>
<div align="center" ><%=copyRightsMess%></div>
```

本例页面效果如图 3.4 所示。

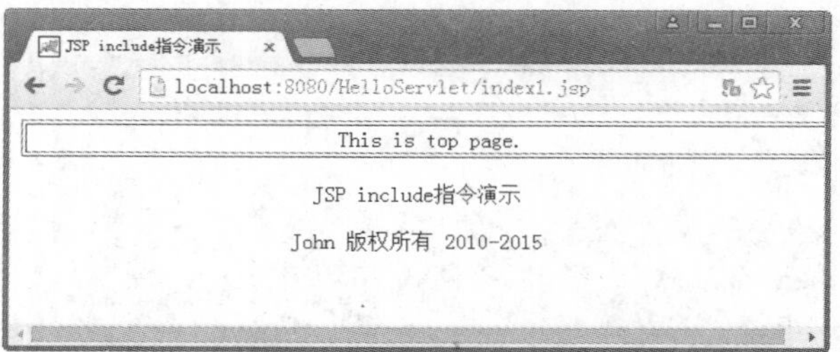

图 3.4 index.jsp 页面的运行结果

include 指令在 JSP 页面被转换成 Servlet 时才将文件导入，这与<jsp:include>动作不同。

3.3.3 与标签相关的 taglib 指令

taglib 指令（又名标签指令）是 JSP 新增的一个指令，用户可以自定义新的标签在页面中执行。taglib 指令的语法如下：

```
<%taglib  uri="tagliburl" prefix="tagPre" %>
```

其中 uri 属性用来表示自定义标签库的存放位置。prefix 属性是为了区分不同标签库的标签名，在页面中引用标签也是以 prefix 开头的。

现在比较流行的 JSTL、EL 标签，看下它们如何在 JSP 中使用。下面的例子是用 JSTL 标签输出 1~10。

JSTL 标签库会在后续的章节详细说明，这里先演示它的基本用法。

【例 3.4】演示 taglib 指令。

```
----------------- index.jsp ----------------
<%@ page language="java" import="java.util.*" pageEncoding="UTF-8"%>
<%@ taglib uri="http://java.sun.com/jstl/core_rt" prefix="c"%>
<!DOCTYPE HTML PUBLIC "-//W3C//DTD HTML 4.01 Transitional//EN">
<html>
  <head>
    <meta content="text/html;charset=UTF-8" http-equiv="Content-Type">
    <title>JSP taglib 指令演示</title>
  </head>
  <body>
      <table>
         <tr>
            <td>输出值</td>
         </tr>
         <c:forEach begin="1" end="10" var="num">
            <tr>
               <td><c:out value="${num}"></c:out></td>
            </tr>
         </c:forEach>
      </table>
  </body>
</html>
```

从上述代码可以看出，JSTL 标签使得 JSP 页面十分的简洁，它不需要定义或初始化对象、方法。但凡事都不能只看表面，应该注重本质，标签的制定在页面显示是如此的简单，但在编制时是个复杂的过程，它通过一定的编程步骤将 JSP 代码和 Java 代码联系起来。标签制定的最大好处就是使得开发者的职责分工更加地明细，标签制定者无需关注业务逻辑的实现，而页面编程人员直接使用标签即可。这样两者就不会冲突，分工明确。下面简单介绍定制标签的过程。

1. 标签库

标签库用于定义标签的 XML 文件。该文件中包含了标签的属性、和标签的名称、JSP 在处理标签时所需的类文件等信息。在使用自定义标签时，必须指明标签库所在目录；在执行标签功能作用时，通过定义标签属性来决定显示的内容。下面看一段标签的源代码，其中包括了标签名的定义、属性定义以及标签对应的 Java 类。

```
<?xml version="1.0" encoding="UTF-8" ?>
<!DOCTYPE taglib
  PUBLIC "-//Sun Microsystems, Inc.//DTD JSP Tag Library 1.2//EN"
```

```
04      "http://java.sun.com/dtd/web-jsptaglibrary_1_3.dtd">
05  <taglib>
06      <tlib-version>1.0</tlib-version>
07      <jsp-version>1.2</jsp-version>
08      <short-name>lms</short-name>
09      <!--  页面引用的 url 地址 -->
10      <uri>/lms-tags</uri>
11
12      <tag>
13          <!--  标签名称 -->
14          <name>page</name>
15          <!--  标签处理类 -->
16          <tag-class>com.common.model.PageTag</tag-class>
17          <body-content>JSP</body-content>
18          <!--  标签描述 -->
19          <description>分页</description>
20          <!--  标签属性 -->
21          <attribute>
22              <!--  标签属性名称 -->
23              <name>object</name>
24              <!--  标签属性是否为必需，true 是必需，false 不是必需 -->
25              <required>true</required>
26              <!--  标签属性对应的处理类 -->
27              <type>com.common.model.PageObject</type>
28              <!--  标签属性描述 -->
29              <description>PageObject 对象</description>
30          </attribute>
31      </tag>
32
33      <tag>
34          <!--  标签名称 -->
35          <name>substring</name>
36          <!--  标签处理类 -->
37          <tag-class>com.gsta.common.tag.SubStringTag</tag-class>
38          <description>substring</description>
39          <attribute>
40              <name>content</name>
41              <required>true</required>
42              <type>java.lang.String</type>
43          </attribute>
44      </tag>
45  </taglib>
```

上述代码中第 6~8 行分别定义标签的版本、JSP 的版本、标签的简称，代码第 12~31 行定义 page 标签，代码第 21~30 行定义 page 标签的属性，在定义属性名称的时候，可以定义该标签的数据类型以及该属性是否是必需的。例如上述代码第 41 行定义的，JSP 页面就会根据其定义在执行的时候检查相关的错误信息。

2. 标签处理类

标签处理类其实就是一个 Java 类，该 Java 类实现标签接口（javax.Servlet.jsp.Tag），当自定义标签被 JSP 处理时即被执行。每个标签中都需要执行的方法是：

```
Public int doStartTag() throws JspException
```

该方法定义了标签被执行时的方法。在标签中定义的属性，那么在 Java 类中就必须有这些属性并且设定其 get/set 方法。例如，在上面标签源码中与 substring 标签对应的标签 Java 类是一个用于返回结果的 Java 类 在 substring 标签定义时, 必须定义其 content 属性并给出 get/set 方法，如下代码所示：

```
private String content;
public void setContent(String content) {
    this.content = content;
}

public void setLength(int length) {
    this.length = length;
}
```

上述代码中，给出了属性 content 的 get/set 方法。

3.4 JSP 动作

上一节介绍了 JSP 的 page、include、taglib 这 3 个指令，这些指令不仅可以帮助 JSP 编程人员编写出复杂的 JSP 页面，而且能简化 JSP 页面代码，为后期维护带来便利。本节介绍 JSP 中另外一组元素：JSP 动作。在 JSP 中一共有 13 个动作：<jsp:include>、<jsp:forward>、<jsp:plugin>、<jsp:param>、<jsp:useBean>、<jsp:getProperty>、<jsp:setProperty>、<jsp:output>、<jsp:attribute>、<jsp:element>、<jsp:body>、<jsp:params>、<jsp:fallback>。其中<jsp:include>和<jsp:forward>用的比较多，<jsp:useBean>、<jsp:getProperty>和<jsp:setProperty>与 JavaBean 相互结合，将在第 8 章详细介绍。本节先介绍其他的 JSP 动作。

3.4.1 <jsp:include>动作

<jsp:include>动作与 include 指令十分相似，它们的作用都是引入文件到目标页面。<jsp:include>语法格式如下：

```
<jsp:include page="relative URL" flush="true"/>
```

例如：

```
<jsp:include page="top.html" flush="true"/>
```

其中 page 属性是<jsp:include>动作的一个必选属性，它指明了需要包含文件的路径，该路径可以是相对的也可以是绝对的。flush 属性用于指定输出缓存是否转移到被导入文件中。如果指定为 true，则包含在被导入文件中；如果指定为 false，则包含在原文件中。

page 属性所指的绝对路径不是文件系统的绝对路径，而是以应用系统目录为根目录的绝对路径。通常 Web 应用下的 WEB-INF 目录就是根目录。一般在设定 URL 时指定相对路径，这样简单些。

使用<jsp:include>动作与 include 指令可以达到相同的界面效果。例如上节中介绍的使用 include 指令演示界面页脚的例子，直接将 include 替换成<jsp:include>动作，它们在页面显示结果中没有什么不同。

使用 jsp:include 包含页脚信息的代码如下：

```
----------------- index_include.jsp -----------------
01  <%@ page language="java" import="java.util.*" pageEncoding="UTF-8"%>
02  <!DOCTYPE HTML PUBLIC "-//W3C//DTD HTML 4.01 Transitional//EN">
03  <html>
04      <head>
05        <meta content="text/html;charset=UTF-8" http-equiv="Content-Type">
06        <title>JSP include 动作演示</title>
07      </head>
08      <body>
09          <jsp:include page="john.html" />
10          <br/>
11          <div align="center">JSP include 指令演示</div>
12          <jsp:include  page="copyRight.jsp"  flush="true"/>
13      </body>
14  </html>
```

上述示例中，代码第 9 行换成 jsp:include 动作。可以看出，两种不同的包含方式达到了相同的效果。

但是，jsp:include 动作与 include 指令还是有些不同的。首先，jsp:include 动作是在页面被访问时导入的，而 include 指令是有 JSP 引擎编译时导入的；其次，include 指令中，被包含的文件会同主页面一块被编译为一个 Servlet 类文件，而 jsp:include 动作包含的文件和主页面会是相对独立的两个文件，在编译时会被编译成两个 Servlet 类文件。因此 jsp:include 在效率上稍微慢些。

【例 3.5】演示 jsp:include 动作。

```
----------------- news.jsp -----------------
01  <%@ page language="java" import="java.util.*" pageEncoding="UTF-8"%>
02   <!DOCTYPE HTML PUBLIC "-//W3C//DTD HTML 4.01 Transitional//EN">
03  <html>
04    <head>
05      <title>新闻摘要</title>
06       <meta http-equiv="pragma" content="no-cache">
07       <meta http-equiv="cache-control" content="no-cache">
08       <meta http-equiv="expires" content="0">
09       <meta http-equiv="keywords" content="keyword1,keyword2,keyword3">
10       <meta http-equiv="description" content="This is my page">
11    </head>
12    <body>
13        <center>
14            <table>
15                <tr>
16                    <th>中华新闻网</th>
17                </tr>
18            </table>
19        </center>
20        <p>最新新闻摘要：
21           <ul>
22               <li><jsp:include page="news1.html" /></li>
23               <li><jsp:include page="news3.html"/></li>
24               <li><jsp:include page="news3.html"/></li>
25           </ul>
26        </p>
27    </body>
28   </html>
```

上述代码中，代码第 22~24 行使用 JSP 动作，页面运行效果如图 3.5 所示。

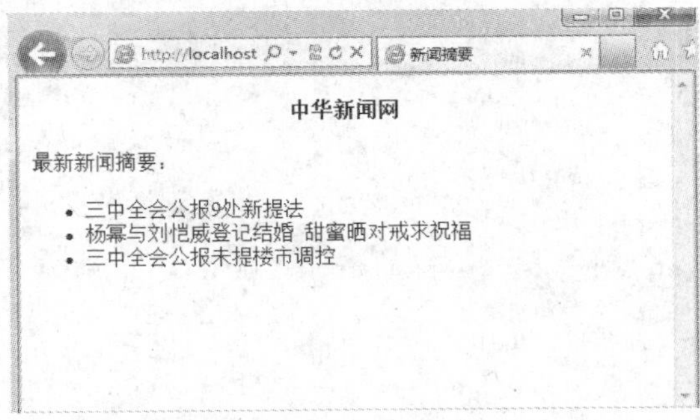

图 3.5　jsp:include 运行结果

3.4.2 <jsp:forward>动作

<jsp:forward>动作的作用是转发请求到另外一个页面中，在请求过程中会连同请求的参数数据一起被转发到目标页面中，目标页面通过 request.getParameter 方法获得参数值进行进一步处理。<jsp:forward>基本语法如下：

```
<jsp:forward page=”relative URL”>
```

<jsp:forward>只有 page 一个属性，其值为相对的 URL。例如：

```
<jsp:forward page=”/error.jsp”/>
```

将跳转到错误页面。

来看一个<jsp:forward>的例子。如果随机数能被 2 整除，则跳转到 even.jsp，否则跳转到 odd.jsp。

【例 3.6】演示 jsp:forward 动作。

```
----------------------- forwardE.jsp--------------------------
<%@ page language="java" import="java.util.*" pageEncoding="utf-8"%>
<%
    String url = "";
    int random = (int)(Math.random()*10);//产生10以内的随机数
    int m = random%2;
    switch(m){
        case 0:
            url = "even.jsp";
            break;
        case 1:
            url = "odd.jsp";
            break;
    }
 %>
<!DOCTYPE HTML PUBLIC "-//W3C//DTD HTML 4.01 Transitional//EN">
<html>
  <head>
    <title>My JSP 'forwardE.jsp' starting page</title>
     <meta http-equiv="pragma" content="no-cache">
     <meta http-equiv="cache-control" content="no-cache">
     <meta http-equiv="expires" content="0">
     <meta http-equiv="keywords" content="keyword1,keyword2,keyword3">
     <meta http-equiv="description" content="This is my page">
  </head>

  <body>
      <jsp:forward page="<%=url %>"/>
  </body>
```

```
29    </html>
```

上述代码中，代码第 2~14 行使用 Java 判断 url 的返回值，其中 even.jsp 中的内容为“能被 2 整除，跳转到偶数页界面”，odd.jsp 中的内容为“不能被 2 整除，跳转到奇数页界面”，本例效果如图 3.6 所示。

从上述代码的运行结果看出，界面已经被重定向到 odd.jsp 页面，但是浏览器中的地址仍显示的是跳转前的地址。这也算是<jsp:forward>动作的一个特点，相对于请求者而言，所看出的响应仍然是原先请求的页面给出的，请求者并不会获得转发后的页面地址。这样相对来说，请求具有隐蔽性。

图 3.6 jsp:forward 运行结果

<jsp:forward>动作中的 url 页面只能是该 Web 应用中的文件，不能是该 Web 应用之外的文件。

3.4.3 <jsp:param>动作

<jsp:param>用来传递参数信息，它经常与其他动作一起使用，比如与<jsp:forward>、<jsp:include>等结合使用传递主页面的参数到目标页面。其基本语法如下：

```
<jsp:param name=" 参数名称" value=" 参数值">
```

例如：

```
<jsp:param name="username" value="李四">
```

【例 3.7】和【例 3.8】分别是<jsp:param>和<jsp:include>、<jsp:forward>结合使用的例子，由主页面和子页面构成。子页面从主页面中获得参数值显示。

【例 3.7】演示<jsp:param>和<jsp:include>结合使用。

```
---------------------- main.jsp---------------------------
<%@ page language="java" import="java.util.*" pageEncoding="utf-8"%>
<%request.setCharacterEncoding("utf-8"); //设定页面传递参数的编码格式%>
<!DOCTYPE HTML PUBLIC "-//W3C//DTD HTML 4.01 Transitional//EN">
<html>
  <head>
   <title>主页面</title>
  </head>

  <body>
```

```
10          <center>
11              <table>
12                  <tr>
13                      <th>主页面</th>
14                  </tr>
15              </table>
16          </center>
17          <jsp:include page="subPage.jsp" >
18              <jsp:param value="John" name="userName"/>
19              <jsp:param value="10086" name="passwd"/>
20              <jsp:param value="北京丰台" name="address"/>
21          </jsp:include>
22      </body>
23  </html>
```

上述代码中，第 17 行引用<jsp:include>动作，代码第 18~20 行使用<jsp:param>参数。

```
----------------------- subPage.jsp--------------------------
<%@ page language="java" import="java.util.*" pageEncoding="utf-8"%>
<%
    String userName= request.getParameter("userName");
    String passwd= request.getParameter("passwd");
    String address= request.getParameter("address");
    System.out.println(address);
 %>
<!DOCTYPE HTML PUBLIC "-//W3C//DTD HTML 4.01 Transitional//EN">
<html>
  <head>
    <title>子页面</title>
  </head>
  <body>
      <center>
          <table>
              <tr>
                  <th>子页面：人员信息</th>
              </tr>
          </table>
      </center>
      <center>
          <table>
              <tr>
                  <td>用户名：<%=userName%></td>
              </tr>
              <tr>
                  <td>密    码：<%=passwd%></td>
              </tr>
```

```
29              <tr>
30                  <td>用户地址：<%=address%></td>
31              </tr>
32          </table>
33      </center>
34   </body>
35  </html>
```

上述代码中，第 22~32 行获取主页面传递的参数值。

所有的页面编码格式要统一，否则在传递中文时会出现乱码问题。request.setCharacterEncoding 方法是设定传递参数的编码格式，在传递中文时要特别注意。在 URL 中包含参数传递时，可能会有中文乱码或者特殊字符问题，需要对这些参数进行编码处理。在 JavaScript（简称 JS）中，通过调用函数 encodeURI 或者 encodeURIComponent 实现编码，在 JSP 脚本中，通过方法 URLEncoder.encode 实现转码。

页面效果如图 3.7 所示。

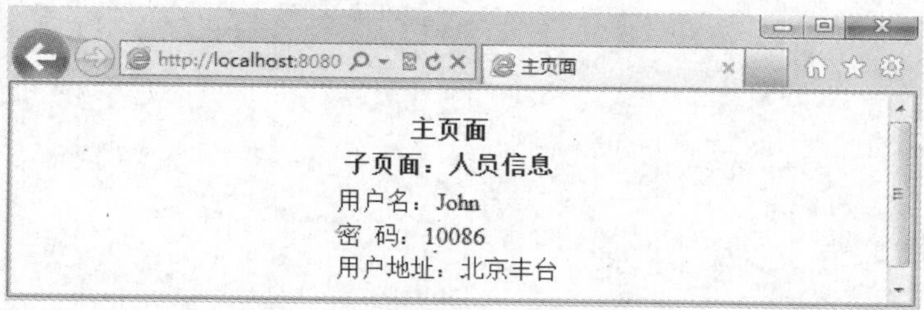

图 3.7　<jsp:param>和<jsp:include>结合使用运行结果

【例 3.8】演示<jsp:param>和<jsp:forward>结合使用。

```
---------------------- main.jsp--------------------------
<%@ page language="java" import="java.util.*" pageEncoding="utf-8"%>
<%request.setCharacterEncoding("utf-8"); //设定页面传递参数的编码格式%>
<!DOCTYPE HTML PUBLIC "-//W3C//DTD HTML 4.01 Transitional//EN">
<html>
  <head>
   <title>主页面</title>
  </head>

  <body>
      <center>
          <table>
              <tr>
                  <th>主页面</th>
              </tr>
          </table>
      </center>
      <jsp:forward page="subPage.jsp" >
          <jsp:param value="Smith" name="userName"/>
          <jsp:param value="10086" name="passwd"/>
```

```
20            <jsp:param value="北京丰台" name="address"/>
21        </jsp:forward>
22    </body>
23  </html>
```

上述代码中，第 2 行设置请求编码，第 17 行使用<jsp:forward>动作，第 18~20 行使用<jsp:param>参数

```
----------------------- subPage.jsp---------------------------
01  <%@ page language="java" import="java.util.*" pageEncoding="utf-8"%>
02  <%
03     String userName= request.getParameter("userName");
04     String passwd= request.getParameter("passwd");
05     String address= request.getParameter("address");
06     System.out.println(address);
07   %>
08  <!DOCTYPE HTML PUBLIC "-//W3C//DTD HTML 4.01 Transitional//EN">
09  <html>
10    <head>
11      <title>子页面</title>
12    </head>
13    <body>
14        <center>
15            <table>
16                <tr>
17                    <th>子页面：人员信息</th>
18                </tr>
19            </table>
20        </center>
21        <center>
22            <table>
23                <tr>
24                    <td>用户名：<%=userName%></td>
25                </tr>
26                <tr>
27                    <td>密    码：<%=passwd%></td>
28                </tr>
29                <tr>
30                    <td>用户地址：<%=address%></td>
31                </tr>
32            </table>
33        </center>
34    </body>
35  </html>
```

上述代码中，第 2~7 行代码获取主页面传递的参数值，第 22~30 行显示获取的参数值，页面效果如图 3.8 所示。

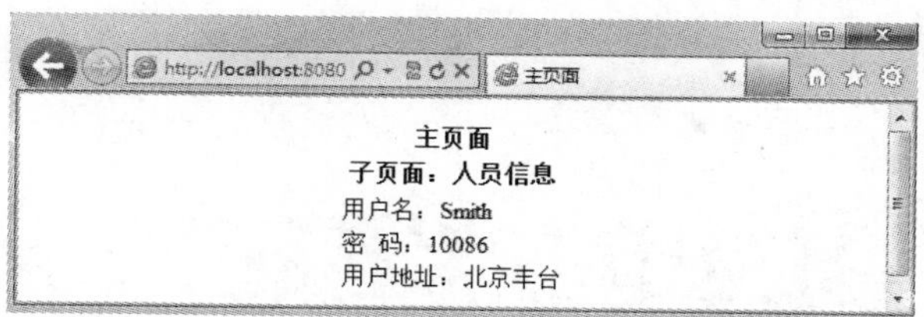

图 3.8　<jsp:param>与<jsp:forward>结合使用运行结果

从上述例子可以看出，<jsp:param>在<jsp:include>、<jsp:forward>中的用法基本一样，只是将<jsp:include>替换成<jsp:forward>动作就可以。

3.5 上机实践

请编写一个 JSP 页面，网页有上部、中间、下部 3 部分，请通过由 1 个主 JSP 页面包含 3 个页面的方式实现页面效果，要求用两种方式实现，效果如图 3.9 所示。

图 3.9　例图

第 4 章

◄ JSP的内置对象 ►

JSP 内置对象的含义是可以直接在 JSP 页面中使用的对象，使用前不需要声明它们。熟悉并了解 JSP 内置对象，可以方便读者更好地操作页面、更好地开发页面、完成更复杂的业务流程。本章要讲解 7 个内置对象 request、response、out、session、application、page、config 的作用和使用方法。

4.1 request 对象

本节介绍 request 对象的作用范围及其常用的方法。用户每访问一个页面，就会产生一个 HTTP 请求。这些请求中一般都包含了请求所需的参数值或者信息，如果将 request 对象看作是客户请求的一个实例，那么这个实例就包含了客户请求的所有数据。因此，可以通过 request 来获取客户端和服务器端的信息，如 IP 地址、传递的参数名和参数值、应用系统名、服务器主机名等等。

4.1.1 request 对象的常用方法

request 对象的常用方法参见表 4.1。

表 4.1 request 对象常用方法

方法	方法说明
getParameter()方法	该方法是取得请求中指定的参数值，它返回的是 String 类型，如果有必要需要将取得的参数值转换为合适的类型
getParameterValues()方法	该方法是将同名称的参数一次性地读入到 String 类型的数组中
getParameterNames()方法	获取参数名称，它返回的是枚举类型
getMethod()方法	获取客户提交信息的方式，是 post 还是 get
getServletPath()方法	获取 JSP 页面文件的目录
getHeader()方法	获取 HTTP 头文件中指定值，例如：accept、user-agent、content-type、content-length 等
getRemoteAddr()方法	获取客户的 IP 地址

（续表）

方法	方法说明
getServerName()方法	获取服务器的名称
getServerPort()方法	获取服务器的端口号
getContextPath()方法	获取项目名称，如果项目为根目录，则得到空的字符串
getHeaders()方法	获取表头信息，它返回的是枚举类型

4.1.2 使用 request 对象接收请求参数

request 对象有两种方法获得请求参数值，一个是 getParameter()，另一个是 getParameterValues()。下面演示使用 getParameter()方法获取页面请求参数。

【例 4.1】获取网页请求参数

getParameter.jsp 是获取页面参数值，代码如下：

```
------------------ getParameter.jsp ----------------
01  <%@ page language="java" import="java.util.*" pageEncoding="UTF-8"%>
02  <!DOCTYPE HTML PUBLIC "-//W3C//DTD HTML 4.01 Transitional//EN">
03  <html>
04    <head>
05      <title>' getParameter.jsp'</title>
06       <meta http-equiv="Content-Type" content="text/html;charset=utf-8">
07    </head>
08    <body >
09        <center>
10            <%
11               String name= request.getParameter("name");
12               String city = request.getParameter("city");
13               if(name!=null&&city!=null)
14               {
15             %>
16               <p>Welcome  <%=name %>,您所在的城市是<%=city %></p>
17              <%
18                }else{
19               %>
20               <p>欢迎访问本页面！</p>
21               <%
22                 }
23                %>
24        </center>
25    </body>
26  </html>
```

上述代码中，第 11、12 行代码分别获得参数 name、city 的参数值。

在访问地址上输入 http://localhost:8080/ch04/getParameter.jsp?name=John&city=Beijing 后

页面效果如图 4.1 所示。

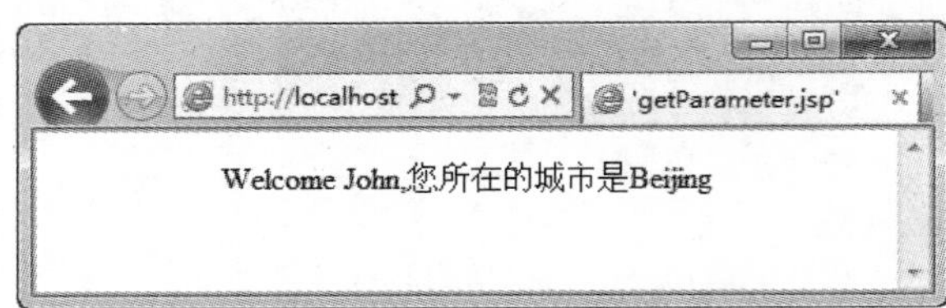

图 4.1　例 4.1 的运行结果

在 URL 参数传递时，在页面地址后使用“？”连接请求参数，参数由“=”相连表示参数名和参数值，一个请求携带多个参数用“&”连接。

实际上，上述传递参数的方法在实际的开发中相比表单提交参数要少用，因为这种提交参数的方法完全将参数暴露出来，不利于隐私保护，所以开发中一般是采取表单提交的方式传递参数。

在表单 form 中，有两个非常重要的属性 action 和 method。属性 action 指明表单提交后的数据跳转到指定页面并处理参数。method 有两个值，分别是 get 和 post，通常指定为 post，因为用 get 时，在表单中设定的参数和参数值将附加到页面地址的末尾以参数的形式提交，这样会暴露参数值，不安全；而指定为 post，提交表单时，表单中的参数将被作为请求头部中的信息发送，不会在目标地址中添加任何参数，这样也就不会暴露出参数值。所有出于安全考虑，一般会选择 post 提交。

【例 4.2】表单 post 提交获取网页请求参数

【程序结构】

ex4_1.jsp 页面将用户填写的内容提交给 getParameter.jsp。页面关系如图 4.2 所示。

图 4.2　例 4.2 的页面关系图

ex4_1.jsp 是用户提交页面，源代码如下：

```
------------------ ex4_1.jsp -----------------
<%@ page language="java" import="java.util.*" pageEncoding="UTF-8"%>
<!DOCTYPE HTML PUBLIC "-//W3C//DTD HTML 4.01 Transitional//EN">
<html>
  <head>
    <title>'ex4_1.jsp'</title>
     <meta http-equiv="Content-Type" content="text/html;charset=utf-8">
  </head>

```

```
09    <body >
10        <center>
11            <form action="getParameter.jsp" method="post">
12                <table>
13                    <tr>
14                        <td>姓名</td>
15                        <td><input type="text" name="name" value=""/></td>
16                    </tr>
17                    <tr>
18                        <td>城市</td>
19                        <td><input type="text" name="city" value=""/></td>
20                    </tr>
21                    <tr>
22                        <td><input type="submit" value="提交"/></td>
23                        <td><input type="reset" value="重置"/></td>
24                    </tr>
25                </table>
26            </form>
27        </center>
28    </body>
29  </html>
```

上述代码中，代码第 11~26 行用表单 form 提交，其中 action 为 getParameter.jsp，methode 方法为 post，代码第 12~25 行用输入框装载姓名和城市。getParameter.jsp 与【例 4.1】中的 getParameter.jsp 代码一样。

在浏览器中访问 ex4_1.jsp 并在“姓名”和“城市”输入框中输入值“John”和“Beijing”，单击“提交”按钮后，可以看到显示的结果与例 4.1 的结果是一样的。

在测试时，读者可以设定 method 的两个值并分别进行测试，查看地址栏的变化情况。

4.1.3　请求中文乱码的处理

在上述例子中，请求信息中的参数值都是用英文，但在实际开发中，请求信息通常是包含中文的。但在上述的例子中，要是使用中文，例如在表单的姓名输入框中输入中文字符，在提交之后显示的值不是中文而是“？”或者其他字符，如图 4.3 所示。

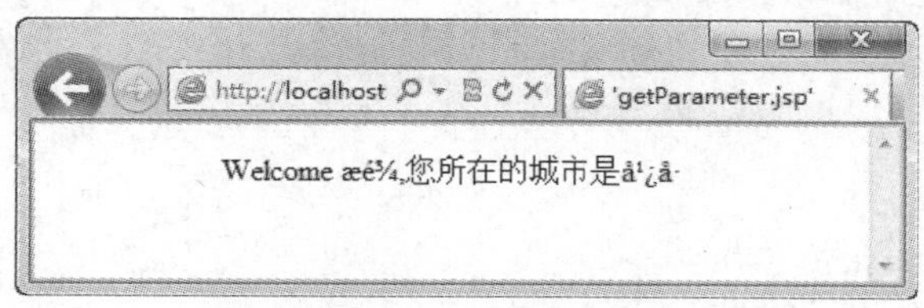

图 4.3　显示奇怪的字符

产生这种原因是由于请求信息所使用的字符集与页面使用的字符集的不同，有 3 种方法可以解决这个问题。

（1）一种方法是在接受请求的页面中规定请求字符编码的代码，例如在例子 4.2 的 getParameter.jsp 页面中添加语句“request.setCharacterEncoding(“utf-8”)”即可。

（2）另一种方法是在取得参数值后，通过转码的方式将参数值转为合适的字符集。例如，将前面的例子中获得参数的代码修改为如下形式：

```
   String name= new
String(request.getParameter("name").getBytes("ISO-8859-1"),"utf-8")
   String city= new
String(request.getParameter("city").getBytes("ISO-8859-1"),"utf-8");
```

通过上述两种方法都可以解决中文乱码问题，但是在代码维护方便带来很大麻烦，因为它不仅增加了代码量，而且移植性也差。

（3）第 3 种方法，通过编写一个 Servlet 过滤器来解决中文乱码问题，并可以通过配置过滤器解决所有请求处理字符集的问题，请求处理页面就不用关心字符集的处理了。第 3 种方法可以有效的防止上述的问题且移植性强。如何编写过滤器将在第 5 章中详细介绍。

4.1.4 获取请求的头部信息

请求的头部信息在实际中也是有其重要作用的，这些信息有时候对服务器的响应特别有用，而且也能查看到服务器中的相关信息。

请求头部信息的相关方法除了 4.1.1 节提到的 getHeaders()外，还有以下方法：

- String getHeader(String name)：获取字符串型的表头信息
- int getIntHeader(String name)：获取整型的表头信息

【例 4.3】获取请求中所有的头名称和对应的值。

getHeaders.jsp 页面用于获取请求中的所有头名称和对应的值，其源代码如下：

```
----------------- getHeaders.jsp -----------------
<%@ page language="java" import="java.util.*" pageEncoding="UTF-8"%>
<!DOCTYPE HTML PUBLIC "-//W3C//DTD HTML 4.01 Transitional//EN">
<html>
  <head>
    <title>'getHeaders.jsp'</title>
     <meta http-equiv="Content-Type" content="text/html;charset=utf-8">
  </head>

  <body >
          <%
              Enumeration<String> enumeration = request.getHeaderNames();
              while(enumeration.hasMoreElements()){
                 String name = enumeration.nextElement();
```

```
14                      String value = request.getHeader(name);
15                      if(value==null||"".equals(value)){
16                          value="空字符串";
17                      }
18              %>
19                  <p>表头名称：<%=name %>  对应的值：<%=value %></p>
20              <%} %>
21    </body>
22  </html>
```

上述代码中，代码第 11~17 行获取请求的 Header 信息，并循环遍历输出。在浏览器中直接输入页面地址，效果如图 4.4 所示。这些头信息的含义说明如下：

- accept：客户端能接收的 MIME 类型
- accept-language：浏览器的首选语言
- user-agent：客户端程序的相关信息，例如：浏览器版本，操作系统类型等
- host：表示服务器的主机名和端口号
- ua-cpu：CPU 类型
- connection：判断客户端是否可以持续性的连接 HTTP
- accept-encoding：指明客户端能够处理的编码类型有哪些
- cookie：会话的信息，每个会话都有一个会话 ID 或者其他信息

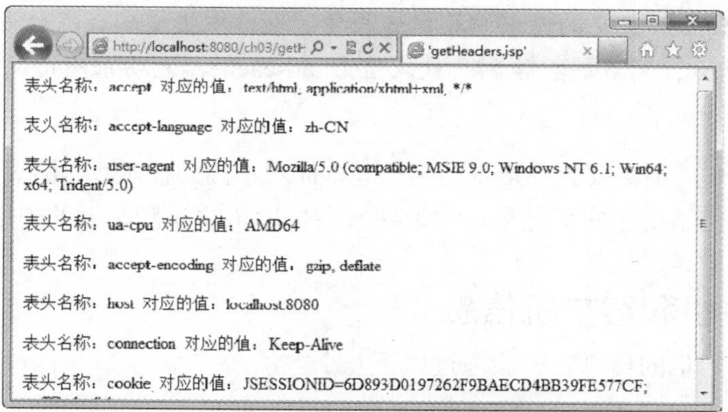

图 4.4　网页的请求头信息

以上是通过直接在地址栏中输入页面地址获得到的头文件信息。如果通过 post 方式跳转的页面，那么头文件信息会稍微多些。例如，在一表单中将 method 的属性指定为 post，然后 action 指定为例 4.3 的跳转页面，则显示结果如图 4.5 所示。

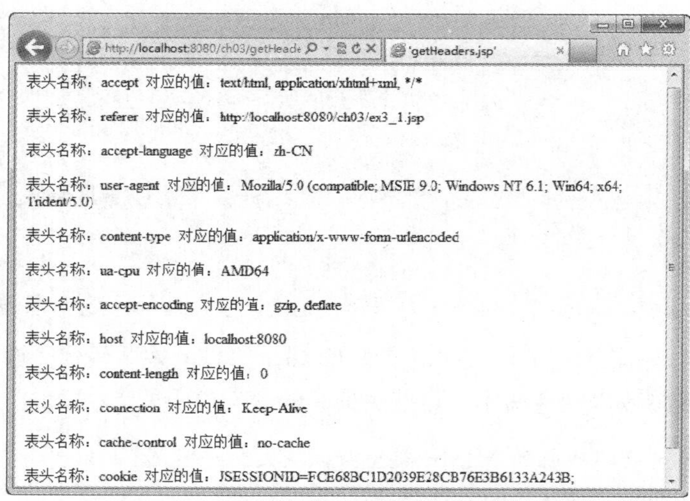

图 4.5　post 方式获取的请求头信息

从结果对比看，增加了以下几个信息：

- referer：向页面发起请求的地址。
- content-type：指明表单的 MIME 编码类型，一般值为 text/plain，multipart/form-data，application/x-www-form-urlencoded。默认为 application/x-www-form-urlencoded，如果表单上传一个文件，则表单中 enctype 设置为 multipart/form-data。
- content-length：提交的请求中数据的长度。表单中提交的数据越多，长度就越大。
- cache-control：网页缓存控制。默认值为 no-cache，表明每次访问都从服务器中获取页面。

在一般的开发中，不需要关注请求的头信息，针对不同的需求，在网页中设置相关的属性值就可以。但当我们了解了它们之后，在解决某些问题上将起到关键的作用。

4.1.5　获取主机和客户机信息

获取主机和客户机的信息，一般通过以下几个方法：

- getRemoteAddr()：获取客户主机 IP
- getRemoteHost()：获取客户主机名称
- getLocalAddr()：获取本地主机 IP
- getLocalHost()：获取本地主机名称
- getServerName ()：获取服务器主机名称
- getServerPort()：获取服务器端口

【例 4.4】获取客户机和主机信息

getHostInfo.jsp 是获取客户机和本地主机信息的页面，其源代码如下：

```
------------------ getHostInfo.jsp -----------------
```

```
01  <%@ page language="java" import="java.util.*" pageEncoding="UTF-8"%>
02  <!DOCTYPE HTML PUBLIC "-//W3C//DTD HTML 4.01 Transitional//EN">
03  <html>
04    <head>
05      <title>'getHostInfo.jsp'</title>
06       <meta http-equiv="Content-Type" content="text/html;charset=utf-8">
07    </head>
08
09    <body >
10        <p>
11                本地机器 IP：<%=request.getLocalAddr() %><br>
12                本地机器名称：<%=request.getLocalName() %><br>
13                本地机器端口：<%=request.getLocalPort() %><br>
14          </p>
15          <p>
16                客户主机 IP：<%=request.getRemoteAddr() %><br>
17                客户主机名称：<%=request.getRemoteHost() %><br>
18                客户主机端口：<%=request.getRemotePort()%><br>
19          </p>
20          <p>
21                服务器 IP：<%=request.getServerName() %><br>
22                服务器端口：<%=request.getServerPort() %><br>
23          </p>
24      </body>
25  </html>
```

上述代码中，第 11~13 行获取本地机器信息，第 16~18 行获取客户主机信息，代码第 21~23 行获取服务器信息。页面效果如图 4.6 所示。

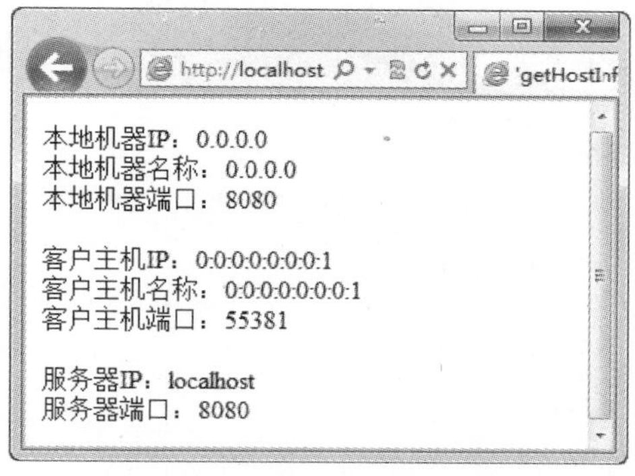

图 4.6　获取主机信息

一般而言，服务器主机名与本地机器 IP 应该是相同的，端口号也应一样。

4.2 response 对象

本节介绍 response 对象的常用方法。当用户访问一个页面时，就会产生一个 HTTP 请求，服务器做出响应时调用的是 response 响应包。response 响应包实现的是接口 javax.servlet.http.HttpServletResponse。

4.2.1 response 对象的常用方法

response 对象的常用方法参见表 4.2。

表 4.2 response 对象的常用方法

方法	方法说明
addHeader(String arg0,String arg1)方法	向页面中添加头和对应的值
addCookie(Cookie arg0)方法	添加 Cookie 信息
sendRedirect(String arg0)方法	实现页面重定向
setStatus(int arg0)方法	设定页面的响应状态代码
setContentType(String arg0)方法	设定页面的 MIME 类型和字符集
setCharacterEncoding(arg0)方法	设定页面响应的编码类型

4.2.2 设置头信息

设置头信息包括设置页面返回的 MIME 类型、返回的字符集、页面中的 meta 等信息。其中设置 MIME 类型和返回的字符集尤为重要，因为它关系到页面的显示是否会出现乱码。有两种方法设置 MIME 类型和返回的字符集，分别如下：

```
response.setContentType(String type)：其中 type 的值为“text/html;charset=utf-8”，当然可以为其他的 MIME 类型和字符集。
```

用 page 指令。基本固定格式为：

```
<%@ page contentType="text/html;charset=utf-8" language="java" %>
```

【例 4.4】 setContentType 的用法示例。

setContentType.jsp 用来设置 MIME 类型和字符集，MIME 设置为“text/html”，字符集设置为“UTF-8”，其源代码如下：

```
------------------setContentType.jsp----------------
<%
    response.setContentType("text/html;charset=UTF-8");//设置字符集和 MIME 类型
    String str = new String("这是测试例子".getBytes("ISO-8859-1"),"utf-8");
%>
<!DOCTYPE HTML PUBLIC "-//W3C//DTD HTML 4.01 Transitional//EN">
```

```
<html>
  <head>
    <title>setContentType.jsp</title>
  </head>

  <body>
      <p>这里是一段中文字符</p>
      </br>
      <%=str %>
  </body>
</html>
```

页面效果如图 4.7 所示。

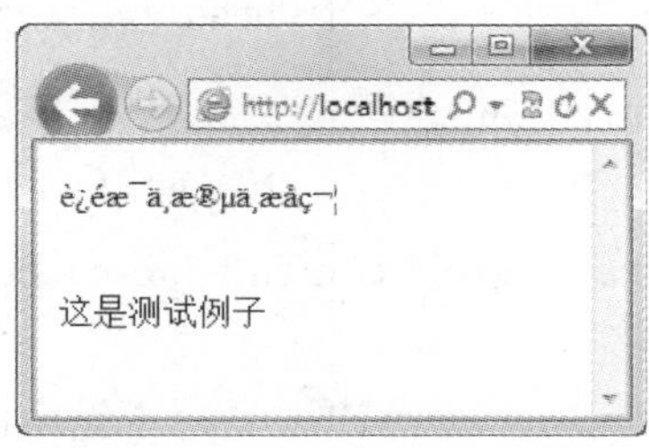

图 4.7　setContentType 运行示例

从运行结果可以看出，脚本中经过转码的文字显示正常，而在 HTML 中的字是乱码。

使用 page 指令设定字符集，在写法和处理上都相对简单些，HTML 中的中文字不需要转码，而且脚本中的中文字也不需要转码。

【例 4.5】 用 page 指令设置页面字符集

```
01  <%@ page language="java" import="java.util.*" pageEncoding="UTF-8"%>
02  <%
03      String str = "这是测试例子";
04   %>
05  <!DOCTYPE HTML PUBLIC "-//W3C//DTD HTML 4.01 Transitional//EN">
06  <html>
07    <head>
08      <title>page 指令 setContentType.jsp</title>
09    </head>
10
11    <body>
12        <p>这里是一段中文字符</p>
13        </br>
14        <%=str %>
15    </body>
16  </html>
```

上述代码中，第 1 行用 page 指令设置页面字符集，页面效果如图 4.8 所示。

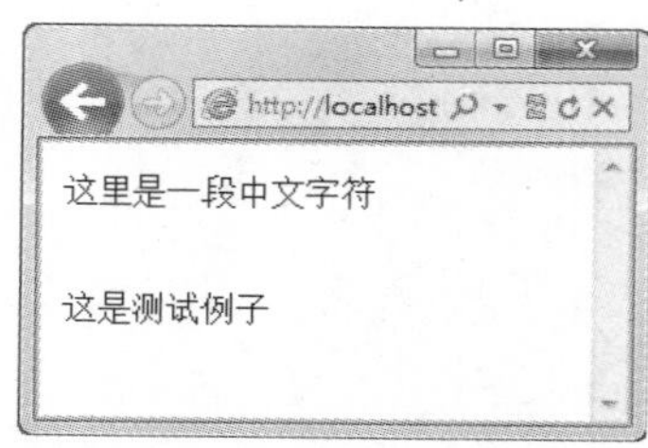

图 4.8　page 指令设置字符集

JSP 中页面设置字符集会破坏 JSP 容器自身的页面编码处理，所以不建议设置。但可以在 Servlet 中设定。用 page 指令设置字符集相对简单，所以在开发中一般都采用这种模式。

设置头 meta 信息是指 HTML 页面中存在于<head></head>之间的信息。例如：

（1）<meta http-equiv="pragma" content="no-cache">：设定禁止浏览器从本地机的缓存中调用页面内容，设定后离开网页就不能从 Cache 中再调用。

（2）<meta http-equiv="cache-control" content="no-cache">：请求和响应遵循的缓存机制策略。

（3）<meta http-equiv="expires" content="0"> ：用于设定网页的到期时间，一旦过期则必须到服务器上重新调用。

（4）<meta http-equiv="keywords" content="keyword1,keyword2,keyword3">：向搜索引擎说明网页的关键字。

（5）<meta http-equiv="description" content="This is my page">：向搜索引擎说明网页的主要内容。

【例 4.6】 设定 meta 头信息

页面 setMeta.jsp 用于设置 meta 中的信息，其源代码如下：

```
-----------------setMeta.jsp----------------
<%@ page language="java" import="java.util.*" pageEncoding="UTF-8"%>
<!DOCTYPE HTML PUBLIC "-//W3C//DTD HTML 4.01 Transitional//EN">
<html>
  <head>
    <title>My JSP 'setMeta.jsp' starting page</title>
  </head>
  <body>
     <center class="aa">
        <p class="bb">
           现在的时间为：<br/>
           <%
               out.print(""+ new Date());
               response.setHeader("refresh","1");
               response.setHeader("description","实时的显示当前时间");
```

```
15                 response.setHeader("keywords","实时,显示,显示当前时间");
16                 response.setHeader("cache-control","no-cache");
17              %>
18            <br/><br/>
19            copyright:2015
20        </center>
21     </body>
22   </html>
```

上述代码中，代码第 12~17 行设置 Header 信息，页面效果如图 4.9 所示。

图 4.9　设置 meta 头信息

从运行结果可以看出，在 response 中设置 meta 信息的效果，但实际开发中是直接在 HTML 标记中写定这些值，而不在响应中设定。

4.2.3　设置页面重定向

重定向是指一个页面在收到一个访问请求后，根据请求的 URL 重新跳转到其他的页面。设置重定向的方法是：

```
response.sendRedirect(String url)
```

其中 url 代表了跳转路径，路径可以是相对路径也可以是绝对路径。

【例 4.7】　设定页面重定向

sendRedirect.jsp 页面中设定了跳转到一个不存在的页面，其源代码如下：

```
------------------sendRedirect.jsp----------------
<%@ page language="java" import="java.util.*" pageEncoding="UTF-8"%>
<%
   response.sendRedirect("sendPageError.jsp");
 %>
<!DOCTYPE HTML PUBLIC "-//W3C//DTD HTML 4.01 Transitional//EN">
<html>
  <head>
    <title>My JSP 'sendRedirect.jsp' starting page</title>
  </head>
```

```

    <body>
      This is my JSP page. <br>
    </body>
  </html>
```

运行结果如图 4.10 所示。

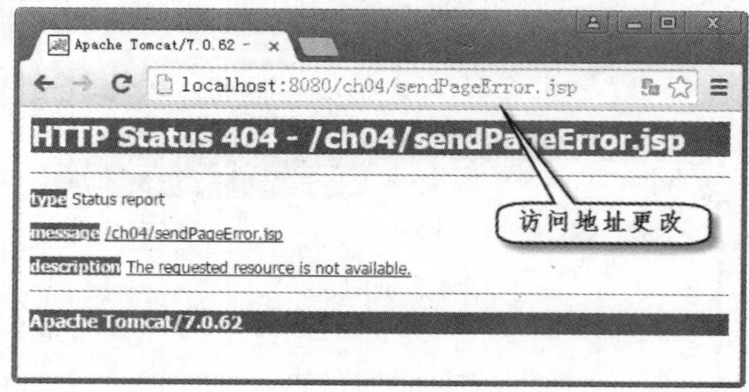

图 4.10　设置页面重定向

从运行结果可以看出跳转页面时，URL 地址改变了且不显示 sendRedirect.jsp 中的页面信息。重定向的运行过程如图 4.11 所示。

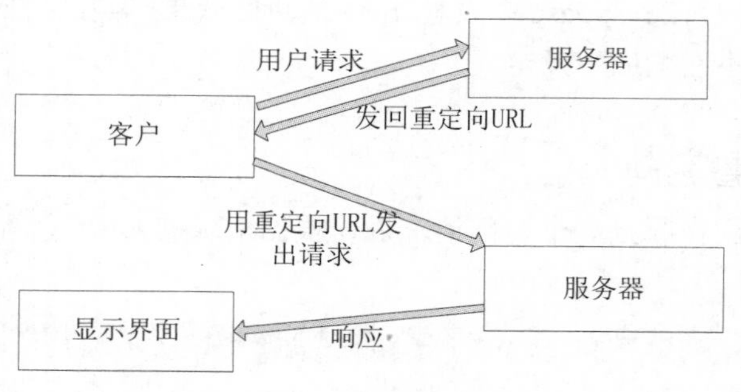

图 4.11　重定向运行过程

4.3 session 对象

在 Web 开发中，session 对象同样占据着极其重要的位置，在开发中它是一个非常重要的对象，可以用来判断是否是同一用户，还可以用来记录客户的连接信息等。HTTP 协议是一种无状态的协议（即不保存连接状态的协议），每次用户请求在接收到服务器的响应后，连接就关闭了，服务器端与客户端的连接就被断开。因此，当用户的浏览器没关闭，这个时候又发起请求，那么网站就应该识别出该用户的情况。这种情况下，session 对象就起到了关键作用。session 相关概念见表 4.3，常用方法见表 4.4。

表 4.3　session 相关概念

名称	说明
会话	当用户打开浏览器连接到一个 Web 应用时或者是某个界面，到关闭浏览器这个过程称为一个会话。其实一打开一个浏览器就意味着打开了一个会话对象
session 对象生命周期	当用户访问某个页面到关闭浏览器，这段时间称之为 session 对象的生命周期。也可以说是会话开始到结束这段时间称为 session 对象的生命周期
session 对象与线程	一个用户一个线程，这样保证了多个用户单击同一页面时的 session 对象唯一性
session 对象与 Cookie 对象	session 对象与 Cookie 对象是一一对应关系。JSP 引擎会将创建好的 session 对象存放在对应的 Cookie 中

表 4.4　session 对象常用方法

方法	说明
void setAttribute(String arg0,String arg1)	将参数名和参数值存放在 session 对象中
Object getAttribute(String arg0)	通过 arg0 中的参数获取参数值
Enumeration getAttribteName()	一个用户一个线程，这样保证了多个用户单击同一页面时的 session 对象唯一性
String getID()	获取 session 对象的 ID 值
void removeAttribute(String arg0)	移除指定 session 中的参数
long getCreateTime()	获取对象创建的时间，返回结果是 long 型的毫秒数
int getMaxInactiveInterval()	获取 session 对象的有效时间
Void setMaxInactiveInterval()	设置 session 对象的有效时间
Boolean isNew()	用于判断是否是一个新客户
Void invalidate()	使 session 对象不合法即失效

4.3.1　获取 session ID

获取 session 对象 ID 可以判断会话是否是同一会话，用户可以通过会话中的信息来进行相关的操作。

【例 4.8】　获取 session ID 值

【程序结构】创建两个 Web 目录并部署应用，页面之间通过超链接联系起来。

假设有 3 个页面：ex4_2.jsp，ex4_3.jsp，ex4_4.jsp。ex4_2.jsp 和 ex4_3.jsp 部署在同一应用下，ex4_4.jsp 在另外一个应用中。ex4_2.jsp 通过表单提交到 ex4_3.jsp，ex4_2.jsp 通过超链接指向 ex4_4.jsp。ex4_3.jsp 通过超链接指向 ex4_2.jsp，ex4_3.jsp 也通过超链接指向 ex4_4.jsp。ex4_4.jsp 通过表单提交到 ex4_2.jsp。它们之间的关系如图 4.12 所示。

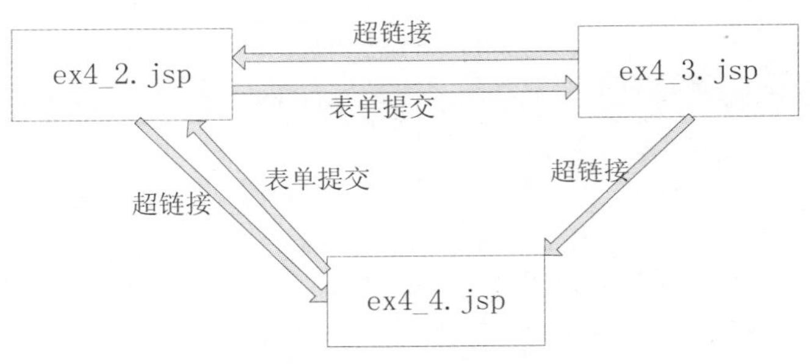

图 4.12　页面之间的关系图

ex4_2.jsp 发送请求和超链接到 ex4_3.jsp 和 ex4_4.jsp，其源代码如下：

```
-----------------ex4_2.jsp----------------
01  <%@ page language="java" import="java.util.*" pageEncoding="UTF-8"%>
02  <!DOCTYPE HTML PUBLIC "-//W3C//DTD HTML 4.01 Transitional//EN">
03  <html>
04    <head>
05      <title>My JSP 'ex4_2.jsp' starting page</title>
06       <meta http-equiv="pragma" content="no-cache">
07       <meta http-equiv="cache-control" content="no-cache">
08    </head>
09    <body>
10    <%
11            String sessionID = session.getId();
12            session.setAttribute("name","John");             //存参数 name
13            String author = (String)session.getAttribute("author");//强制转为String 类型
14            long time = session.getCreationTime();
15            Date date = new Date(time);
16     %>
17     <center>
18            <p>您访问的是 ex4_2.jsp 页面</br>
19              <%=author %>, 您的 session 对象 ID 为: </br>
20               <%=sessionID%></br>
21               session 对象创建时间: <%=date %>
22               </br>
23            </p>
24            <form action="ex4_4.jsp" method="post">
25                <input type="submit" value="转向 ex4_3.jsp"/>
26            </form>
27            <a href="../ch04/ex4_4.jsp">欢迎到 ex4_4.jsp 页面</a>
28     </center>
29    </body>
30  </html>
```

上述代码中，第 10~15 行获取 session 中 author 参数值，并设置当前日期，第 19~20 行显示获取的参数值，页面效果如图 4.13 所示。

图 4.13　ex4_2.jsp 运行结果

从图 4.13 中可以看出，页面输出了 session 对象的 ID 值和创建的时间，但是没有获得从页面 ex4_4.jsp 传来的参数值 author。

ex4_4.jsp 超链接到 ex4_4.jsp 和 ex4_4.jsp，其源代码如下：

```
-----------------ex4_3.jsp-----------------
<%@ page language="java" import="java.util.*" pageEncoding="UTF-8"%>
<!DOCTYPE HTML PUBLIC "-//W3C//DTD HTML 4.01 Transitional//EN">
<html>
  <head>
    <title>My JSP 'ex4_3.jsp' starting page</title>
     <meta http-equiv="pragma" content="no-cache">
     <meta http-equiv="cache-control" content="no-cache">
  </head>
  <body>
  <%
        String sessionID = session.getId();
        String name = (String)session.getAttribute("name");//强制转为String 类型
        long time = session.getCreationTime();
        Date date = new Date(time);
   %>
   <center>
        <p>您访问的是 ex4_3.jsp 页面</br>
           <%=name %>,您的 session 对象 ID 为：</br>
           <%=sessionID%></br>
           session 对象创建时间：<%=date %>
           </br>
        </p>
        <a href="ex4_2.jsp"><%=name %>,欢迎到 ex4_2.jsp 页面</a></br>
        <a href="../ch04/ex4_4.jsp"><%=name %>,欢迎到 ex4_4.jsp 页面</a>
```

```
    </center>
  </body>
</html>
```

运行结果如图 4.14 所示。从图中可以看出，页面 ex4_3.jsp 获得的 session 对象 ID 和创建时间与 ex4_2.jsp 中的 session 对象 ID 和创建时间是一样的，且获得到了参数 name 的值。

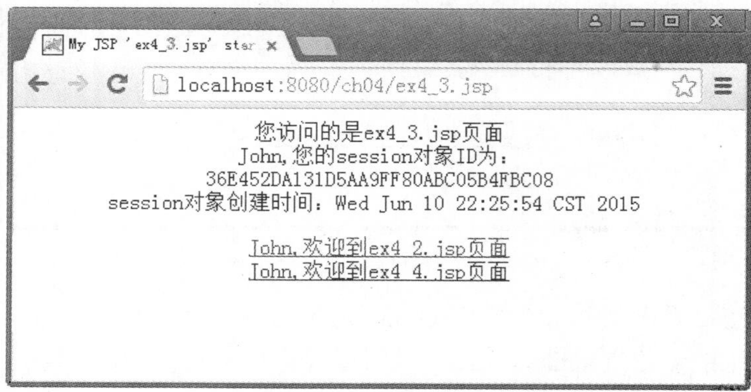

图 4.14　ex4_4.jsp 运行结果

ex4_4.jsp 发送请求和超链接到 ex4_2.jsp，其源代码如下：

```
------------------ex4_4.jsp-----------------
<%@ page language="java" import="java.util.*" pageEncoding="UTF-8"%>
<!DOCTYPE HTML PUBLIC "-//W3C//DTD HTML 4.01 Transitional//EN">
<html>
  <head>
    <title>My JSP 'ex4_4.jsp' starting page</title>
     <meta http-equiv="pragma" content="no-cache">
     <meta http-equiv="cache-control" content="no-cache">
  </head>
  <body>
  <%
         String sessionID = session.getId();
         String name = (String)session.getAttribute("name");//强制转为String类型
         session.setAttribute("author","Smith");            //存参数author
         long time = session.getCreationTime();
         Date date = new Date(time);
   %>
   <center>
         <p>您访问的是 ex4_4.jsp 页面</br>
            <%=name %>,您的 session 对象 ID 为：</br>
            <%=sessionID%></br>
            session 对象创建时间：<%=date %>
            </br>
```

```
23            </p>
24            <form action="../ch04/ex4_2.jsp" method="post">
25                <input type="submit" value="转向 ex4_2.jsp"/>
26            </form>
27        </center>
28      </body>
29    </html>
```

上述代码中，第 11~15 行获取 session 中的用户名参数，第 18~21 行显示参数值，页面效果如图 4.15 所示。

图 4.15　ex4_4.jsp 运行结果

从图 4.15 中可以看出，页面输出了 session 对象的 ID 值和创建时间，但是它和页面 ex4_2.jsp、ex4_3.jsp 页面的 session 对象的 ID 值不同，并且页面没有获得从 ex4_2.jsp 传来的参数值 name。

从以上的结果可以看出：一个 Web 应用的 session 对象的 ID 值是唯一的，并且两个应用之间的参数用 session 对象是获取不到值的。

读者务必要理解 session 对象的生命周期。

4.3.2　登录用户信息的保存

【例 4.8】登录用户信息的保存

【程序结构】login.jsp 是用户登录界面，validate.jsp 页面是验证用户合法性界面，class.jsp 页面是登录成功显示班级管理界面，logout.jsp 是退出登录界面。它们之间的关系如图 4.16 所示。

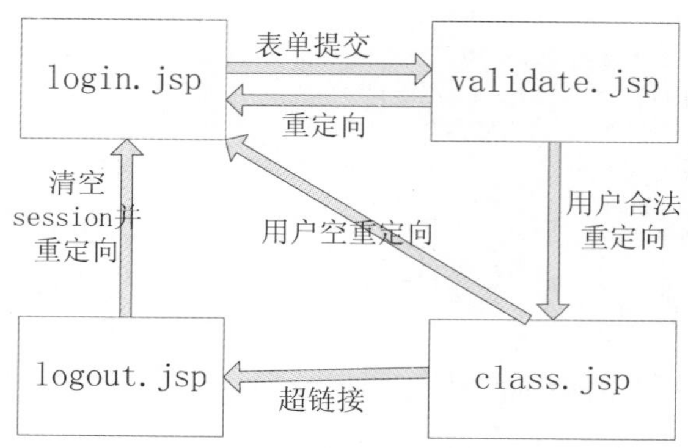

图 4.16　例 4.8 页面之间的关系

login.jsp 登录页面，其源代码如下：

```
-----------------login.jsp----------------
<%@ page language="java" import="java.util.*" pageEncoding="UTF-8"%>
<%
String path = request.getContextPath();
String basePath =
request.getScheme()+"://"+request.getServerName()+":"+request.getServerPort()+path+"/";
%>
<!DOCTYPE HTML PUBLIC "-//W3C//DTD HTML 4.01 Transitional//EN">
<html>
  <head>
    <base href="<%=basePath%>">
    <title>用户登录</title>
     <meta http-equiv="pragma" content="no-cache">
     <meta http-equiv="cache-control" content="no-cache">
     <meta http-equiv="expires" content="0">
     <meta http-equiv="keywords" content="keyword1,keyword2,keyword3">
     <meta http-equiv="description" content="This is my page">
  </head>

  <body>
     <center>
            <font size="8">用户登录</font>
            <hr/>
            <form action="validate.jsp" method="post">
                  用户名称：<input type="text" name="username"/>
            <br/>
                  用户密码：<input type="password" name="password"/>
            <br/>
            <input type="submit" value="登录"/>
```

```
29              </form>
30          </center>
31
32      </body>
33  </html>
```

上述代码中，第 23~29 行用 form 表单提交登录信息。

validate.jsp 是验证页面，其源代码如下：

```
-----------------validate.jsp----------------
<%@ page language="java" import="java.util.*" pageEncoding="UTF-8"%>
<!DOCTYPE HTML PUBLIC "-//W3C//DTD HTML 4.01 Transitional//EN">
<html>
  <head>
    <title>success.jsp</title>
     <meta http-equiv="pragma" content="no-cache">
     <meta http-equiv="cache-control" content="no-cache">
     <meta http-equiv="expires" content="0">
     <meta http-equiv="keywords" content="keyword1,keyword2,keyword3">
     <meta http-equiv="description" content="This is my page">
  </head>
   <%!
        //声明一个用户集合,模拟从数据库中取出用户集
        Map<String, String> map =new HashMap<String, String>();
        //声明验证的标识
        boolean flag = false;
   %>
     <%
         //向集合添加数据
        map.put("John","123456");
        map.put("Smith","222222");
        map.put("Bob","333333");
        map.put("Bruth","666666");
      %>
     <%!
        //声明验证方法
        boolean validate(String username,String password){
            String passwd = map.get(username);
            if(passwd!=null&&passwd.equals(password)){
                return true;
            }else{
              return false;
            }
        }
      %>
      <%
```

```
37              //获得页面提交的用户名跟密码
38              String username = request.getParameter("username");
39              String password = request.getParameter("password");
40 if(username==null||username==""||password==null||password==""){
41                  response.sendRedirect("login.jsp");
42              }
43              flag = validate(username,password);
44              if(flag){
45                  //保存在 session 对象中
46                  session.setAttribute("username",username);
47                  session.setAttribute("password",password);
48                  response.sendRedirect("class.jsp");
49              }
50          %>
51      <body>
52          <center>
53              <font size="6">用户登录</font>
54          </center>
55          <br/>
56          <center>
57              <%if(!flag){ %>
58                  <a href="login.jsp">重新登录系统</a>
59              <%} %>
60          </center>
61      </body>
62  </html>
```

上述代码中，第 18~24 行设置用户集合；代码第 26~35 行声明验证用户合法性方法；第 38~49 行获取参数值，并判断其合法性，合法的保存在 session 中，否则跳转到 login.jsp 中。

class.jsp 是班级管理页面，其源代码如下：

```
-----------------class.jsp----------------
<%@ page language="java" import="java.util.*" pageEncoding="UTF-8"%>
<%
    String name = (String)session.getAttribute("username");
    if(name==null){
        response.sendRedirect("login.jsp");
    }
%>

<!DOCTYPE HTML PUBLIC "-//W3C//DTD HTML 4.01 Transitional//EN">
<html>
  <head>
    <title>My JSP 'score.jsp' starting page</title>
    <meta http-equiv="pragma" content="no-cache">
```

```
14        <meta http-equiv="cache-control" content="no-cache">
15        <meta http-equiv="expires" content="0">
16        <meta http-equiv="keywords" content="keyword1,keyword2,keyword3">
17        <meta http-equiv="description" content="This is my page">
18    </head>
19
20    <body>
21        <center>
22            <font size="5">班级管理</font>
23        <hr/>
24        <h3>学生：<%=name %></h3>
25        <table>
26            <tr>
27                <td>
28                    <a href="addClass.jsp">班级录入</a>
29                </td>
30                <td>
31                    <a href="modifyClass.jsp">班级修改</a>
32                </td>
33                <td>
34                    <a href="queryClass.jsp">班级查询</a>
35                </td>
36                <td>
37                    <a href="delClass.jsp">班级删除</a>
38                </td>
39            </tr>
40        </table>
41        <a href="logout.jsp">退出登录</a>
42        </center>
43    </body>
44 </html>
```

上述代码中，第 28~37 行列出班级的各项方法。logout.jsp 是退出页面，其源代码如下：

```
-----------------logout.jsp----------------
<%@ page language="java" import="java.util.*" pageEncoding="UTF-8"%>
<%
  String username = (String)session.getAttribute("username");
  session.removeAttribute("John");
  session.invalidate();
  response.sendRedirect("login.jsp");
%>

<!DOCTYPE HTML PUBLIC "-//W3C//DTD HTML 4.01 Transitional//EN">
<html>
  <head>
```

```
12        <title>My JSP 'logout.jsp' starting page</title>
13      </head>
14      <body>
15      </body>
16    </html>
```

【代码解析】

login.jsp 页面中第 23 行代码 form 表单用 post 方式提交，action 指向 success.jsp 页面。Validate.jsp 页面中第 12~17 行代码是声明 map 集合和成功标识 flag，第 18~24 行向 map 中添加数据，第 25~34 行是验证的方法，第 36~50 行是获得页面传递的参数以及验证学生是否存在，如果存在则存放在 session 中并重定向到 class.jsp。第 57~59 行验证失败则显示重新登录。logout.jsp 页面第 2~6 行获得 session 中的 name 参数值并移除该学生，session 失效。

运行结果如图 4.17 所示。

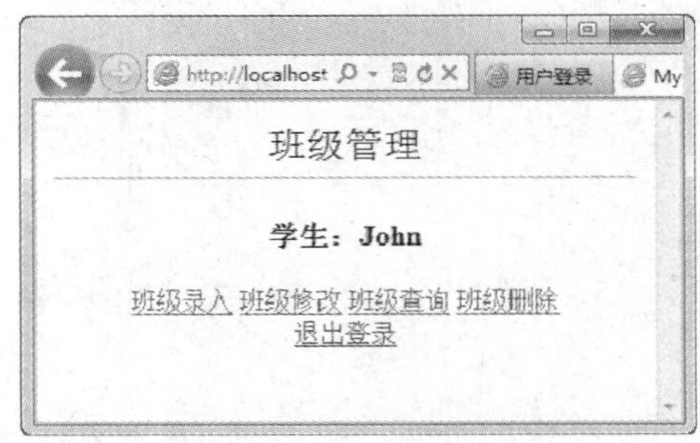

图 4.17　例 4.8 运行结果

该例子用户退出，只是用简单的 session.invalidate 方法，在开发中经常是结合 struts 框架一起使用。

4.4 application 对象

application 对象实现接口 javax.servlet.ServletContext。它的生命周期是从 application 对象创建到应用服务器关闭，也就是说当服务器关闭 application 对象才消失。可以将它视为 Web 应用的全局变量，当服务器运行时有效，如果关闭服务器，其中保存的信息也都消失了。

4.4.1 application 对象的常用方法

application 对象的常用方法参见表 4.5。

表 4.5 application 对象的常用方法

方法	方法说明
getAttribute(String arg0)方法	获得存放在 application 中的含有关键字为 arg0 的对象
setAttribute(String arg0,Object obj)方法	将关键字 arg0 的指定的对象 obj 放进 application 对象中
Enumeration getAttributeNames()方法	获取 application 中所以参数的名字，返回值是枚举类型
removeAttribute(String arg0)方法	移除 application 对象中 arg0 指定的参数值
getServletInfo()方法	获取 Servlet 的当前版本信息
getContext(String arg0)方法	获取 arg0 指定路径的 context 内容
getRealPath(String arg0)方法	获取指定文件的实际路径
getMimeType(String arg0)方法	获取指定的文件格式

4.4.2 获取指定页面的路径

【例 4.9】 获取指定页的实际路径、相对路径和当前应用程序路径

application.jsp 是指定页输出其所在的实际路径和相对路径，其源代码如下：

```
------------------application.jsp-----------------
<%@ page language="java" import="java.util.*" pageEncoding="UTF-8"%>
<!DOCTYPE HTML PUBLIC "-//W3C//DTD HTML 4.01 Transitional//EN">
<html>
  <head>
    <title>My JSP 'application.jsp' starting page</title>
  </head>
  <body>
      <h3>指定页的实际路径、相对路径和当前应用程序路径</h3>
      <hr/>
      <table border="1" bordercolor="black">
          <tr>
              <td>当前服务器的名称和版本</td>
              <td><%=application.getServerInfo() %></td>
          </tr>
          <tr>
              <td>页面 application.jsp 的实际路径</td>
              <td><%=application.getRealPath("application.jsp") %></td>
          </tr>
          <tr>
              <td>页面 application.jsp 的 URL</td>
              <td><%=application.getResource("application.jsp") %></td>
          </tr>
          <tr>
              <td>当前 Web 程序的路径</td>
              <td><%=application.getContextPath() %></td>
          </tr>
      </table>
```

```
28     </body>
29   </html>
```

上述代码中，第 11~25 行分别输出 application 中指定页的实际路径、相对路径和当前应用程序路径等信息。页面效果如图 4.18 所示。

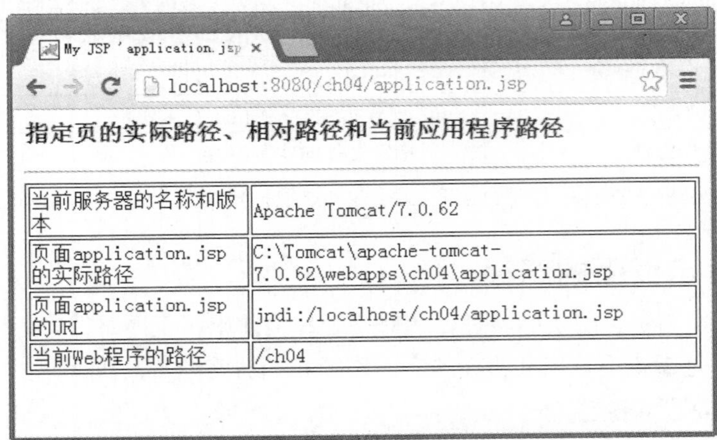

图 4.18　application.jsp 运行结果

4.4.3　设计一个网站计数器

application 对象还可以保存访问网站的人数，也就是常说的网站计数器，下面通过一个例子来演示。

【例 4.10】网站计数器

applicationCount.jsp 是网站计数器页面，其源代码如下：

```
----------------- applicationCount.jsp----------------
<%@ page language="java" import="java.util.*" pageEncoding="UTF-8"%>
<%
    Integer count =(Integer) application.getAttribute("count");
    if(count==null){
     count=1;
    }else{
        count++;
    }
    application.setAttribute("count",count);
 %>

<!DOCTYPE HTML PUBLIC "-//W3C//DTD HTML 4.01 Transitional//EN">
<html>
  <head>
    <title>网站计数器</title>
  </head>
  <body>
```

```
18          欢迎访问本网站，您是第<%=count %>位访问客户！
19    </body>
20  </html>
```

【代码解析】

第 02~09 行代码从页面中获得计数值，如果为空则设定初始值为 1，如果有则加 1。程序运行结果如图 4.19 所示。

图 4.19　网站计数器

application 对象在 Web 应用的运行时是一直存在于服务器中的，因此保存这种全局变量相对来说比较占用资源，因此不被推荐使用。在实际的开发中，一般都是让对象存在于必要的时间中，否则当访问量加剧时，会造成内存不足等情况。

4.5　out 对象

out 对象是继承 javax.servlet.jsp.JspWriter 类的一个输出流对象。它包含有很多 IO 流中的方法和特性。最常用的方法就是输出内容到 HTML 中。

4.5.1　out 对象的常用方法

out 对象的常用方法参见表 4.6。

表 4.6　out对象的常用方法

方法	方法说明
append(char c)方法	将字符添加到输出流中
clear()方法	清空页面缓存中的内容
close()方法	关闭网页流的输出
flush 方法	网页流的刷新
println()方法	将内容直接打印在 HTML 标记中
write()方法	与 println()方法相似。区别在于 print()方法可以输出各种类型的数据而 write 方法只能输字符相关的数据，例如：字符、字符数组、字符串等

4.5.2　out 对象的使用示例

out 对象中的方法相对比较简单而且也比较少用，下面就通过几个简单的例子来演示。

【例 4.11】演示 out 对象的 println 方法

outprintln.jsp 是直接在 Java 脚本中描述 HTML 标记，其源代码如下：

```
----------------- outprintln.jsp----------------
01  <%@ page language="java" import="java.util.*" pageEncoding="UTF-8"%>
02  <!DOCTYPE HTML PUBLIC "-//W3C//DTD HTML 4.01 Transitional//EN">
03  <html>
04    <head>
05      <title>演示 out 对象</title>
06    </head>
07
08    <body>
09       <center>
10           <hr>
11           <h4>以下就是一个表格</h4>
12           <%
13           out.println("<table border='2'>");
14           out.println("<tr>");
15           out.println("<td width='60'>"+"姓名"+"</td>");
16           out.println("<td width='40'>"+"性别"+"</td>");
17           out.println("<td width='80'>"+"出生年月"+"</td>");
18           out.println("<td width='60'>"+"城市"+"</td>");
19           out.println("</tr>");
20           out.println("<tr>");
21           out.println("<td width='60'>"+"Smith"+"</td>");
22           out.println("<td width='60'>"+"Male"+"</td>");
23           out.println("<td width='60'>"+"1984.8"+"</td>");
24           out.println("<td width='60'>"+"NerYork"+"</td>");
25           out.println("</tr>");
26           out.println("</table>");
27           %>
28       </center>
29    </body>
30  </html>
```

上述代码中，第 13~26 行用 out 对象输出 HTML 格式。页面效果如图 4.20 所示。

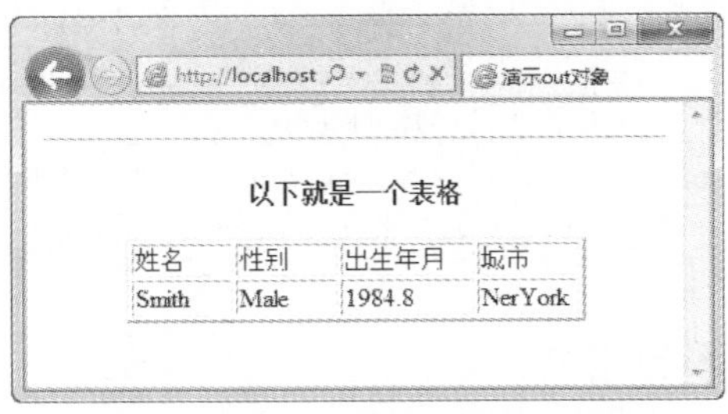

图 4.20　outprintln.jsp 页面的运行结果

现在开发中很少用这种方法输出 HTML 标记，因为比较繁琐而且容易出错。

【例 4.12】演示 out 对象的 clear 方法

outclear.jsp 页面是清空缓冲区中的内容，其源代码如下：

```
----------------- outclear.jsp-----------------
01  <%@ page language="java" import="java.util.*" pageEncoding="UTF-8"%>
02  <!DOCTYPE HTML PUBLIC "-//W3C//DTD HTML 4.01 Transitional//EN">
03  <html>
04    <head>
05      <title>演示 out 对象 clear 方法</title>
06    </head>
07
08    <body>
09       <h4>这是 HTML 中的内容</h4>
10       <%
11          out.print("<h4>这是 out 对象输出的信息</h4>");
12          out.clear();
13        %>
14        <center><h4>这是 HTML 中的信息</h4></center>
15    </body>
16  </html>
```

上述代码中，第 12 行清除缓冲区中的内容。页面效果如图 4.21 所示。

图 4.21　outclear.jsp 页面运行结果

从运行结果看出，若在 JSP 页面中调用 clear 方法，那么以前向客户输出流中写入的数据都将被 clear。

在调用 clear 之前，不要调用 flush 方法，否则会抛出 IO 异常。

4.6 page 对象

page 对象的实质是 java.lang.Object 对象，它代表转译后的 Servlet。page 对象指当前的 JSP 页面本身，在实际开发中并不常用到。

4.6.1 page 对象的常用方法

page 对象的常用方法参见表 4.7。

表 4.7 page对象常用方法

方法	方法说明
getClass()方法	返回当时被转译的 Servlet 类
hashCode()方法	返回此时被转译的 Servlet 类的哈希代码
toString()方法	将此时被转译的 Servlet 类转换成字符串
equals(Object obj)方法	比较此时的对象是否与指定的对象相等
clone()方法	将此事的对象复制到指定的对象中
copy(Object obj)方法	对指定对象进行克隆

4.6.2 page 对象的使用示例

下面就通过简单的例子来演示 page 中的方法。

【例 4.13】演示输出 page 对象的 toString()方法和 hashCode()方法

page.jsp 页面演示其 toString 方法和 hashCode 方法，其源代码如下：

```
----------------- page.jsp----------------
01  <%@ page language="java" import="java.util.*" pageEncoding="UTF-8"%>
02  <!DOCTYPE HTML PUBLIC "-//W3C//DTD HTML 4.01 Transitional//EN">
03  <html>
04    <head>
05      <title>演示 page 对象</title>
06    </head>
07
08    <body>
09       <%
10           int code = page.hashCode();//hashcode
11           String str = page.toString();
12           out.println("page 对象的 hash 码:"+code);
13           out.println("page 对象的值:"+str);
14        %>
15    </body>
16  </html>
```

上述代码中，第 10~13 行获取 page 对象中的值，并输出在页面中，页面效果如图 4.22 所示。

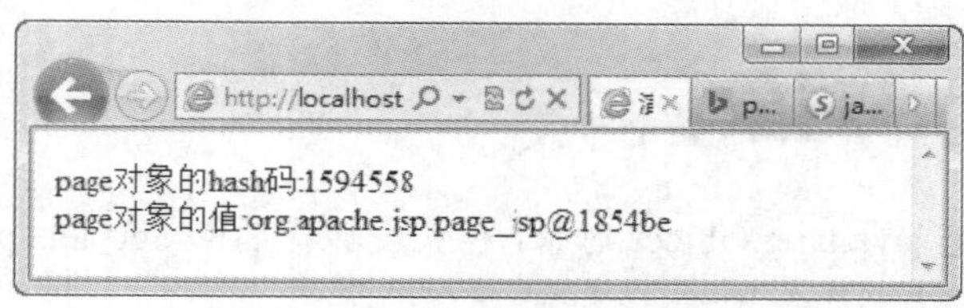

图 4.22 page.jsp 运行结果

4.7 config 对象

config 对象是实现了 javax.servlet.ServletConfig 接口，它一般是在页面初始化时传递参数用的。

4.7.1 config 对象的常用方法

config 对象的常用方法参见表 4.8。

表 4.8　config 对象的常用方法

方法	方法说明
getInitParameter(String arg0)方法	获得指定的初始值
getServletName()方法	获得 servlet 名字
getServletContext()方法	获得 servletContext 值
equals(Object obj)方法	比较此时的对象是否与指定的对象相等
getInitParameterNames()方法	获得初始化值的枚举值
toString()方法	获得此对象的值

4.7.2 config 对象的使用示例

下面就通过简单的例子来演示 config 中的方法。

【例 4.13】演示输出 config 对象的 getInitParameter()方法

config.jsp 页面演示其 getInitParameter()方法，在此假设 WEB-INF 文件夹下面有 web.xml 文件，内容如下：

```
<?xml version="1.0" encoding="UTF-8"?>
<web-app version="2.5"
    xmlns="http://java.sun.com/xml/ns/javaee"
    xmlns:xsi="http://www.w4.org/2001/XMLSchema-instance"
    xsi:schemaLocation="http://java.sun.com/xml/ns/javaee
    http://java.sun.com/xml/ns/javaee/web-app_2_5.xsd">
  <welcome-file-list>
    <welcome-file>index.jsp</welcome-file>
  </welcome-file-list>

  <servlet>
     <servlet-name>
         jspconfigdemo
      </servlet-name>
      <jsp-file>/config.jsp</jsp-file>
      <init-param>
```

```
17              <param-name>url</param-name>
18              <param-value>http://www.baidu.com</param-value>
19          </init-param>
20      </servlet>
21      <servlet-mapping>
22          <servlet-name>
23              jspconfigdemo
24          </servlet-name>
25          <url-pattern>/config.jsp</url-pattern>
26      </servlet-mapping>
27  </web-app>
```

上述代码中，第 11~26 行在 web.xml 中配置 Servlet，包括其初始化参数等信息。config.jsp 页面显示配置内容，其源代码如下：

```
01  <%@ page language="java" import="java.util.*" pageEncoding="UTF-8"%>
02  <!DOCTYPE HTML PUBLIC "-//W3C//DTD HTML 4.01 Transitional//EN">
03  <html>
04    <head>
05      <title>演示 config 对象</title>
06    </head>
07  
08    <body>
09       <%
10           String url = config.getInitParameter("url");
11           String str = config.toString();
12           out.print("page 对象的 initParameter 方法："+url+"</br>");
13           out.print("page 对象的 toString 方法："+str);
14       %>
15    </body>
16  </html>
```

上述代码中，第 10 行用 config 对象获取初始化参数值信息，页面效果如图 4.23 所示。

图 4.23　config.jsp 页面的运行结果

一般而言很少在页面中使用 config 对象，因为 JSP 页面实质是 Servlet。

4.8　上机实践

1. 请用 session 对象统计访问页面的客户数。

2. 请编写一个会员注册程序，要求用户在注册页面填写信息后，提交请求，接收注册请求的页面可以显示出信息。

第 5 章

◂ Servlet技术的应用 ▸

在 Web 应用技术中，Servlet 是另一个重要的技术。Servlet 是用 Java 类编写的服务端程序，与平台架构、协议无关。JSP 的底子其实是 Servlet，因为所有的 JSP 页面传回服务端时都要转为 Servlet 进行编译、运行。由于 JSP 编写 HTML 页面直观且易调试，所以 JSP 逐步取代 Servlet 在开发页面中的作用。

本章首先讲解 Servlet 的原理、生命周期、部署的方法等，让读者了解如何编写一个 Servlet，如何完成一个动作流程。然后介绍 Servlet 的一些进阶 API、过滤器和监听器。过滤器和监听器是 Servlet 规范里的两个高级特性，过滤器的作用是通过对 request、response 修改实现特定的功能，例如：请求数据字符编码、IP 地址过滤、异常过滤、用户身份认证等。监听器的作用是用于监听 Web 程序中正在执行的程序，根据发生地事件做出特定的响应。合理利用这两个特性，能够轻松地解决某些 Web 特殊问题。

5.1 Servlet 是什么

本节首先介绍 Servlet 的基本概念，Servlet 是用 Java 类编写的服务端应用程序，顾名思义，它通常是在服务端运行的程序，打开浏览器即可调用。它可以被看作是位于客户端和服务器端的一个中间层，负责接收和请求客户端用户的响应。

Servlet 使用了很多 Web 服务器都支持的 API，可调用和扩展 Java 中提供的大量程序设计接口、类、方法等功能。

Servlet 可以提供以下功能：

（1）对客户端发送的数据进行读取和拦截

客户端在发送一个请求时，一般而言都会携带一些数据（例如：URL 中的参数、页面中的表单、Ajax 提交的参数等），当一个 Servlet 接收到这些请求时，Java Servlet 中的类通过所提供的方法就能得到这些参数（例如：方法 request.getParameterName(name)获得参数名为 name 的参数值），也正因为这个原因，Servlet 可以对发送请求做拦截作用，它将在某些请求前先做个预处理分析，来判断客户端是否可以做这些请求（例如：检查访问权限、设定程序的字符集、检查用户角色等），当 Servlet 具有如上功能时，一般称之为拦截器。

（2）读取客户端请求的隐含数据

客户端请求的数据可以分为隐含数据和显式数据：隐含数据一般是不直接跟随于 URL 中，它存在于请求的来源、缓存数据（Cookie）、客户端类型等；显示数据显然是用户可以直观看到的（例如表单数据和 URL 参数）。Servlet 不但可以处理显示数据而且可以处理隐含数据，是个多面手。

（3）运行结果或者生成结果

当一个 Web 应用程序对客户端发出的请求做出响应时，一般需要很多中间过程才能得到结果。那么 Servlet 就担起了这个中间角色功能，它协调各组件各部分完成相应的功能，根据不同的请求做出相应的响应请求显示结果。

（4）发送响应的数据

Servlet 在对客户端做出响应并经过处理得出结果后，就会对客户端发送响应的数据，以便让客户端获取请求的结果数据。在 Web 应用程序中，Servlet 的这个作用相当的突出，无论现有的技术多么的突出，都是基于这个作用点出发的。

综上所述，Servlet 的程序运行顺序大致如图 5.1 所示。

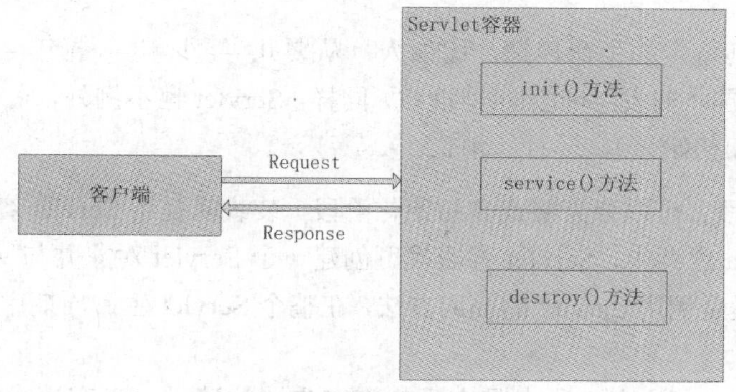

图 5.1 Servlet 运行顺序

5.2 Servlet 的技术特点

上一节介绍了 Servlet 的概念和运行顺序，让读者对 Servlet 有一个整体的印象和概念，并了解其运行的顺序等等，那么本节介绍 Servlet 的一些技术特点，让读者了解 Servlet 的优点。

Servlet 在开发中带来的优点就是能及时地响应和处理 Web 端的请求，使得一个不懂网页的 Java 开发人员也能编写出 Web 应用程序，只是在开发/修改一个 Web 程序时比较麻烦，因为代码的可读性比较差，也比较难以维护。但是它却有以下优点：

1. 高效率

Servlet 本身就是一个 Java 类，在运行的时候在同一个 Java 虚拟机中，可以快速地响应客户端的请求并生成结果。在 Web 服务器中处理一个请求使用的都是线程而非进程，也就是说在性能开销方面就小很多，无需大量的启动进程时间，在高并发量访问时，一个进程可以有多个线程，并发时线程在 CPU 使用的开销代价要远小于进程的开销。

2. 简单方便

开发一个 Web 程序时，从开发顺序上说比较简单，首先定义一个 Servlet 类，然后在系统（web.xml）中配置程序，继而发布程序，这样一个 Web 程序就可以了。在开发的过程中，系统提供了大量的实用工具方法，可以处理复杂的 HTML 表单数据、处理 cookie、跟踪网页会话等。

5.3 Servlet 的生命周期

一个生命都有其特定的生命周期，比如人：从婴儿—>少年—>青年—>壮年—>老人；再有开发项目：从立项—>开发—>运维—>消亡，同样，Servlet 也不例外，它有 3 个阶段分别是：初始化（包括装载和初始化）、运行、消亡。

（1）初始化阶段：可以分为装载和初始化阶段。装载就是由 Servlet 容器装载一个 Servlet 类，把它装载到 Java 内存中，Servlet 容器就可创建一个 Servlet 对象并与 web.xml 中的配置对应起来；初始化阶段是调用 Servlet 的 ini()方法，在整个 Servlet 生命周期中 init()方法只会被调用一次。

（2）运行阶段：在这个阶段中是实际响应客户端的请求。当有请求时，Servlet 会创建 HttpServletRequest 和 HttpServletResponse 对象，然后调用 service(HttpServletRequest request, HttpServletResponse response)方法。serivce()方法通过 request 对象获得请求对象的信息并加以处理，再由 response 对象给客户端做出响应。

（3）消亡阶段：当 Servlet 应用被终止后，Servlet 容器会调用 destory()方法对 Servlet 对象进行销毁动作。在消亡的过程中，Servlet 容器将释放被它所占的资源，例如：关闭流、关闭数据库连接等。同样在整个 Servlet 生命周期中 destroy()方法也只被调用一次。

下面通过编写一个 Servlet 类来说明下它的生命周期，完整的代码如下：

```
package com.etch.edu;

import java.io.IOException;
import java.io.PrintWriter;

```

```
import javax.servlet.ServletException;
import javax.servlet.ServletRequest;
import javax.servlet.ServletResponse;
import javax.servlet.http.HttpServlet;
import javax.servlet.http.HttpServletRequest;
import javax.servlet.http.HttpServletResponse;

public class HelloServlet extends HttpServlet {

    //序列化  可以自动生成或者自行定义
    private static final long serialVersionUID = 1L;

    public void init() throws ServletException{
        System.out.println("初始化 init 方法");
    }

      public void service(ServletRequest request,
            ServletResponse response)  throws   ServletException,
            IOException{
            System.out.println("调用 public service 方法");
            response.setContentType("text/html;charset=gbk");
            PrintWriter out = response.getWriter();
            out.println("收到 service 请求");
    }
    protected void service(HttpServletRequest request,
          HttpServletResponse response) throws ServletException{
        System.out.println("调用 protected  service 方法");
    }

    public void doGet(HttpServletRequest request,
            HttpServletResponse response)throws ServletException,
    IOException{
        System.out.println("调用 doGet()方法");
       //设置响应的页面类别跟页面编码
        response.setContentType("text/html;charset=gbk");
       PrintWriter out = response.getWriter();
       out.println("收到 HelloServlet doGet()请求");
    }

    public void doPost(HttpServletRequest request,
               HttpServletResponse response)throws ServletException,
    IOException{
        System.out.println("调用 doPost()方法");
        //设置响应的页面类别跟页面编码
        response.setContentType("text/html;charset=gbk");
```

```
            PrintWriter out = response.getWriter();
            out.println("收到 HelloServlet  doPost()请求");
        }

        public void destory(){
            System.out.println("调用 destory()方法");
        }
    }
```

可以看出，上述例子各个方法都只是执行打印的功能，它们只是为了说明一个 Servlet 的生命周期的执行过程。

完成上述 Servlet 编译后，还得配置下 web.xml，具体配置如下：

```
<servlet>
     <servlet-name>helloServlet</servlet-name>
     <servlet-class>com.etch.edu.HelloServlet</servlet-class>
 </servlet>
 <servlet-mapping>
          <servlet-name>helloServlet</servlet-name>
        <url-pattern>/HelloServlet</url-pattern>
 </servlet-mapping>
```

配置 servlet-name 时，需要区分英文大小写。

除了在 web.xml 配置 Servlet，在 Servlet 3.0 中还可以通过直接注入的方式进行配置，其代码如下：

```
package com.etch.edu;

import java.io.IOException;
import java.io.PrintWriter;

import javax.servlet.ServletException;
import javax.servlet.ServletRequest;
import javax.servlet.ServletResponse;
import javax.servlet.http.HttpServlet;
import javax.servlet.http.HttpServletRequest;
import javax.servlet.http.HttpServletResponse;
@WebServlet(
        urlPatterns = { "/HelloServlet " },
        name = " helloServlet "
)
public class HelloServlet extends HttpServlet {

    //序列化  可以自动生成或者自行定义
    private static final long serialVersionUID = 1L;

```

```
21      public void init() throws ServletException{
22          System.out.println("初始化 init 方法");
23      }
24
25      public void doGet(HttpServletRequest request,
26          ......
27      }
28          ......
29  }
```

如上述代码中，第 12~15 行就是用注入声明的方式表示这是一个 Servlet 类，@ WebServlet 中的参数见表 5.1 所示。从上述配置可以看出，这种方法比较简单也是现在主流的开发形式，在随后的章节中还会继续介绍和使用这种方法。如果用注入的方式，那么 web.xml 就不用配置 Servlet。

表 5.1　@WebServlet 主要属性列表

属性名	描述
name	指定 Servlet 的 name 属性，等价于 <servlet-name>；如果没有指定，则该 Servlet 的取值即为类的全名
urlPatterns	指定 Servlet 的 URL 匹配模式；等价于<url-pattern> 标签
loadOnStartup	指定 Servlet 的加载顺序，等价于<load-on-startup> 标签。当值为 0 或者大于 0 时，表示容器在应用启动时就加载并初始化这个 servlet；当值小于 0 或者没有指定时，则表示容器在该 servlet 被选择时才会去加载；正数的值越小，该 servlet 的优先级越高，应用启动时就越先加载；当值相同时，容器就会自己选择顺序来加载

配置完 web.xml 之后，可以通过以下步骤查看 Servlet 的生命周期执行过程。

（1）启动 Tomcat，将项目工程放在 webapps 文件夹下。

（2）在浏览器地址中输入与该 Servlet 配置相对应的 URL，在浏览器中看到“收到 service 请求”内容，在控制台中会输出：

```
初始化 init 方法
调用 public service 方法
```

（3）再在浏览器中输入 URL，在浏览器中看到“收到 service 请求”内容，在控制台中会输出：

```
调用 public service 方法
```

通过上述过程，验证了 Servlet 的生命周期过程。图 5.2 进一步说明了生命周期的不同阶段。

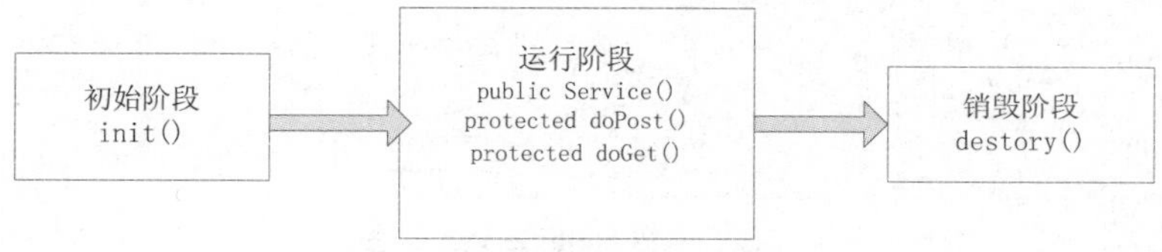

图 5.2　Servlet 生命周期

从程序的运行结果还可以知道，当重写了 service()方法之后，doPost()方法和 doGet()方法是不会被处理的，由 service()来管理转向对应的方法。

5.4 编写和部署 Servlet

上一节叙述了 Servlet 的生命周期过程，本节介绍如何编写一个 Servlet 类和部署 Servlet 工程，编写和部署 Servlet 是开发一个 Web 工程的基础，是读者必须掌握的内容。

5.4.1 编写 Servlet 类

通过上节的讲述，知道了一个 Servlet 的生命周期全过程，以及执行的过程顺序，本小节讲述如何编写一个简单的 Servlet 类和部署。

编写 Servlet 的开发工具本书以 MyEclipse 为例。在 MyEclipse 的主界面中选择 File|New|Web Project 命令，然后在 Project Name 文本框中输入自己定制的工程名字，创建一个 Servlet 工程，如图 5.3 所示。单击 Next 按钮直至完成。完成后出现编辑页面，一个 Servlet 工程初始化的结构如图 5.4 所示。其中 src 是放置源码包的，下面存放工程源码，默认新建一个 index.jsp，并在 WEB-INF 文件夹下生成 web.xml，存放 Servlet 的配置信息。

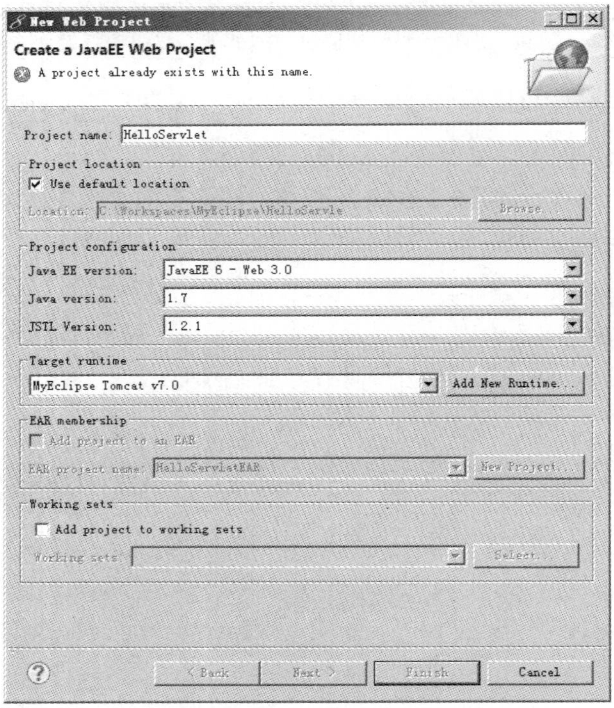

图 5.3 创建 HelloServlet 工程

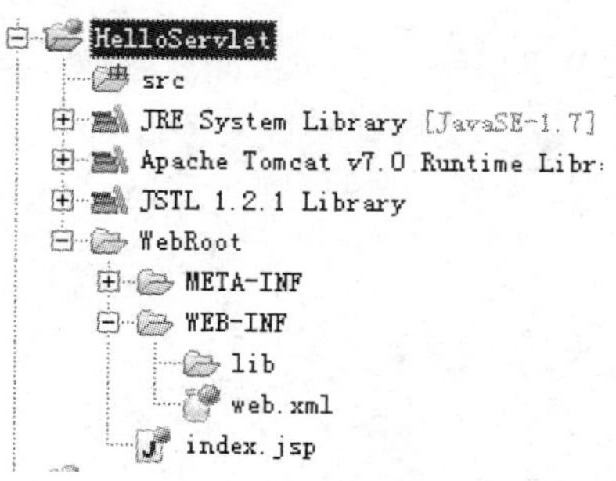

图 5.4　Servlet 工程目录

为了目录结构更加有层次感，新建包 com.yxtech（右击 src，选择 New|Package），然后右键单击该文件，选择 File|New|Class 命令，然后在 Name 中输入自己定制的 Servlet 程序文件的名字，创建 Servlet 的对话框如图 5.5 所示。

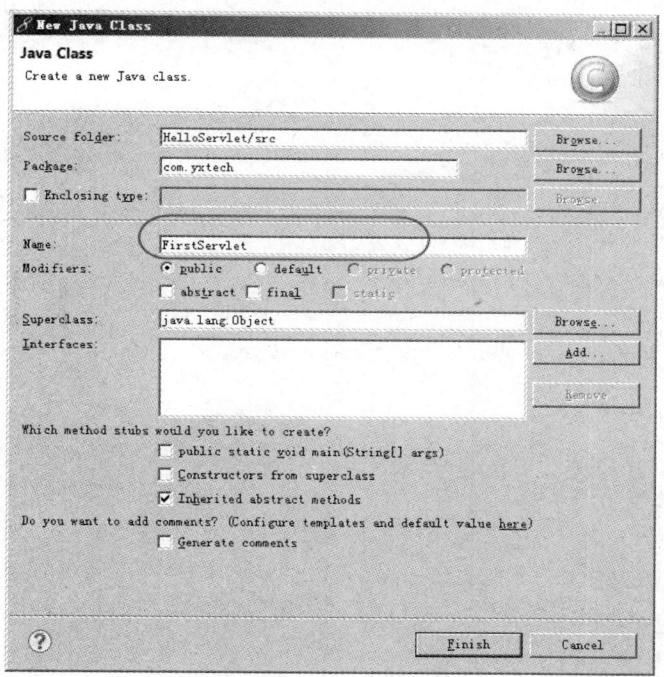

图 5.5　创建一个 Servlet 类

单击 Finish 按钮后出现主编辑界面。在编写 Servlet 类时，需要继承 HttpServlet 类，这样才可以开发具体的功能，假设想在页面中输出 0~10 的循环，那么程序如下：

```
package com.yxtech;

import java.io.IOException;
import java.io.PrintWriter;

```

```
import javax.servlet.ServletException;
import javax.servlet.http.HttpServlet;
import javax.servlet.http.HttpServletRequest;
import javax.servlet.http.HttpServletResponse;

public class FirstServlet extends HttpServlet {

    //序列化  可以自动生成或者自行定义
    private static final long serialVersionUID = 1L;

    public void init() throws ServletException{
    System.out.println("初始化  init()");
    }

    public void doGet(HttpServletRequest request,
                    HttpServletResponse response)
                throws ServletException,IOException
    {
        System.out.println("调用 doGet()方法");
       //设置响应的页面类别跟页面编码
        response.setContentType("text/html;charset=gbk");
       PrintWriter out = response.getWriter();
       out.println("<html>");
       out.println("<head><title>测试0到10循环结果</title>");
       out.println("<body>");
       out.println("开始执行。。。。。");
       int count=0;
       for(int i=0;i<=10;i++){
        count+=i;
       }
       out.println("程序执行结果："+count);
       out.println("</body>");
       out.println("</html>");
       out.flush();
       out.close();
    }
   public void doPost(HttpServletRequest request,
                   HttpServletResponse response)
               throws ServletException,IOException{
```

```
            doGet(request,response);
        }
        public void destory(){
            System.out.println("调用 destory()方法");
        }
}
```

接下来，在 WebRoot 目录下的 WEB-INF 文件夹下的 web.xml 中配置 FirstServlet 类文件。在 MyEclipse 中双击 web.xml 文件，在<web-app>中新增标签<servlet>和</servlet>以及<servlet-mapping>和</servlet-mapping>，具体的编辑内容如下：

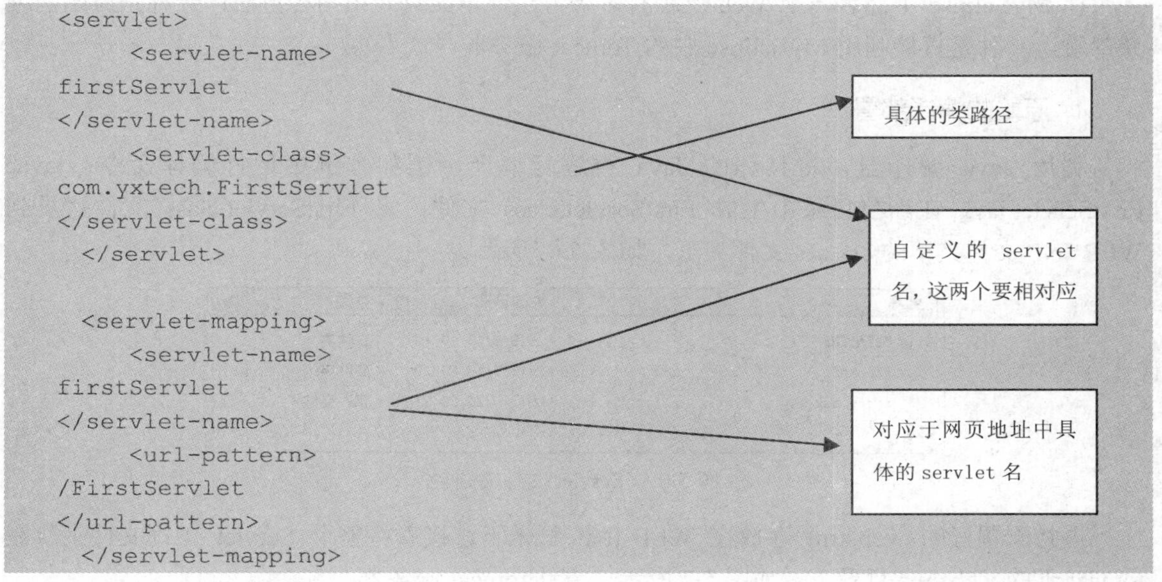

```
<servlet>
    <servlet-name>
firstServlet
</servlet-name>
    <servlet-class>
com.yxtech.FirstServlet
</servlet-class>
  </servlet>

  <servlet-mapping>
    <servlet-name>
firstServlet
</servlet-name>
    <url-pattern>
/FirstServlet
</url-pattern>
  </servlet-mapping>
```


现在分析下创建 Servlet 的步骤：

（1）引入相应的包，例如：java.servlet 包或者 java.servlet.http 包，这两个包的区别在于前者是与协议无关的，后者是与 HTTP 协议相关的。在平时开发的过程中，一般都是继承 HttpServlet 类，因为它封装了很多基于 HTTP 协议下的 Servlet 功能，基本够用，当然要想自己开发一个协议还可以继承 GenericServlet 类。FirstServlet 中有两个处理请求的方法，一个是 doGet()方法，一个是 doPost()方法，doGet()响应 http Get 请求，doPost()响应 http Post 请求。

（2）创建一个扩展类，例如本例中的 FirstServlet 类。

（3）override 一个 doGet()或者 doPost()方法。例如本例中重构了 doGet()方法，在该方法中完成处理请求，并输出到 HTML 页面中。

（4）配置 web.xml，在 web.xml 中<servlet>和<servlet-mapping>是一对，它们之间通过<servlet-name>相连，标签属性<servlet-class>是具体的 servlet 类路径，而<url-pattern>标签属性是地址栏中要指定的 servlet 名，例如：

本例地址 http://localhost:8080/HelloServlet/FirstServlet。

地址和端口　　工程名　　url-patter 中配置的 servlet 名

5.4.2　部署 Servlet 类

在 MyEclipse 中部署一个 Web 工程有多种方法，本文只介绍两种方法：一种是原始的编译部署，一种是直接利用 MyEclipse 中的 Tomcat 服务器。

1. 原始的编译部署

利用 Java 编译器，将具体的 Java 类进行编译，例如本例中运行编译命令：javac FirstServlet.java 在当前目录中生成 FirstServlet.class 文件，将 FirstServlet.class 文件拷贝到 WEB-INF 文件夹下的 classes 文件夹下，如图 5.6 所示。

名称	修改日期	类型
META-INF	2015/5/13 20:19	文件夹
WEB-INF	2015/5/13 20:19	文件夹
index.jsp	2013/9/10 20:57	JSP 文件

图 5.6　Servlet 文件夹路径

再将配置好的 web.xml 存放在 WEB-INF 目录下，接着将整个 Servlet 工程文件夹放在 tomcat 下的 webapps 目录下，如图 5.7 所示，启动 tomcat 服务器。

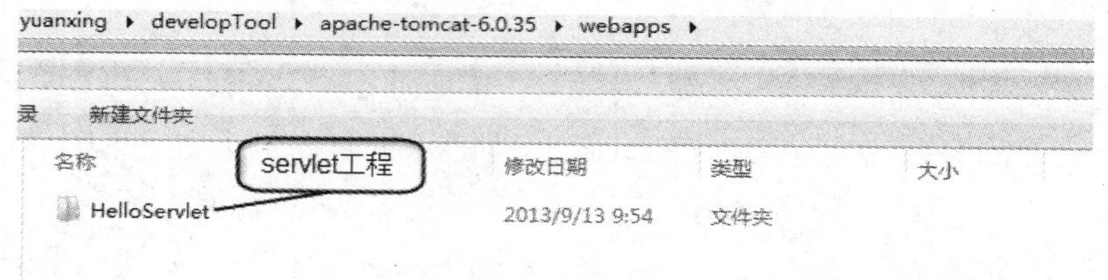

图 5.7　Servlet 放置目录

2. 利用 MyEclipse 中 Tomcat 服务器

单击 Servlet 工程，选择部署图标弹出选择框，单击 Add 按钮，弹出选择框，在 Server 列表中选择 Tomcat 7.x，单击 Finish 按钮，完成部署如图 5.8 所示。

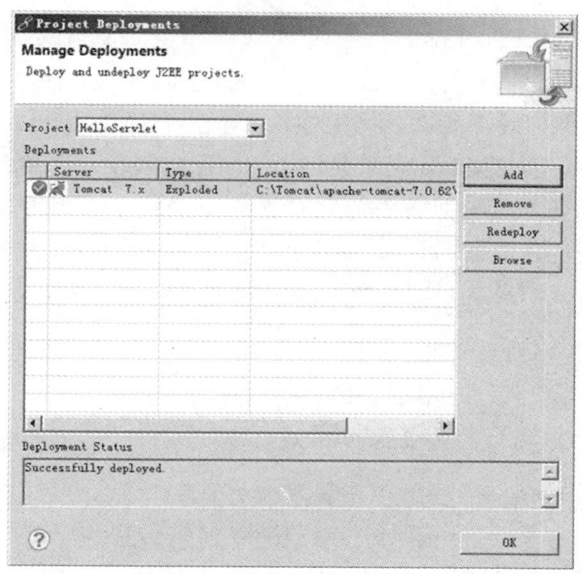

图 5.8　MyEclipse 中部署 Servlet 工程

单击按钮 选择 Tomcat 7.x，单击 Start 按钮，启动 Tomcat 服务器，如图 5.9 所示。

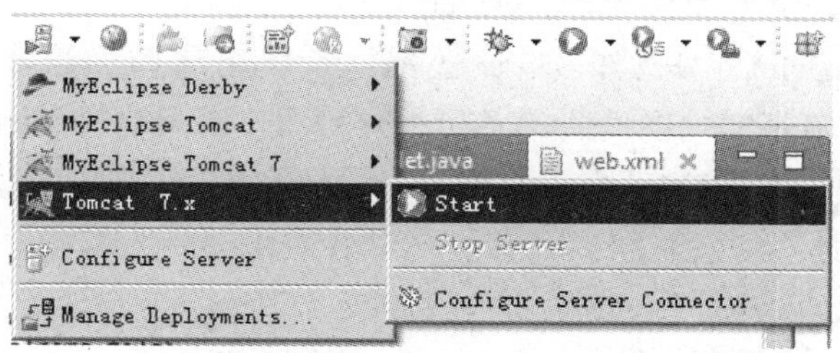

图 5.9　启动 MyEclipse Tomcat 服务器

打开 IE 浏览器，在地址栏中输入地址：

http://localhost:8080/HelloServlet/FirstServlet，然后按回车键。在页面中显示信息“开始执行……程序执行结果：55”。恭喜你，一个 Servlet 工程部署并运行成功。

5.5 Servlet 与 JSP 的比较

上一节介绍了如何编写和部署 Servlet 程序，让读者了解编写一个 Servlet 的基本步骤，使得读者能编写出简单的 Servlet 类。本节介绍 Servlet 与 JSP 之间的区别与联系，包括它们之间的内在联系是什么。其实从本质上讲，Servlet 与 JSP 是一样的，因为 JSP 页面最后运行时会被转换成一个 Servlet。但是从开发者的视角看，运用这两个技术还是有些区别的，这些区别也限制了在开发中的选择。

JSP 与 Servlet 的主要区别如下：

（1）Servlet 是 Java 代码，JSP 是页面代码（HTML 和 JSP 表达式）

编写 Servlet 就是编写 Java 代码，所以应用 Java 中的规范去编写 Servlet 类就可以。但是想在客户端中响应出结果，就必须在代码中加入大量的 HTML 代码，可想而知当要想得到一个比较美观、复杂的界面，HTML 代码量会相当的多而且繁琐。而 JSP 是以 HTML 代码为主，在页面中适当的嵌入 Java 代码来处理业务上的逻辑。显然，JSP 会比 Servlet 较易编写和美观。

基于这点的差异，也是选择 Servlet 或者 JSP 技术的考量标准之一。如果业务中主要是页面为主，则选择 JSP 技术；反之，则选择 Servlet 技术，它适合服务器端开发。

（2）Servlet 运行速率快过 JSP

因为 Servlet 本身就是一个 Java 类，编译的时候直接被转为 class 类文件。而 JSP 需要先被编译为 Java 类，而后再运行。所以 Servlet 运行速率较快。

（3）Servlet 需要手动编译，JSP 由服务器自动编译

Servlet 类要被编译成为 class 文件后，手动的复制到 Web 应用程序目录下。而 JSP 页面部署到 Web 应用则简单很多，只需要将 JSP 页面拷贝到指定的目录下，当它第一次被访问时，Web 服务器自动将 JSP 代码转换为 Java 代码并自动编译。

（4）编辑 HTML 工具不支持编辑 Servlet

当下有很多制作网页的工具例如：Dreamweaver、golive 等供选择，利用这些工具可以快速地开发网页，编写出复杂的界面而且具备可视化效果，极大地提高网页开发的效率。但对于大多数的网页工具而言，在 HTML 页面中加入 Java 代码是可以的，但是要在 Java 代码中加入 HTML 代码却不行，而且不能及时纠错。因此，初学者应用不同的编辑器工具编写 JSP 和 Servlet 也是可以理解的。

了解 Servlet 与 JSP 的差别之后，在编写 Web 应用程序时，就要根据当前的需要权衡使用 JSP 或者 Servlet。开发者要尽量使 JSP 和 Servlet 都发挥出最大的作用，又便于日后的代码维护工作。一般而言，Servlet 多数负责对客户端的请求进行处理和调用 JavaBean；由 JavaBean 负责提供可复用的数据以及数据的访问数据等。而 JSP 页面主要负责页面的展示，将动态的数据展现给客户。这也就是开发者提出的简易 MVC 模式。这样分工大大减少了 JSP 页面中 Java 程序和 HTML 代码的耦合度，对维护工作具有重大的意义。

5.6 Servlet 进阶 API

在编写完一个 Servlet 类后，通常需要在 web.xml 中进行相关的配置，这样 Web 容器才能读取 Servlet 设置的信息包括其类地址、初始化等。对于每个 Servlet 的配置，Web 都会生成与之相对应的的 ServletConfig 对象，从 ServletConfig 对象中可以得到 Servlet 初始化参数。

本节将介绍 ServletConfig 与 GenericServlet 的关系，如何使用 ServletConfig 和 ServletContext 对象来获取 Servlet 初始化参数。

5.6.1　Servlet、ServletConfig 与 GenericServlet

在 Web 容器启动后，通过加载 web.xml 文件读取 Servlet 的配置信息，并实例化 Servlet 类，并且为每个 Servlet 配置信息产生唯一一个 ServletConfig 对象。在运行 Servlet 时，调用 Serlvet 接口的 init()方法，将生产的 ServletConfig 作为参数传入 Servlet 中。流程如图 5.10 所示。

初始化init()方法
创建ServletConfig对象
并获取参数值
Web容器
HttpServlet
创建destroy()方法

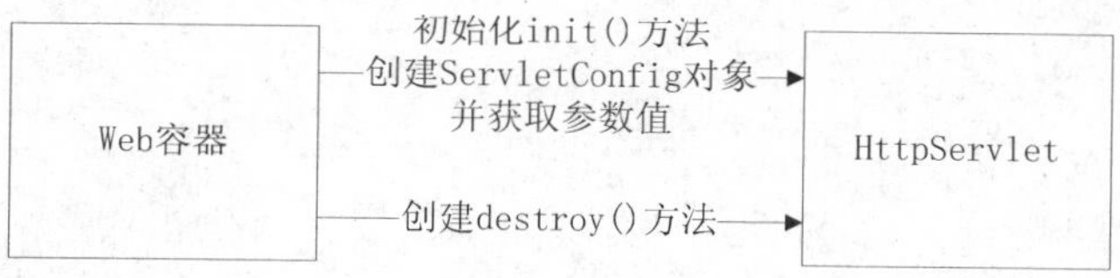

图 5.10　创建 Servlet 与 ServletConfig 示意图

初始化方法只会被调用一次，即容器在启动时，实例化 Servlet 和创建 ServletConfig 对象，且 Servlet 与 ServletConfig 是一一对应关系，之后就直接执行 service()方法。

从 Java Servlet Api 中，可以得知 GenericServlet 类是同时实现了 Servlet、ServletConfig 两个接口，如图 5.11 所示。

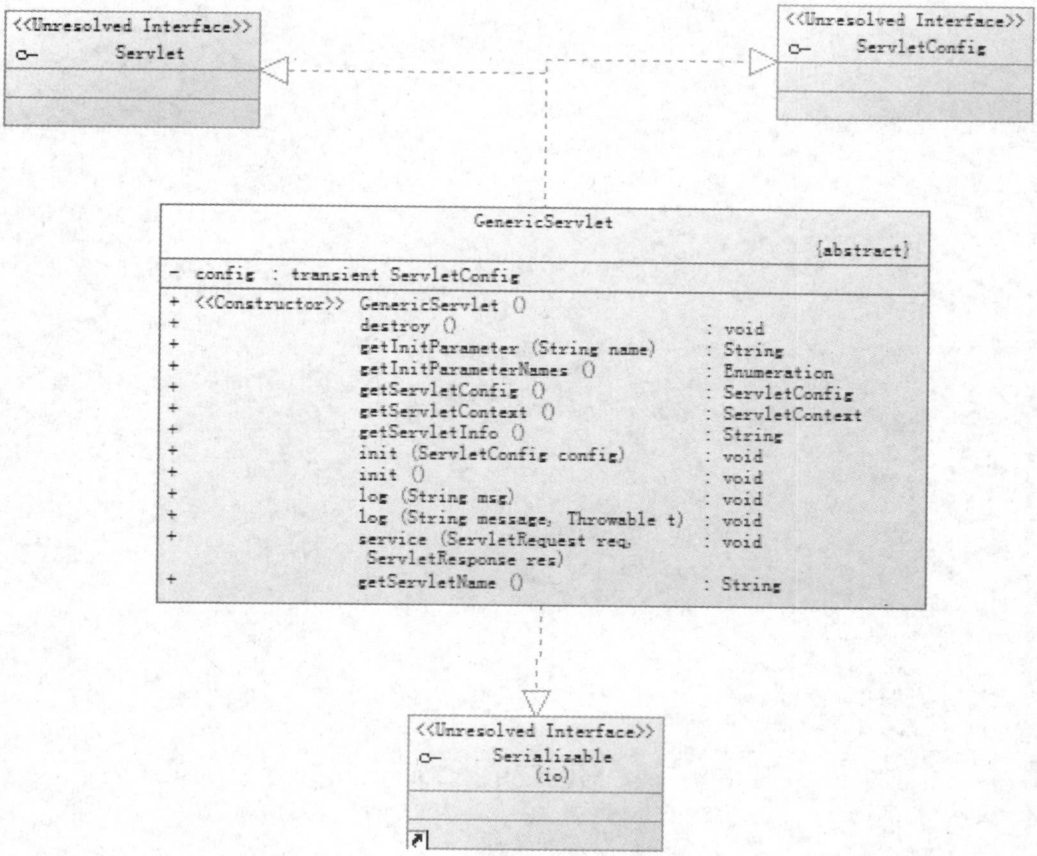

<<Unresolved Interface>>
Servlet
<<Unresolved Interface>>
ServletConfig
GenericServlet
{abstract}
- config : transient ServletConfig
+ <<Constructor>> GenericServlet ()
+ destroy () : void
+ getInitParameter (String name) : String
+ getInitParameterNames () : Enumeration
+ getServletConfig () : ServletConfig
+ getServletContext () : ServletContext
+ getServletInfo () : String
+ init (ServletConfig config) : void
+ init () : void
+ log (String msg) : void
+ log (String message, Throwable t) : void
+ service (ServletRequest req, ServletResponse res) : void
+ getServletName () : String
<<Unresolved Interface>>
Serializable
(io)

图 5.11　GenericServlet 类图

这个类的存在使得编写 Servlet 更加方便。它提供了一个简单的方法，这个方法用来执行有关 Servlet 生命周期的方法以及在初始化时对 ServletConfig 对象和 ServletContext 对象进行说明。

先来看下 GenericServlet 类的源代码：

```
----------------------GenericServlet.java----------------------------
package javax.servlet;

import java.io.IOException;
import java.util.Enumeration;

public abstract class GenericServlet
    implements Servlet, ServletConfig, java.io.Serializable
{

    private transient ServletConfig config;
    //GenericServlet 默认的构造方法
    public GenericServlet() { }
    //GenericServlet 默认的销毁方法
    public void destroy() {
     log("destroy");
    }
    //获得初始化参数方法
    public String getInitParameter(String name) {
     return getServletConfig().getInitParameter(name);
    }
    //获得参数名称，并返回枚举类型
    public Enumeration getInitParameterNames() {
     return getServletConfig().getInitParameterNames();
    }
    //获得 ServletConfig 对象
    public ServletConfig getServletConfig() {
     return config;
    }
    //获得 ServletContext 对象
    public ServletContext getServletContext() {
     return getServletConfig().getServletContext();
    }

    public String getServletInfo() {
     return "";
    }
    //GenericServlet 初始化方法
    public void init(ServletConfig config) throws ServletException {
     this.config = config;
     log("init");
     this.init();
    }
    //GenericServlet 的空初始化方法
    public void init() throws ServletException {

    }

    public void log(String msg) {
```

```
49         getServletContext().log(getServletName() + ": "+ msg);
50        }
51
52        public void log(String message, Throwable t) {
53         getServletContext().log(getServletName() + ": " + message, t);
54        }
55
56        public abstract void service(ServletRequest req, ServletResponse res)
57         throws ServletException, IOException;
58        //GenericServlet 获得 Servlet 名称
59        public String getServletName() {
60           return config.getServletName();
61        }
62    }
```

分析上述代码第 38~41 行，得知在 Servlet 调用 init()方法是，GenericServlet 类将 ServletConfig 封装了。

从代码第 18~32 行中，得知 GenericServlet 类还定义获取 ServletConfig 对象的方法，当编写 Servlet 类时就可以通过这些方法来获取所要得配置信息，而不用重新 New 出 ServletConfig 对象。

5.6.2　使用 ServletConfig

上节介绍过，当容器初始化 Servlet 时，会为 Servlet 创建惟一的 ServletConfig 对象。Web 容器读取 web.xml 文件，将初始化参数传给 ServletConfig，而 ServletConfig 作为对象参数传递到 init()方法中。

在 ServletConfig 接口中，各方法说明见表 5.2。

表 5.2　ServletConfig接口方法说明

方法	描述
getServletName()方法	该方法返回一个 servlet 实例的名称
getServletContext()方法	返回一个 ServletContext 对象的引用
getInitParameter(String name)方法	返回一个由参数 String name 决定的初始化变量的值，如果该变量不存在，返回 null
getInitParameterNames()方法	返回一个存储所有初始化变量的枚举类型。如果 servlet 没有初始化变量，返回一个空枚举类型

从 Servlet 3.0 开始，配置 Servlet 允许注入的方式进行配置，而不仅仅在 web.xml 中配置。因此配置 Servlet 的形式可以有如下形式：

```
@WebServlet(
 urlPatterns = { "/servletConfigDemo.do" },
 loadOnStartup = 1,
 name = "ServletConfigDemo",
 displayName = "demo",
```

```
    initParams = {
        @WebInitParam(name = "success", value = "success.html"),
        @WebInitParam(name = "error", value = "error.html")
    }
)
```

它等价于：

```
<servlet>
<display-name>ss</display-name>
<!—Servlet 名称-->
<servlet-name>ServletConfigDemo</servlet-name>
<!—Servlet 类路径-->
    <servlet-class>com.eshore.ServletConfigDemo</servlet-class>
<load-on-startup>-1</load-on-startup>
<!—初始化参数-->
    <init-param>
        <param-name>success</param-name>
        <param-value>success.html</param-value>
    </init-param>
    <init-param>
        <param-name>error</param-name>
        <param-value>error.html</param-value>
    </init-param>
</servlet>
<servlet-mapping>
 <!—Servlet 映射-->
    <servlet-name>ServletConfigDemo</servlet-name>
    <url-pattern>/servletConfigDemo.do</url-pattern>
</servlet-mapping>
```

上述的两个配置过程是等价的，在 Servlet 3.0 中可以同时兼容，在 Servlet 2.0 中只能在 web.xml 中配置。@WebServlet 和@WebInitParam 的主要属性分别参见表 5.3 和表 5.4 所示。

表 5.3　@WebServlet 主要属性列表

属性名	描述
name	指定 Servlet 的 name 属性，等价于<servlet-name>。如果没有指定，则该 Servlet 的取值即为类的全名
value	该属性与 urlPatterns 属性等价，但两个属性不能同时使用
urlPatterns	指定 Servlet 的 URL 匹配模式，等价于<url-pattern>标签
loadOnStartup	指定 Servlet 的加载顺序，等价于<load-on-startup>标签。当值为 0 或者大于 0 时，表示容器在应用启动时就加载并初始化这个 Servlet；当值小于 0 或者没有指定时，则表示容器在该 Servlet 被选择时才会去加载；正数的值越小，该 Servlet 的优先级越高，应用启动时就越先加载；当值相同时，容器就会自己选择顺序来加载
initParams	指定 Servlet 初始化参数，等价于<init-param>标签
asyncSupported	声明 Servlet 是否支持异步操作模式，等价于<async-supported>标签。该属性在 Servlet 3.0 才有

表 5.4　@WebInitParam 主要属性列表

属性名	描述
name	指定参数的名字，等价于<param-name>
value	指定参数的值，等价于<param-value>
description	参数的描述，等价于<description>

以下用例子说明 ServletConfig 的用法以及其自身方法的使用。

【例 5.1】利用初始化信息设定跳转信息

编写 ServletConfigDemo 类，输出其初始化参数，当用户成功登录时跳转到成功页面，反之跳转到错误页面。源代码如下：

```
---------------------- ServletConfigDemo.java---------------------------
@WebServlet(
    urlPatterns = { "/servletConfigDemo.do" },
    loadOnStartup = 1,
    name = "ServletConfigDemo",
    displayName = "demo",
    initParams = {
        @WebInitParam(name = "success", value = "success.html"),
        @WebInitParam(name = "error", value = "error.html")
    }
)
public class ServletConfigDemo extends HttpServlet {

    public void doPost(HttpServletRequest request, HttpServletResponse
    response)
            throws ServletException, IOException {
        //获取 ServletConfig 对象
        ServletConfig config = getServletConfig();

        //1.getInitParameter(name)方法
        String success = config.getInitParameter("success");
        String error = config.getInitParameter("error");

        System.out.println("success-----------"+success);
        System.out.println("error-----------"+error);

        //2.getInitParameterNames 方法
        Enumeration enumeration = config.getInitParameterNames();
        while(enumeration.hasMoreElements()){
            String name = (String) enumeration.nextElement();
            String value = config.getInitParameter(name);
            System.out.println("name-----------"+name);
            System.out.println("value-----------"+value);
```

```
32          }
33          //3.getServletContext 方法
34          ServletContext servletContext = config.getServletContext();
35          System.out.println("servletContext-----------"+servletContext);
36          //5.getServletName 方法
37          String servletName = config.getServletName();
38          System.out.println("servletName-----------"+servletName);
39
40           request.setCharacterEncoding("UTF-8");
41           response.setContentType("text/html;charset=UTF-8");
42           String userId = request.getParameter("userId");
43           String passwd = request.getParameter("passwd");
44
45           //判断是否是 linl 用户且密码相符。
46           if(userId!=null&&"linl".equals(userId)
47                  &&passwd!=null&&"123456".equals(passwd)){
48               //获得 sessioin 对象
49               HttpSession session = request.getSession();
50               //设置 user 参数
51               session.setAttribute("user", userId);
52               //跳转页面
53               RequestDispatcher dispatcher = request.
54                      getRequestDispatcher(success);
55               dispatcher.forward(request, response);
56           }else{
57               RequestDispatcher dispatcher = request.
58                      getRequestDispatcher(error);
59               dispatcher.forward(request, response);
60           }
61      }
62      //以下省略 init()、destroy()等方法
63  }
```

上述代码中第 1~10 行，是用注入的方式配置初始化参数，该 Servlet 的名称为 ServletConfigDemo，url 路径为 servletConfigDemo.do（所以在浏览器里不要忘记输入后面的 do）。代码第 16~38 行演示 ServletConfig 类方法的使用，第 40~60 行是具体的执行动作，用户验证成功跳转成功页面，否则跳转失败页面。

使用注入方式配置 Servlet 时，服务器必须要支持 Servlet 3.0 才可以，否则运行 Servlet 时会报错。

5.6.3 使用 ServletContext

ServletContext 对象是 Servlet 中全局的存储信息，当服务器启动时，Web 容器为 Web 应用

创建惟一的 ServletContext 对象，应用内的 Servlet 共享同一个 ServletContext。可以认为在 ServletContext 中存放着共享数据，应用内的 Servlet 可以通过 ServletContext 对象提供的方法获取共享数据。ServletContext 对象只有在 web 应用被关闭的时候才销毁。

ServletContext 接口中定义了运行 Servlet 应用程序环境信息，可以用来获取请求资源的 URL、设置与存储全局属性、Web 应用程序初始化参数。ServletContext 中常见的方法见表 5.5 所示。

表 5.5　ServletContext 常用方法

方法	描述
getRealPath(String path)	获取给定的虚拟路径所对应的真实路径名
getResource(String uripath);	返回由 path 指定的资源路径对应的一个 URL 对象
getResourceAsStream(String uripath);	返回一个指定位置资源的 InputStream。返回的 InputStream 可以是任意类型和长度的。使用时指定路径必须以“/”开头，表示相对于应用程序环境根目录
getRequestDispatcher(String uripath);	返回一个特定 URL 的 RequestDispatcher 对象，否则就返回一个空值
getResourcePaths(String path)	返回一个存储 web-app 中所指资源路径的 Set（集合），如果是个目录信息，会以“/”作结尾
getServerInfo	获取服务器的名字和版本号

以下用例子说明 ServletContext 的用法。

【例 5.2】说明 ServletContext 的用法

编写 ServletContextDemo 类，输出其初始化参数，并用 getResourceAsStream 方法输出指定文件内容到页面中。源代码如下：

```
----------------------- ServletContextDemo.java---------------------------
@WebServlet(
    urlPatterns = { "/servletContextDemo.do" },
    loadOnStartup = 0,
    name = "ServletContextDemo",
    displayName = "demo",
    initParams = {
        @WebInitParam(name = "dir", value = "/dir"),
        @WebInitParam(name = "success", value = "success.html"),
        @WebInitParam(name = "resourcePath", value = "/dir/test.txt")
    }
)
public class ServletContextDemo extends HttpServlet {

    public void doPost(HttpServletRequest request, HttpServletResponse
    response)
            throws ServletException, IOException {
```

```
16             String dir = getInitParameter("dir");
17             String success = getInitParameter("success");
18             String resourcePath = getInitParameter("resourcePath");
19             //获取 ServletContext 对象
20             ServletContext context =  getServletContext();
21             //getRealPath 获得真实路径
22             String path = context.getRealPath(success);
23             System.out.println("path 真实路径-----"+path);
24             //getResourcePaths 获得指定路径的内容
25             Set set = context.getResourcePaths(dir);
26             for(Object str:set){
27                 System.out.println("文件内容-----"+(String)str);
28             }
29             //获得服务器版本
30             String serverInfo = context.getServerInfo();
31             System.out.println("获得服务器版本-----"+serverInfo);
32             //getResourceAsStream 获得资源文件内容
33             InputStream in = context.getResourceAsStream(resourcePath);
34             OutputStream out = response.getOutputStream();
35             byte[] buffer = new byte[1024];
36             while(in.read(buffer)!=-1){
37                 out.write(buffer);
38             }
39             in.close();
40             out.close();
41         }
42 
43         //以下省略 init()、destroy()等方法
44     }
```

上述代码中第 20 行的 getServletContext()获得 ServletContext 对象，代码第 22~30 行利用 ServletContext 对象获取指定文件的真实路径、文件夹内容和服务器版本号，代码第 33~38 行，用 getResourceAsStream 方法获取文件流，将文件内容输出到页面中。上述代码运行结果如图 5.12 和图 5.13 所示。

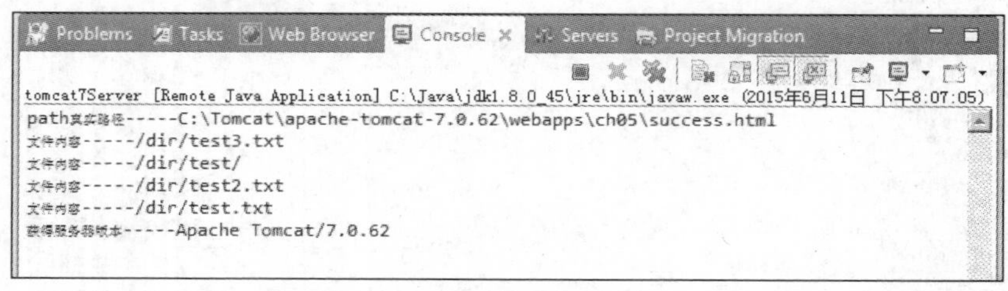

图 5.12　后台输出

图 5.13　页面输出流

以“/”作为开头时有时称为环境相对路径。在 Servlet 中，若是环境相对路径，则直接委托给 ServletContext 的 getRequestDispatcher()。

5.7 应用程序事件、监听器

何谓监听？顾名思义就是监视行为。在 Web 系统中，所谓的监听器就是应用监听事件来监听请求中的行为而创建的一组类。HttpServletRequest、HttpSession、ServletContext 对象在 Web 容器中遵循生成、运行、销毁这样的生命周期。当进行相关的监听配置，Web 容器就会调用监听器上的方法，做出对应的事件处理，了解它运行的情况或者运行别的程序。各监听器接口和事件类见表 5.6 所示。

表 5.6　监听接口和事件类

类别	监听接口	监听事件
与 ServletContext 相关	ServletContextListener	ServletContexEvent
	ServletContexAttributeListener	ServletContexAttributeEvent
与 HttpSession 相关	HttpSessionListener	HttpSessionEvent
	HttpSessionActivationListener	
	HttpSessionAttributeListener	HttpSessionAttributeEvent
	HttpSessionBindingListener	
与 ServletRequest 相关	ServletRequestListener	ServletRequestEvent
	ServletRequestAttributeListener	ServletRequestAttributeEvent

使用监听器需要实现相应的监听接口。在触发监听事件时，应用服务器会自动调用监听方法。开发人员不需要关心应用服务器如何调用，只需要实现这些方法就行。

5.7.1 ServletContext 事件、监听器

如上所述，与 ServletContext 有关的监听器有两个 ServletContextListener 与 ServletContext AttributeListener。

1. ServletContextListener

ServletContextListener 被称为“ServletContext 生命周期监听器”，可以用来监听 Web 程序初始化或者结束时响应的动作事件。

ServletContextListener 接口的类是 javax.servlet.ServletContextListener，该接口提供两个监听方法。

（1）public void contextInitialized (ServletContextEvent sce)：该方法用于通知监听器，已经加载 Web 应用和初始化参数。

（2）public void contextDestroyed (ServletContextEvent sce)：该方法用于通知监听器，Web 应用即将关闭。

在 Web 应用程序启动时，会自动开始监听，首先调用的是 contextInitialized()方法，并传入 ServletContextEvent 参数，它封装了 ServletContext 对象，可以通过 ServletContextEvent 的 getServletContext()方法取得 ServletContext 对象，通过 getInitParameter()方法取得初始化参数。在 Web 应用关闭时，会自动调用 contextDestroyed()方法，同样会传入 ServletContextEvent 参数。在 contextInitialized()中可以实现应用程序资源的准备事件，在 contextDestroyed()中可以实现对资源的释放。例如，可以在 contextInitialized()方法中实现 Web 应用的数据库连接、读取应用程序设置等；在 contextDestroyed()中设置数据库资源的释放。

在 Web 中，实现 ServletContextListener 的步骤如下：

首先，编写一监听类并实现 ServletContextListener 接口；

进行相关的配置如下：

```
<listener>
    <listener-class>
      com.eshore.MyServletContextListener
    </listener-class>
</listener>
```

或者用注入的方式注入监听类：

```
@WebListener
public class MyServletContextListener implements ServletContextListener{
}
```

需要初始化参数则需要在 web.xml 进行配置，例如：

```
<context-param>
     <param-name>user_name</param-name>
     <param-value>linlong</param-value>
 </context-param>
```

@WebListener 也是 Servlet 3.0 才有的，因为它没有设置初始化参数的属性，所以也需要在 web.xml 中设定。

【例 5.3】说明 ServletContextListener 的用法

编写 MyServletContextListener 类，并用 Log4j 将输出日志写到指定文件。

```
------------------------MyServletContextListener.java------------------------
01   import javax.servlet.ServletContext;
02   import javax.servlet.ServletContextEvent;
03   import javax.servlet.ServletContextListener;
04   import javax.servlet.annotation.WebListener;
05
06   import org.apache.log4j.Logger;
07
08
09   @WebListener
10   public class MyServletContextListener implements ServletContextListener{
11
12       private static Logger log = 
Logger.getLogger("MyServletContextListener");
13
14       public void contextInitialized(ServletContextEvent sce) {
15           //通过 ServletContextEvent 获得 ServletContext 对象
16           ServletContext context = sce.getServletContext();
17           String name = context.getInitParameter("user_name");
18           log.debug("初始化参数 name 的值: "+name);
19           log.debug("Tomcat 正在启动中......");
20       }
21
22       public void contextDestroyed(ServletContextEvent sce) {
23           log.debug("Tomcat 正在关闭中......");
24       }
25   }
```

上述代码中，第 16 行获取 ServletContext 对象，第 17 行获取初始化值。部署应用后，在 tomcat 的 log 文件夹下 log.log 的文件中，输出了初始化参数值和相关的信息，如图 5.14 所示。

本例中使用了 Log4j 日志管理，在 log4j.properties 中配置输出的日志文件。例如 log4j.appender.D.File = E:/downloads/apache-tomcat-5.0.53/logs/log.log。

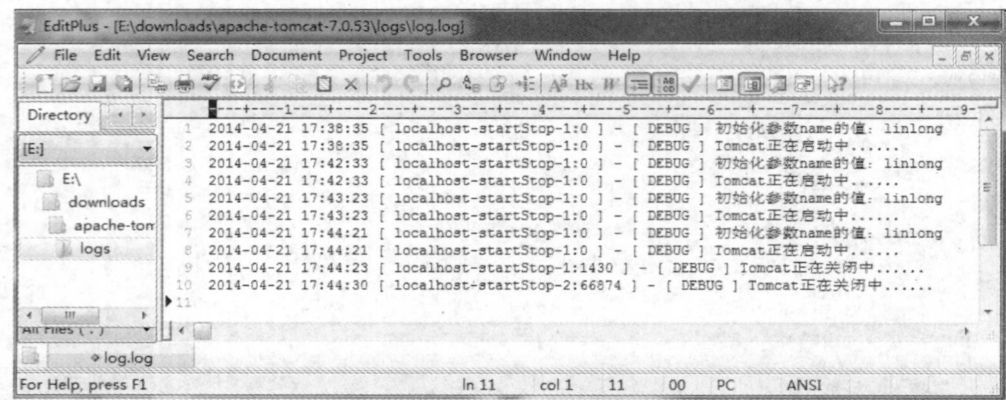

图 5.14　ServletContextListener 效果图

2. ServletContextAttributeListener

ServletContextAttributeListener 被称为“ServletContext 属性监听器”，可以用来监听 Application 属性的添加、移除或者替换时响应的动作事件。

ServletContextAttributeListener 接口的类是 javax.servlet.ServletContextAttributeListener，该接口提供 3 个监听方法。

（1）public void attributeAdded(ServletContextAttributeEvent scab)：该方法用于通知监听器，有对象或者属性被添加到 Application 中。

（2）public void attributeRemoved(ServletContextAttributeEvent scab)：该方法用于通知监听器，有对象或者属性被移除到 Application 中。

（3）public void attributeReplaced (ServletContextAttributeEvent scab)：该方法用于通知监听器，有对象或者属性被更改到 Application 中。

当 ServletContext 中添加属性、移除属性或者更改属性时，与其相对应的方法就会被调用。同样，在 Web 应用程序中，实现 ServletContextAttributeListener 方法同样有两种方法，形式如下：

注入的方式注入监听类：

```
@WebListener
public class MyServletContextAttributeListener implements
ServletContextAttributeListener{
}
```

或者在 web.xml 中配置：

```
<listener>
    <listener-class>
      com.eshore. MyServletContextAttributeListener
    </listener-class>
</listener>
```

5.7.2　HttpSession 事件监听器

从表 5.5 中，可以发现与 HttpSession 有关的监听器有 4 个：HttpSessionListener、HttpSessionAttributeListener、HttpSessionBindingListener、HttpSessionActivationListener。

1. HttpSessionListener

HttpSessionListener 是“HttpSession 生命周期监听器”，可以用来监听 HttpSession 对象初始化或者结束时响应的动作事件。

HttpSessionListener 接口的类是 javax.servlet.http.HttpSessionListener，该接口提供两个监听方法。

（1）public void sessionCreated (HttpSessionEvent se)：该方法用于通知监听器，产生了新的会话。

（2）public void sessionDestroyed(HttpSessionEvent se)：该方法用于通知监听器，已经消除一个会话。

在 HttpSession 对象初始化或者结束前，会自动调用 sessionCreated() 方法和 sessionDestroyed()方法，并传入 HttpSessionEvent 参数，它封装了 HttpSession 对象，可以通过 HttpSessionEvent 的 getSession()方法取得 HttpSession 对象。

在 Web 应用程序中，实现 HttpSessionListener 方法同样有两种方法，形式如下：

注入的方式注入监听类：

```
@WebListener
public class MyHttpSessionListener implements HttpSessionListener{
}
```

或者在 web.xml 中配置：

```
<listener>
    <listener-class>
      com.eshore. MyHttpSessionListener
    </listener-class>
</listener>
```

【例 5.4】说明 HttpSessionListener 的用法，利用 HttpSessionListener 记录在线人数

首先编写用于登录的 Servlet，源代码如下：

```
---------------------- Login.java-------------------------
import java.io.IOException;
import java.util.HashMap;
import java.util.Map;

import javax.servlet.ServletException;
import javax.servlet.annotation.WebInitParam;
import javax.servlet.annotation.WebServlet;
```

```
import javax.servlet.http.HttpServlet;
import javax.servlet.http.HttpServletRequest;
import javax.servlet.http.HttpServletResponse;

@WebServlet(
        urlPatterns = { "/Login.do" },
        loadOnStartup = 0,
        name = "Login",
        displayName = "demo",
        initParams = {
            @WebInitParam(name = "success", value = "success.jsp")
        }
)
public class Login extends HttpServlet {

    Map<String, String> users;
    //在构造方法中，初始化用户值
    public Login() {
        users = new HashMap<String, String>();
        users.put("zhangsan", "123456");
        users.put("lisi", "123456");
        users.put("wangwu", "123456");
        users.put("zhaoliu", "123456");
    }

    public void doGet(HttpServletRequest request, HttpServletResponse
    response)
            throws ServletException, IOException {
        doPost(request, response);
    }
    //重写doPost()方法
    public void doPost(HttpServletRequest request, HttpServletResponse
    response)
            throws ServletException, IOException {
        request.setCharacterEncoding("UTF-8");
        String userId = request.getParameter("userId") ;   //获取userId
        String passwd = request.getParameter("passwd");    //获取passwd
        //匹配用户名跟密码，如果一致则记录数加1
        if (users.containsKey(userId) && users.get(userId).
        equals(passwd)) {
            request.getSession().setAttribute("user", userId);
            request.getSession().setAttribute("count",
                    MyHttpSessionListener.getCount());
        }
        String success = getInitParameter("success");
```

```
            //获取初始化参数 success 值
49          response.sendRedirect(success);                 //跳转页面
50      }
51  }
```

代码第 11~19 行以注入的方式编写 Servlet，代码第 37~48 行从页面获取用户名和密码，如果用户验证通过，取得 HttpSession 实例并设置用户属性。如果想在程序中加上在线人数的功能，则可以实现 HttpSessionListener 接口。例如：

```
-----------------------
MyHttpSessionListener.java--------------------------
01  import javax.servlet.annotation.WebListener;
02  import javax.servlet.http.HttpSessionEvent;
03  import javax.servlet.http.HttpSessionListener;
04  @WebListener
05  public class MyHttpSessionListener implements HttpSessionListener{
06
07      private static int count;       //统计数
08
09      public static int getCount() {
10          return count;
11      }
12     //在 Session 开始时，统计数加1
13      public void sessionCreated(HttpSessionEvent se) {
14          MyHttpSessionListener.count++;
15      }
16     //在 Session 销毁时，统计数减1
17      public void sessionDestroyed(HttpSessionEvent se) {
18          MyHttpSessionListener.count--;
19      }
20
21  }
```

上述代码，以注入的方式注入 HttpSessionListener 监听 HttpSession 对象。代码第 14 行显示每一次 HttpSession 创建的时候，count 都会递增，而销毁 HttpSession 的时候，count 都会递减。

成功显示的页面 success.jsp，源代码如下：

```
-----------------------success.jsp---------------------------
<%@ page language="java" import="java.util.*" pageEncoding="UTF-8"%>
<%
String path = request.getContextPath();
%>
<!DOCTYPE HTML PUBLIC "-//W3C//DTD HTML 5.01 Transitional//EN">
<html>
  <head>
```

```
08      <title>登录成功界面</title>
09    </head>
10    <body>
11        <h3>目前在线人数为：${sessionScope.count}</h3>
12        <h4>欢迎您：${sessionScope.user}</h4>
13        <a href="<%=path%>/logout.do?userId=${sessionScope.user}">注销</a>
14    </body>
15  </html>
```

上述代码中，第 11 行显示在线人数，第 13 行退出用户登录。运行效果如图 5.15 所示。

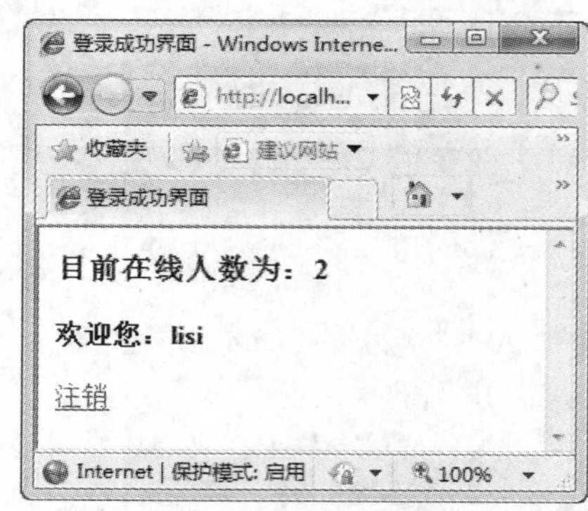

图 5.15　HttpSessionListener 效果图

2. HttpSessionAttributeListener

HttpSessionAttributeListener 是“HttpSession 属性改变监听器”，可以用来监听 HttpSession 对象加入属性、移除属性或者替换属性时响应的动作事件。

HttpSessionAttributeListener 接口的类是 javax.servlet.http.HttpSessionAttributeListener，该接口提供 3 个监听方法。

（1）public void attributeAdded(HttpSessionBindingEvent se)：该方法用于通知监听器，已经在 Session 中添加一个对象或者变量。

（2）public void attributeRemoved(HttpSessionBindingEvent se)：该方法用于通知监听器，已经在 Session 中移除一个对象或者变量。

（3）public void attributeReplaced (HttpSessionBindingEvent se)：该方法用于通知监听器，已经在 Session 中替换一个对象或者变量。

当对 session 范围的对象或者变量进行操作时，Web 容器会自动调用实现接口类相对应的方法。HttpSessionBindingEvent 是一个对象，可以利用其 getName()方法得到操作对象或者变量的名称，利用 getValue()方法得到操作对象或者变量的值。

在 Web 应用程序中，实现 HttpSessionAttributeListener 方法同样有两种方法，形式如下：

注入的方式注入监听类：

```
@WebListener
```

```
public class MyHttpSessionAttributeListener implements
HttpSessionAttributeListener{
}
```

或者在 web.xml 中配置：

```
<listener>
    <listener-class>
      com.eshore. MyHttpSessionAttributeListener
    </listener-class>
</listener>
```

3. HttpSessionBindingListener

HttpSessionBindingListener 是“HttpSession 对象绑定监听器”，可以用来监听 HttpSession 中，设置成 HttpSession 属性或者从 HttpSession 中移除时，得到 session 的通知。

HttpSessionBindingListener 接口的类是 javax.servlet.http. HttpSessionBindingListener，该接口提供两个监听方法。

（1）public void valueBound(HttpSessionBindingEvent event)：该方法用于通知监听器，已经绑定一个 session 范围的对象或者变量。

（2）public void valueUnbound (HttpSessionBindingEvent event)：该方法用于通知监听器，已经解绑一个 session 范围的对象或者变量。

参数 HttpSessionBindingEvent 是一个对象，可以通过 getSession()方法得到当前用户的 session，getName()方法得到操作的对象或者变量名称，getValue()方法得到操作的对象或者变量值。

在 Web 应用程序中，实现 HttpSessionBindingListener 接口，不需要注入或者在 web.xml 中配置，只需将设置成 session 范围的属性实现 HttpSessionBindingListener 接口就行。

4. HttpSessionActivationListener

HttpSessionActivationListener 是“HttpSession 对象转移监听器”，可以用来实现它对同一会话在不同的 JVM 中转移，例如，在负载均衡中，Web 的集群服务器中的 JVM 位于网络中的多台机器中。当 session 要从一个 JVM 转移至另一 JVM 时，必须先在原来的 JVM 上序列化所有的属性对象，若属性对象实现 HttpSessionActivationListener，就调用 sessionWillPassivate()方法，而转移后，就会调用 sessionDidActivate()方法。

HttpSessionActivationListener 接口的类是 javax.servlet.http. HttpSessionActivationListener，该接口提供两个监听方法。

（1）public void sessionDidActivate(HttpSessionEvent se)：该方法用于通知监听器，该会话已变为有效状态。

（2）public void sessionWillPassivate (HttpSessionEvent se)：该方法用于通知监听器，该会话已变为无效状态。

5.7.3 HttpServletRequest 事件、监听器

从表 5.5 中，可以发现与 HttpServletRequest 有关的监听器有两个：ServletRequestListener、ServletRequestAttributeListener。

1. ServletRequestListener

ServletRequestListener 是“Request 生命周期监听器”，可以用来监听 Reuqest 对象初始化或者结束时响应的动作事件。

ServletRequestListener 接口的类是 javax.servlet.ServletRequestListener，该接口提供两个监听方法。

（1）public void requestInitialized(ServletRequestEvent arg0)：该方法用于通知监听器，产生了新的 request 对象。

（2）public void requestDestroyed (ServletRequestEvent arg0)：该方法用于通知监听器，已经消除一个 request 对象。

在 Request 对象初始化或者结束前，会自动调用 requestInitialized ()方法和 requestDestroyed ()方法，并传入 ServletRequestEvent 参数，它封装了 ServletRequest 对象，可以通过 ServletRequestEvent 的 getServletContext()方法取得 Servlet 上下文对象，getServletRequest()方法得到请求对象。

在 Web 应用程序中，实现 ServletRequestListener 方法有两种方法，形式如下：

（1）注入的方式注入监听类

```
@WebListener
public class MyServletRequestListener implements ServletRequestListener{
}
```

（2）或在 web.xml 中配置

```
<listener>
    <listener-class>
      com.eshore. MyServletRequestListener
    </listener-class>
</listener>
```

2. ServletRequestAttributeListener

ServletRequestAttributeListener 是“Request 属性改变监听器”，可以用来监听 Request 对象加入属性、移除属性或者替换属性时响应的动作事件。

ServletRequestAttributeListener 接口的类是 javax.servlet.http. ServletRequestAttributeListener，该接口提供 3 个监听方法。

（1）public void attributeAdded(ServletRequestAttributeEvent arg0)：该方法用于通知监听器，已经在 Request 中添加一个对象或者变量。

（2）public void attributeRemoved(ServletRequestAttributeEvent arg0)：该方法用于通知监听器，已经在 Request 中移除一个对象或者变量。

（3）public void attributeReplaced(ServletRequestAttributeEvent arg0)：该方法用于通知监听器，已经在 Request 中替换一个对象或者变量。

当对 request 范围的对象或者变量进行操作时，Web 容器会自动调用实现接口类相对应的方法。ServletRequestAttributeEvent 是一个对象，可以利用其 getName()方法得到操作对象或者变量的名称，利用 getValue()方法得到操作对象或者变量的值。

在 Web 应用程序中，实现 ServletRequestAttributeListener 方法同样有两种方法，形式如下：

注入的方式注入监听类：

```
@WebListener
public class MyServletRequestAttributeListener  implements
HttpSessionAttributeListener{
}
```

或者在 web.xml 中配置：

```
<listener>
    <listener-class>
      com.eshore. MyServletRequestAttributeListener
    </listener-class>
</listener>
```

【例 5.5】说明 HttpServletRequest 的监听器示例

编写 MyRequestListener 监听类，它同时实现 ServletRequestListener 监听器与 ServletRequestAttributeListener 监听器，并用 log4j 将日志输出到指定文件。其源代码如下：

```
----------------------- MyRequestListener.java---------------------------
import javax.servlet.ServletRequestAttributeEvent;
import javax.servlet.ServletRequestAttributeListener;
import javax.servlet.ServletRequestEvent;
import javax.servlet.ServletRequestListener;
import javax.servlet.annotation.WebListener;

import org.apache.log4j.Logger;

@WebListener
public class MyRequestListener implements ServletRequestListener,
        ServletRequestAttributeListener {
    private static Logger log = Logger.getLogger("MyRequestListener");
    public void requestDestroyed(ServletRequestEvent arg0) {
        log.debug("一个请求消亡");
    }
    //request 初始化，日志新增记录
```

```
17        public void requestInitialized(ServletRequestEvent arg0) {
18            log.debug("产生一个新的请求");
19        }
20        //新增一个 request 属性，日志新增属性记录
21        public void attributeAdded(ServletRequestAttributeEvent arg0) {
22            log.debug("加入一个 request 范围的属性，名称为："+
23               arg0.getName()+",其值为："+arg0.getValue());
24        }
25        //移除一个 request 属性，日志移除属性记录
26        public void attributeRemoved(ServletRequestAttributeEvent arg0) {
27            log.debug("移除一个 request 范围的属性，名称为："+arg0.getName());
28        }
29        //修改一个 request 属性，日志修改属性记录
30        public void attributeReplaced(ServletRequestAttributeEvent arg0) {
31            log.debug("修改一个 request 范围的属性，名称为："+
32                arg0.getName()+",修改前的值为："+arg0.getValue());
33        }
34    }
```

上述代码中，第 9 行用注入的方式注入 requestListener 监听器，第 12 行获得 Logger 对象，代码第 13~33 行分配实现监听方法并实现相应的动作。然后编写 request 页面，用来请求。requestListener.jsp 页面源代码如下：

```
---------------------- requestListener.jsp-------------------------
01  <%@ page language="java" import="java.util.*" pageEncoding="UTF-8"%>
02  <%@taglib prefix="c" uri="http://java.sun.com/jsp/jstl/core" %>
03  <%
04  String path = request.getContextPath();
05  %>
06
07  <!DOCTYPE HTML PUBLIC "-//W3C//DTD HTML 5.01 Transitional//EN">
08  <html>
09    <head>
10      <title>使用 RequestListener 监听器</title>
11    </head>
12
13    <body>
14         使用 RequestListener 监听器<br/>
15         <c:set value="zhangsan" var="username" scope="request"/>
16         姓名为：<c:out value="${requestScope.username}"/>
17        <c:remove var="username" scope="request"/>
18    </body>
19  </html>
```

上述代码中，第 2 行代码引入 JSTL 标签库，代码第 15 行设定一个 request 参数 username，代码第 17 行移除一个 request 参数 username。运行上述代码，在页面输入页面路径，在日志文

件中打印信息如图 5.16 所示。

```
2014-04-23 11:09:57 [ localhost-startStop-1:0 ] - [ DEBUG ] 初始化参数name的值: linlong
2014-04-23 11:09:57 [ localhost-startStop-1:0 ] - [ DEBUG ] Tomcat正在启动中......
2014-04-23 11:10:43 [ http-apr-8080-exec-2:45451 ] - [ DEBUG ] 产生一个新的请求
2014-04-23 11:10:43 [ http-apr-8080-exec-2:45467 ] - [ DEBUG ] 修改一个request范围的属性, 名称为: org.apache.catalina.ASYNC_SUPPORTED,修改前的值为: true
2014-04-23 11:10:45 [ http-apr-8080-exec-2:47932 ] - [ DEBUG ] 加入一个request范围的属性, 名称为: username,其值为: zhangsan
2014-04-23 11:10:45 [ http-apr-8080-exec-2:47948 ] - [ DEBUG ] 移除一个request范围的属性, 名称为: username
2014-04-23 11:10:45 [ http-apr-8080-exec-2:47948 ] - [ DEBUG ] 一个请求消亡
```

图 5.16　HttpServletRequest 的监听器效果图

5.8 过滤器

上一节介绍了 Servlet 监听器，使读者了解了什么是监听器以及如何使用监听器，本节将介绍 Servlet 的另一个高级特性过滤器，以及如何编写和部署过滤器。

5.8.1　过滤器的概念

何为过滤器？顾名思义它的作用就是阻挡某些事件的发生。在 Web 应用程序中，过滤器是介于 Servlet 之前，即可以拦截过滤浏览器的请求，也可以改变对浏览器的响应。它在服务器端与客户端起到了一个中间组件的作用，对二者之间的数据信息进行过滤，其处理过程如图 5.17 所示。由图中可以看出，当客户端浏览器发起一个请求时，服务器端的过滤器将检查请求数据中的内容，它可改变这些内容或者重新设置报头信息，再转发给服务器上被请求的目标资源，处理完毕后再向客户端响应处理结果。

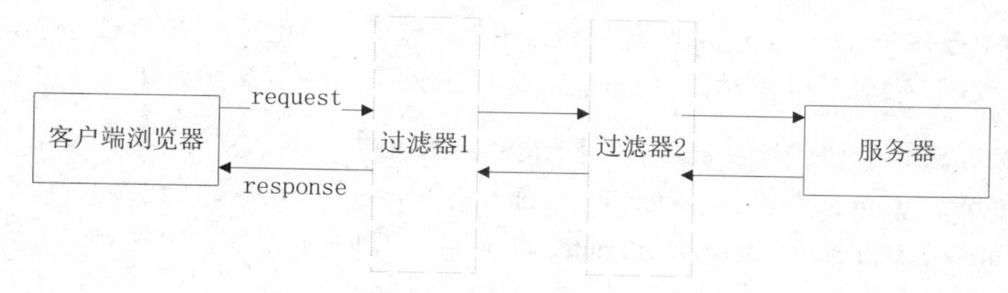

图 5.17　Filter 处理过程

一个 Web 应用程序，可以有多个过滤器，组成一个过滤器链，如经常使用过滤器完成字符编码的设定和验证用户的合法性。过滤器链中的每个过滤器都各司其职地处理并转发数据。

一般而言，在 Web 开发中，经常用 Filter 来进行以下操作。

- 对用户请求进行身份认证
- 对用户发送的数据进行过滤或者替换
- 转换图像数据格式
- 数据压缩
- 数据加密

- XML 数据的转换
- 修改请求数据的字符集

5.8.2 实现与设置过滤器

在 Servlet 中要实现过滤器，必须实现 Filter 接口，并用注入的方式或者在 web.xml 中定义过滤器，让 Web 容器知道该加载哪些过滤器。

1. Filter 接口

Filter 接口的类是 javax.servlet.Filter，该接口有 3 个方法。

（1）public void init(FilterConfig filterConifg)：该方法用来初始化过滤器。filterConifg 参数是一个 FilterConfig 对象。利用该对象可以得到过滤器中初始化配置参数信息。

（2）public void doFilter(ServletRequest request, ServletResponse response,FilterChain chain)：该方法是过滤器中主要实现过滤的方法。当客户端请求目标资源时，Web 应用程序就会调用与此目标资源相关的 doFilter()方法，在该方法中，实现对请求和响应的数据处理。参数 request 表示客户端的请求，response 表示对应请求的响应，chain 是过滤器链对象。在该方法中特定操作完成后，可调用 FilterChain 对象的 doFilter(request,response)将请求传给过滤链中的下一个过滤器，也可以直接返回响应内容，还可以将目标重定向。

（3）public void destroy()：该方法用于释放过滤器中使用的资源。

2. FilterConfig 接口

过滤器中还有 FilterConfig 接口，该接口用于在过滤器初始化时 Web 容器向过滤器传送初始化配置参数，并传入过滤器对象的 init()方法中。FilterConfig 接口中有 4 个方法可以调用。

（1）public String getFilterName()：用于得到过滤器的名字。

（2）public String getInitParameter(String name)：得到过滤器中初始化参数值。

（3）public Enumeration getInitParameterNames()：得到过滤器配置中的所有初始化参数名字的枚举类型。

（4）public ServletContext getServlet ()：得到 Servlet 上下文文件对象。

3. 设置过滤器

实现过滤器有两种方法：注入或者在 web.xml 中配置。其形式如下。

注入方式：

```
@WebFilter(
description = "demo",
filterName = "myfilter",
servletNames = { "*.do" },
urlPatterns = { "/*" },
```

```
    initParams = {
        @WebInitParam(name = "param", value = "paramvalue")
    },
    dispatcherTypes = { DispatcherType.REQUEST }
)
```

它等价于 web.xml 配置如下：

```
<filter>
    <description>demo</description>
  <!--过滤器名称-->
    <filter-name>myfilter</filter-name>
 <!--过滤器类-->
    <filter-class>com.eshore.MyFilter</filter-class>
 <!--过滤器初始化参数-->
    <init-param>
       <param-name>param</param-name>
       <param-value>paramvalue</param-value>
    </init-param>
  </filter>
  <!--过滤器映射配置-->
  <filter-mapping>
     <filter-name>myfilter</filter-name>
     <servlet-name>*.do</servlet-name>
     <url-pattern>/*</url-pattern>
     <dispatcher>REQUEST</dispatcher>
  </filter-mapping>
```

上述的两个配置过程是等价的，在 Servlet 3.0 中可以同时兼容，在 Servlet 2.0 中只能在 web.xml 中配置。@ WebFilter 的主要属性见表 5.7 所示，@ WebInitParam 的主要属性参见表 5.4 所示。

表 5.7　@WebFilter的主要属性列表

属性名	描述
value	该属性与 urlPatterns 属性等价，但两个属性不能同时使用
urlPatterns	指定 Filter 的 URL 匹配模式，等价于<url-pattern>标签
filterName	指定 Filter 的 name 属性，等价于<filter-name>标签
servletNames	指定 Filter 的 Servlet 过滤对象，等价于<servlet-name>标签。当与 urlPatterns 同时存在时，则 Web 容器先比对 urlPatterns 中的 URL，再比对 servletNames 中的配置
dispatcherTypes	指定 Filter 的过滤时间，等价于<dispatcher>标签，其值有 FORWARD、INCLUDE、REQUEST、ERROR、ASYNC 等。默认值是 REQUEST
asyncSupported	声明 Servlet 是否支持异步操作模式，等价于<async-supported>标签。该属性在 Servlet 3.0 才有
initParams	设置过滤器的初始参数

5.8.3 请求封装器

请求封装器是指利用 HttpServletRequestWrapper 类将请求中的内容进行统一修改，例如修改请求字符编码、替换字符、权限验证等。

下面通过例子来说明请求封装器的实现。

【例 5.6】实现编码过滤器

通常在 Web 中实现编码过滤器为如下形式：

```
---------------------- EncodingFilter.java---------------------------
@WebFilter(
        description = "字符编码过滤器",
        filterName = "encodingFilter",
        urlPatterns = { "/*" },
        initParams = {
            @WebInitParam(name = "ENCODING", value = "UTF-8")
        }
    )
public class EncodingFilter implements Filter{
    private static Logger log = Logger.getLogger("EncodingFilter");
    private String encoding="";
    private String filterName="";

    public void init(FilterConfig filterConfig) throws ServletException {
        //通过 filterConfig 获得初始化中编码值
        encoding = filterConfig.getInitParameter("ENCODING");
        filterName = filterConfig.getFilterName();
        if(encoding==null||"".equals(encoding)){
            encoding="UTF-8";
        }
        log.debug("获得编码值");
    }

    public void doFilter(ServletRequest request, ServletResponse
response,
            FilterChain chain) throws IOException, ServletException {
        //分别对请求和响应进行编码设置
        request.setCharacterEncoding(encoding);
        response.setCharacterEncoding(encoding);
        HttpServletRequest req = (HttpServletRequest)request;
        log.debug("请求被"+filterName+"过滤");
        //传输给过滤器链过滤
```

```
32            chain.doFilter(req, response);
33            log.debug("响应被"+filterName+"过滤");
34        }
35
36        public void destroy() {
37            log.debug("请求销毁");
38        }
39    }
```

上述代码中通过注入的方式注入过滤器，并用 log4j 将日志写到文件中，代码第 16~18 行取得过滤器中的初始化值，代码第 27~32 行设置编码并传输给过滤器链。代码第 01~08 行等价于在 web.xml 中的配置如下：

```
<filter>
     <description>字符编码过滤器</description>
     <filter-name>encodingFilter</filter-name>
     <filter-class>com.eshore.EncodingFilter</filter-class>
     <init-param>
        <param-name>ENCODING</param-name>
        <param-value>UTF-8</param-value>
     </init-param>
</filter>

<filter-mapping>
      <filter-name>encodingFilter</filter-name>
      <url-pattern>/*</url-pattern>
</filter-mapping>
```

上述编码过滤器对于 post 请求，是没有问题的，但是对于 get 请求获取中文参数时还是会出现乱码问题。这是因为 post 方式请求时，参数是在请求数据包的消息体中；而 get 请求，参数存放在请求数据包的 URI 字段中。而 request.setCharacterEncoding(encoding);只对消息体中的数据有作用，对 URI 字段中的参数不起作用。基于这种情况，想到请求包装器包装请求，将字符编码转换的工作添加到 getParameter()方法中，这样就可以对请求的参数进行统一转换。

编写请求包装类 RequestEncodingWrapper 并继承 HttpServletRequestWrapper，其源代码如下：

```
   -----------------------
RequestEncodingWrapper.java---------------------------
import java.io.UnsupportedEncodingException;

import javax.servlet.http.HttpServletRequest;
import javax.servlet.http.HttpServletRequestWrapper;

```

```
06  public class RequestEncodingWrapper extends HttpServletRequestWrapper{
07      private String encoding="";
08      public RequestEncodingWrapper(HttpServletRequest request) {
09          //必须调用父类构造方法
10          super(request);
11      }
12      public RequestEncodingWrapper(HttpServletRequest request,String
encoding) {
13          //必须调用父类构造方法
14          super(request);
15          this.encoding = encoding;
16      }
17      //重新定义 getParameter 方法
18      public String getParameter(String name){
19          String value = getRequest().getParameter(name);
20          try {
21              //将参数值进行编码转换
22              if(value!=null&&!"".equals(value))
23                  value = new String(value.trim().getBytes("ISO-8859-1"),
encoding);
24          } catch (UnsupportedEncodingException e) {
25              // TODO Auto-generated catch block
26              e.printStackTrace();
27          }
28          return value;
29      }
30  }
```

上述代码中，RequestEncodingWrapper 类继承了 HttpServletRequestWrapper，第 12~16 行自定义构造方法并实现父类的构造方法，HttpServletRequest 将通过此构造方法传入，如果要取得被封装的 HttpServletRequest，则可以调用 getRequest()方法，代码第 19~23 行，重构 getParameter()方法，在此方法中，从 HttpServletRequest 对象上取得请求参数值进行编码转换。这时过滤器 EncodingFilter 的 doFilter 方法更改如下：

```
public void doFilter(ServletRequest request, ServletResponse response,
        FilterChain chain) throws IOException, ServletException {
    //分别对请求和响应进行编码设置
    HttpServletRequest req = (HttpServletRequest)request;
    log.debug("请求被"+filterName+"过滤");
    //如果请求的方法是 get，则用请求包装器，否则直接设定编码
    if("GET".equals(req.getMethod())){
        req = new RequestEncodingWrapper(req,encoding);
    }else{
        request.setCharacterEncoding(encoding);
    }
```

```
        response.setCharacterEncoding(encoding);
        //传输给过滤器链过滤
        chain.doFilter(req, response);
        log.debug("响应被"+filterName+"过滤");
    }
```

请求参数的编码设置是通过过滤器初始化参数来设置的，并在过滤器中的 init()中设置，过滤器仅在 GET 方法请求时创建 RequestEncodingWrapper 实例，POST 方法则通过直接设定编码方式实现，最后调用 FilterChain 的 doFilter()方法传入实例。

5.8.4　响应封装器

响应封装器是指利用 HttpServletResponseWrapper 类将响应中的内容进行统一修改，例如压缩输出内容，替换输出内容等。有些时候需要对网站的输出内容进行控制，一般有两种方法：

一是在保存在数据库前对不合法的内容进行替换；

二是在输出端进行替换。能想到要是对每一个 Servlet 进行输出控制，任务量就非常大而且繁琐。

读者会想到利用过滤器对 Servlet 进行统一处理，但是因为 HttpServletResponse 不能缓存输出内容，所以需要自定义一个具备缓存功能的 response。下面通过两个例子说明响应封装器的实现。

【例 5.7】实现内容替换过滤器

首先，编写一个响应的封装器 ResponseReplaceWrapper，用它来缓存 response 中的内容。其源代码如下：

```
------------------------ResponseReplaceWrapper.java----------------------------
import java.io.CharArrayWriter;
import java.io.IOException;
import java.io.PrintWriter;
import javax.servlet.http.HttpServletResponse;
import javax.servlet.http.HttpServletResponseWrapper;

public class ResponseReplaceWrapper extends HttpServletResponseWrapper{

    private CharArrayWriter charWriter = new CharArrayWriter();
    public ResponseReplaceWrapper(HttpServletResponse response) {
        //必须调用父类构造方法
        super(response);
    }

    public PrintWriter getWriter() throws IOException{
        //返回字符数组 Writer，缓存内容
```

```
17          return new PrintWriter(charWriter);
18      }
19
20      public CharArrayWriter getCharWriter() {
21          return charWriter;
22      }
23  }
```

上述代码中第 11 行调用父类的构造方法，代码第 15~17 行重构 getWriter()方法并用字符数组缓存输出内容。

然后，编写内容过滤器 ReplaceFilter，其源代码如下：

```
---------------------- ReplaceFilter.java--------------------------
......
@WebFilter(
    description = "内容替换过滤器",
    filterName = "replaceFilter",
    urlPatterns = { "/*" },
    initParams = {
        @WebInitParam(name = "filePath", value = "replace_ZH.properties")
    }
)
public class ReplaceFilter implements Filter{

    private Properties propert = new Properties();
    public void init(FilterConfig filterConfig) throws ServletException {
        //通过 filterConfig 获得初始化文件名
        String filePath = filterConfig.getInitParameter("filePath");
        try {
            //导入资源文件
            propert.load(ReplaceFilter.class.getClassLoader()
                .getResourceAsStream(filePath));
        } catch (FileNotFoundException e) {
            e.printStackTrace();
        } catch (IOException e) {
            e.printStackTrace();
        }
    }
    public void doFilter(ServletRequest request, ServletResponse response,
            FilterChain chain) throws IOException, ServletException {
        HttpServletResponse res = (HttpServletResponse)response;
        //实例化响应器包装类
        ResponseReplaceWrapper resp = new ResponseReplaceWrapper(res);
        chain.doFilter(request, resp);
        //缓存输出字符
```

```
33          String outString = resp.getCharWriter().toString();
34          //循环替换不合法的字符
35          for(Object o:propert.keySet()){
36              String key = (String) o;
37              outString = outString.replace(key, propert.getProperty(key));
38          }
39          //用原先的 HttpServletResponse 输出字符
40          PrintWriter out = res.getWriter();
41          out.write(outString);
42      }
43      public void destroy() {
44
45      }
46  }
```

上述代码中第 2~9 行，用注入的方式注入内容过滤器并设定文件名称，代码第 13~25 行，在 init 初始化过滤器方法中，用类反射加载内容过滤文件内容，代码第 30 行，实例化响应包装类 ResponseReplaceWrapper 进行缓存输出内容，代码第 35~38 行替换不合法的字符，代码第 41 行用原先的 HttpServletResponse 将替换后的内容输出到浏览器中。

本例中 ResponseReplaceWrapper 只是一个假的 response。它不负责输出到客户端只负责将输出内容缓存起来。输出到客户端的还是通过原先的 response 完成。

replace_ZH.properties 配置文件的内容如下：

```
\u8272\u60c5=****
\u60c5\u8272=****
\u8d4c\u535a=****
```

replace_ZH.properties\文件时经过 native2ascii 编码而得到的。native2ascii 命令在 JDK 中 bin 目录下。实现命令为：native2ascii –encoding utf-8 源文件转换文件。

最后，编写一个测试的 Servlet，其源代码如下：

```
------------------------ TestServlet.java----------------------------
@WebServlet(
    urlPatterns = { "/Test.do" },
    loadOnStartup = 0,
    name = "testServlet"
)
public class TestServlet extends HttpServlet {
    public void doGet(HttpServletRequest request, HttpServletResponse
    response)
            throws ServletException, IOException {
        doPost(request, response);
    }
```

```
11
12      public void doPost(HttpServletRequest request, HttpServletResponse
        response)
13              throws ServletException, IOException {
14          response.setContentType("text/html;charset=utf-8");
15          PrintWriter out = response.getWriter();
16          out .println("<!DOCTYPE HTML PUBLIC \"-//W3C//DTD
17           HTML 5.01 Transitional//EN\">");
18          out.println("<HTML>");
19          out.println("  <HEAD><TITLE>测试内容输出过滤</TITLE></HEAD>");
20          out.println("  <BODY>");
21          out.println("dfdasf <br/>色情  <br/>情色  <br/>赌博");
22          out.println("  </BODY>");
23          out.println("</HTML>");
24          out.flush();
25          out.close();
26      }
27  }
```

上述测试的 Servlet 很简单，代码第 1~4 行用注入的方式声明一个 Servlet，代码第 14~24 直接用 response 直接输出内容。上述代码运行效果图如图 5.18 所示。

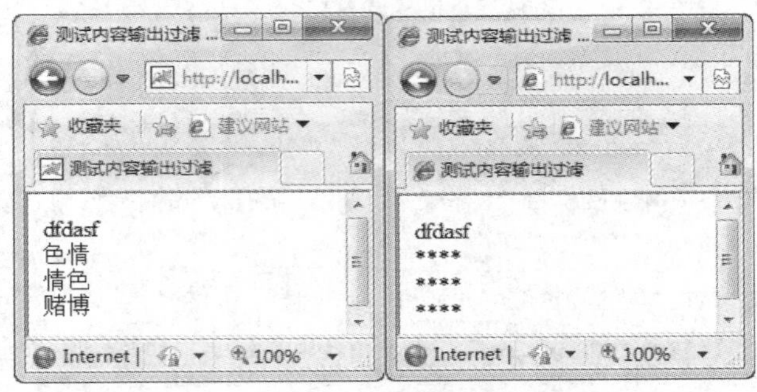

图 5.18　内容过滤效果图

上面例子介绍了内容替换的过滤器，使读者了解如何编写一个自定义的响应封装器和如何使用。下面介绍页面内容压缩过滤器，因为网站常使用 GZIP 压缩算法对网页内容进行压缩，然后传给浏览器，这样可以减少数据传输量和提高响应速度。当浏览器接收到 GZIP 压缩数据后会自动解压并正确显示。

【例 5.8】实现 GZIP 压缩过滤器

首先，编写一个自定义的 ServletOutputStream 类使它具有压缩功能，这里压缩的功能采用 GZIP 算法，这是现在主流浏览器都可以接受的压缩格式，应用 JDK 自带的 GZIPOutputStream 类来完成，其源代码如下：

```
----------------------- GZIPResponseStream.java----------------------------
import java.io.*;
```

```
import java.util.zip.GZIPOutputStream;
import javax.servlet.*;
import javax.servlet.http.*;

public class GZIPResponseStream extends ServletOutputStream {
    //将压缩后的数据存放在 ByteArrayOutputStream 对象中
  protected ByteArrayOutputStream bArrayOutputStream = null;
    //JDK 中自带的 GZIP 压缩类
  protected GZIPOutputStream gzipOutputStream = null;
  protected boolean closed = false;
  protected HttpServletResponse response = null;//原先的 response
  protected ServletOutputStream outputStream = null;//response 中的输出流
  //构造方法，初始化定义值
  public GZIPResponseStream(HttpServletResponse response) throws IOException {
    super();
    closed = false;
    this.response = response;
    this.outputStream = response.getOutputStream();
    bArrayOutputStream = new ByteArrayOutputStream();
    gzipOutputStream = new GZIPOutputStream(bArrayOutputStream);
  }
  //执行压缩，并将数据输出到浏览器
  public void close() throws IOException {
    if (closed) {
      throw new IOException("This output stream has already been closed");
    }
    //执行压缩，必须要调用这个方法
    gzipOutputStream.finish();
    //将压缩后的数据输出到浏览器中
    byte[] bytes = bArrayOutputStream.toByteArray();
    //设置压缩算法为 gzip，浏览器会自动解压数据
    response.addHeader("Content-Length",Integer.toString(bytes.length));
    response.addHeader("Content-Encoding", "gzip");
    //输出到浏览器
    outputStream.write(bytes);
    outputStream.flush();
    outputStream.close();
    closed = true;
  }

  public void flush() throws IOException {
    if (closed) {
```

```
44          throw new IOException("不能刷新关闭的流！");
45        }
46        gzipOutputStream.flush();
47      }
48      //重写 write 方法，如果流关闭则抛异常
49      public void write(int b) throws IOException {
50        if (closed) {
51          throw new IOException("输出流关闭中！");
52        }
53        gzipOutputStream.write((byte)b);
54      }
55      //重写 write 方法
56      public void write(byte b[]) throws IOException {
57        write(b, 0, b.length);
58      }
59      //重写 write 方法，如果流关闭则抛异常
60      public void write(byte b[], int off, int len) throws IOException {
61        if (closed) {
62          throw new IOException("输出流关闭中！");
63        }
64        gzipOutputStream.write(b, off, len);
65      }
66
67      public boolean closed() {
68        return (this.closed);
69      }
70
71    }
```

上述代码中，GZIPResponseStream 继承 ServletOutputStream，使用时传入原始的响应对象，它的主要功能是将 response 中的数据用 ByteArrayOutputStream 缓存起来，然后用 GZIPOutputStream 类将数据压缩，并输出到客户端浏览器中。代码第 15~22 行初始化自定义参数，代码第 24~40 行是执行压缩方法，并将缓存中的数据输出到浏览器中。代码第 49~65 行都是重构输出方法，用 GZIPOutputStream 输出。

然后，自定义 response 包装类 GZIPResponseWrapper，它只对输出的内容进行压缩，不进行将内容输出到客户端的操作。因为 response 要处理的不单单是字符内容还有处理压缩的内容即二进制内容，所以它需要重写 getOutputStream()和 getWriter()方法，其源代码如下：

```
----------------------- GZIPResponseWrapper.java---------------------------
import java.io.*;
import java.util.*;
import javax.servlet.*;
import javax.servlet.http.*;
```

```

public class GZIPResponseWrapper extends HttpServletResponseWrapper {
  //原始的 response
  private HttpServletResponse response = null;
  //自定义的 outputStream,对数据压缩，并且输出
  private ServletOutputStream outputStream = null;
  //自定义 PrintWriter,将内容输出到 ServletOutputStream
  private PrintWriter printWriter = null;

  public GZIPResponseWrapper(HttpServletResponse response) {
    super(response);
    this.response = response;
  }

  //用 GZIPResponseStream 创建输出流
  public ServletOutputStream createOutputStream() throws IOException {
    return (new GZIPResponseStream(response));
  }
  //执行这个方法对数据进行 GZIP 压缩，并输出到浏览器中
  public void finishResponse() {
    try {
      if (printWriter != null) {
       printWriter.close();
      } else {
        if (outputStream != null) {
         outputStream.close();
        }
      }
    } catch (IOException e) {}
  }
  //刷新 ServletOutputStream 输出流
  public void flushBuffer() throws IOException {
      outputStream.flush();
  }
  //覆盖 getOutputStream 方法，处理二进制内容
  public ServletOutputStream getOutputStream() throws IOException {
    if (printWriter != null) {
      throw new IllegalStateException("getWriter() has already been
     called!");
    }
    //如果 outputStream 为空，则创建流
    if (outputStream == null)
     outputStream = createOutputStream();
    return (outputStream);
  }
```

```
49    //处理 getWriter 方法，处理字符内容
50    public PrintWriter getWriter() throws IOException {
51      if (printWriter != null) {
52        return (printWriter);
53      }
54
55      if (outputStream != null) {
56        throw new IllegalStateException("getOutputStream() has already been
           called!");
57      }
58      outputStream = createOutputStream();
59      //通过 outputStream 获得 printWriter 方法
60      printWriter = new PrintWriter(new OutputStreamWriter(outputStream,
        "UTF-8"));
61      return (printWriter);
62      }
63      //压缩后数据长度有变化，所以不用重写该方法
64      public void setContentLength(int length) {}
65  }
```

在上述代码中，第 24~34 行执行数据压缩的方法，getOutputStream()方法和 getWriter()方法都实现 GZIPResponseStream 实例，这样就使得它们都具有压缩功能。代码第 42 行和第 56 行，是因为同一个 Servlet 请求中，getWriter()和 getOutputStream()只能调用一个，两个同时调用必须抛出 IllegalStateException 异常。

接着，编写压缩过滤器类 GZIPFilter，过滤器类中通过检查 Accept-Encoding 标头是否包含 gzip 字符来要判断浏览器是否支持 GZIP 压缩算法，如果支持，则进行 GZIP 压缩数据。否则直接输出。其源代码如下：

```
----------------------- GZIPFilter.java--------------------------
import java.io.*;
import javax.servlet.*;
import javax.servlet.annotation.WebFilter;
import javax.servlet.annotation.WebInitParam;
import javax.servlet.http.*;

@WebFilter(
    description = "内容替换过滤器",
    filterName = "gzipFilter",
    urlPatterns = { "/*" }
)
public class GZIPFilter implements Filter {

    public void doFilter(ServletRequest req, ServletResponse res,
            FilterChain chain) throws IOException, ServletException {
```

```
16          if (req instanceof HttpServletRequest) {
17              HttpServletRequest request = (HttpServletRequest) req;
18              HttpServletResponse response = (HttpServletResponse) res;
19              //依据浏览器的 Header 的信息，判断支持的编码方式
20              String ae = request.getHeader("Accept-Encoding");
21              //如果浏览器支持 gzip 格式，则使用 gzip 压缩数据
22              if (ae != null && ae.toLowerCase().indexOf("gzip") != -1) {
23                  GZIPResponseWrapper wrappedResponse =
24              new GZIPResponseWrapper(response);
25                  chain.doFilter(req, wrappedResponse);
26                  //输出压缩数据
27                  wrappedResponse.finishResponse();
28                  return;
29              }
30              chain.doFilter(req, res);
31          }
32      }
33  }
```

上述代码中，代码第 22 行判断浏览器标头是否包含 gzip 字符，代码第 23~24 行传入压缩响应对象 GZIPResponseWrapper。代码第 27 行输出压缩后的数据到浏览器中。

最后，写个测试的 GzipServlet 测试压缩结果。其源代码如下：

```
------------------------GzipServlet.java---------------------------
@WebServlet(
        urlPatterns = { "/gzip.action" },
        loadOnStartup = 0,
        name = "gzipServlet"
    )
public class GzipServlet extends HttpServlet{

    public void doGet(HttpServletRequest request, HttpServletResponse
    response)
            throws ServletException, IOException {
        doPost(request, response);
    }
    public void doPost(HttpServletRequest request, HttpServletResponse
    response)
            throws ServletException, IOException {
        response.setContentType("text/html;charset=utf-8");
        response.setCharacterEncoding("UTF-8");
        PrintWriter out = response.getWriter();
       String[] urls = {//设置一组 URL 地址
            "http://localhost:8080/ch07/11.png",
            "http://code.jquery.com/ui/1.10.3/jquery-ui.js",
```

```
20                "http://localhost:8080/ch07/login.jsp"
21                };
22            out.println("<!DOCTYPE HTML PUBLIC \"-//W3C//DTD HTML 5.01
23            Transitional//EN\">");
24            out.println("<HTML>");
25            out.println("  <HEAD><TITLE>测试内容压缩输出过滤</TITLE></HEAD>");
26            out.println("  <BODY>");
27
28            for(String url:urls){
29                //模拟一个浏览器
30              URLConnection connGzip = new URL(url).openConnection();
31              //模拟实质浏览器的表头信息支持 gzip 压缩格式
32                connGzip.setRequestProperty("Accept-Encoding", "gzip");
33                int lengthGzip = connGzip.getContentLength();//获取压缩后的长度
34                //模拟另一个浏览器
35                URLConnection connCommon = new URL(url).openConnection();
36               int lengthCommon = connCommon.getContentLength();//获取压缩前的长度
37           double rate = new Double(lengthGzip)/lengthCommon;//计算压缩比率
38           out.println("<table border=\"1\" cellpadding=\"2\"
             cellspacing=\"1\">");
39           out.println("<tr>");
40           out.println("<td colspan=\"3\">网址: "+url+"</td>");
41           out.println("</tr>");
42           out.println("<tr>");
43           out.println("<td>压缩后数据: "+lengthGzip+"byte</td>");
44           out.println("<td>压缩前数据: "+lengthCommon+"byte</td>");
45           out.println("<td>压缩率  :
46          "+NumberFormat.getPercentInstance().format(1-rate)+"</td>");
47           out.println("</tr>");
48           out.println("</table>");
49           }
50            out.println("  </BODY>");
51            out.println("</HTML>");
52            out.flush();
53            out.close();
54       }
55   }
```

上述代码中，用注入的方式声明 Servlet，代码第 17~21 行存放一组 URL 地址，代码第 29~33 行模拟两个浏览器，一个支持 gzip 压缩算法，一个不支持，代码第 38~48 行输出结果到浏览器中。运行效果如图 5.19 所示。

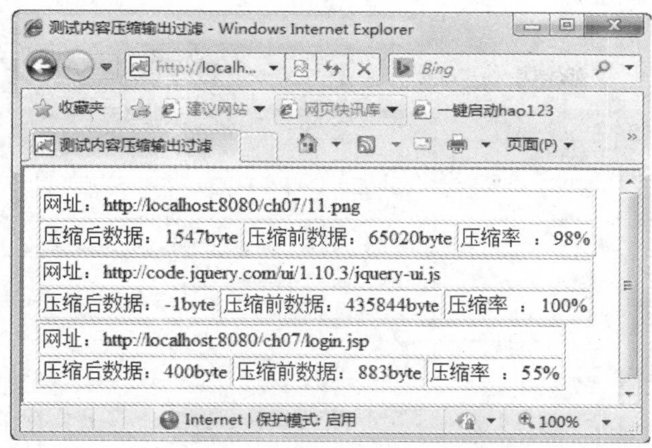

图 5.19 压缩过滤器效果图

5.9 异步处理

在 Servlet 2.0 中，一个普通的 Servlet 工作流程大致是：

- 首先，Servlet 接收请求，对数据进行处理；
- 然后，调用业务接口方法，完成业务处理；
- 最后，将结果返回到客户端。

Servlet 中最耗时的是在第二步的业务处理，因为它会做些数据库操作或者其他的跨网络调用等，在处理业务的过程中，该线程占用的资源不会被释放，这有可能造成性能的瓶颈。

异步处理是 Servlet 3.0 中新增的一个特性，它可以先释放容器被占用的资源，将请求交给另一个异步线程来执行，业务方法执行完成后再生成响应数据。

5.9.1 AsyncContext 简介

在 Servlet 3.0 中，ServletRequest 提供了两个方法启动 AsyncContext。分别如下：

```
AsyncContext startAsync()
```

与

```
AsyncContext startAsync(ServletRequest servletRequest,
ServletResponse servletResponse)
```

上述两个方法都能得到 AsyncContext 接口实现对象。当一个 Servlet 调用了 startAsync() 方法之后，该 Servlet 的响应就会被延迟，并释放容器分配的线程。AsyncContext 接口的主要方法见表 5.8 所示。

表 5.8　AsyncContext 接口的主要方法

方法	描述
void addListener(AsyncListener listener)	添加 AsyncListener 监听器
complete()	响应完成
dispatch()	指定 URL 进行响应完成
getRequest()	获取 servlet 请求对象
getResponse ()	获取 servlet 响应对象
setTimeout(long timeout)	设置超时时间
start(java.lang.Runnable run)	异步启动线程

在 Servlet 3.0 中，有两种方式实现 Servlet 支持异步处理：注入声明和 web.xml。其形式分别如下所示：

```
@WebServlet(
  asyncSupported=true,
  urlPatterns={"/asyncdemo.do"},
  name="myAsyncServlet"
)
public class MyAsyncServlet extends HttpServlet{
......
}
```

web.xml 中配置如下：

```
<servlet>
    <servlet-name>myAsyncServlet</servlet-name>
    <!—异步 Servlet 类路径->
<servlet-class>com.eshore.MyAsyncServlet</servlet-class>
 <!—异步支持属性-->
    <async-supported>true</async-supported>
  </servlet>
```

如果支持异步处理的 Servlet 前面有 Filter，则 Filter 也需要支持异步处理。

【例 5.9】演示异步处理 Servlet

编写一个支持异步通信的 Servlet 类，对于每个请求，该 Servlet 会取得异步信息 AsyncContext，同时释放占用的内存，延迟响应。然后启动 AsyncRequest 对象定义的线程，在这线程中做业务处理，等业务处理完成，输出页面信息并调用 AsyncContext 的 complete()方法表示异步完成。异步处理 Servlet 的源代码如下：

```
---------------------- MyAsyncServlet.java--------------------------
@WebServlet(
    asyncSupported = true,
    urlPatterns = { "/asyncdemo.do" },
```

```
    name = "myAsyncServlet"
)
public class MyAsyncServlet extends HttpServlet {
    SimpleDateFormat sdf = new SimpleDateFormat("yyyy-MM-dd HH:mm:ss");
    public void doGet(HttpServletRequest request, HttpServletResponse
    response)
            throws ServletException, IOException {
        doPost(request, response);
    }

    public void doPost(HttpServletRequest request, HttpServletResponse
    response)
            throws ServletException, IOException {
        response.setContentType("text/html;charset=UTF-8");
        PrintWriter out = response.getWriter();
        out.println("开始时间：" + sdf.format(new Date()) + "  ");
        out.flush();
        //在子线程中执行业务调用，并由其负责输出响应，主线程退出
        AsyncContext asyncContext = request.startAsync(request,response);
        asyncContext.setTimeout(900000000);//设置最大的超时时间
        new Thread(new Executor(asyncContext)).start();
        out.println("结束时间：" + sdf.format(new Date())+ "  ");
        out.flush();
    }
    //内部类
    public class Executor implements Runnable {
        private AsyncContext ctx = null;
        public Executor(AsyncContext ctx){
            this.ctx = ctx;
        }
        public void run(){
            try {
                //等待二十秒钟，以模拟业务方法的执行
                Thread.sleep(20000);
                PrintWriter out = ctx.getResponse().getWriter();
                out.println("业务处理完毕的时间：" + sdf.format(new Date()) +
".");

                out.flush();
                ctx.complete();
            } catch (Exception e) {
                e.printStackTrace();
            }
        }
    }
}
```

上述代码中，第 2 行表示该 Servlet 支持异步处理。代码第 20 行获得 AsyncContext 对象。第 21 行设置 AsyncContext 超时时间，其默认的超时时间是 10 秒。第 27 行启动一线程 Executor 类，第 35 行等待 20 秒钟用以模拟业务处理，第 36 行输出响应结果。第 39 行表示 AsyncContext 类处理完成，运行效果如图 5.20 所示。

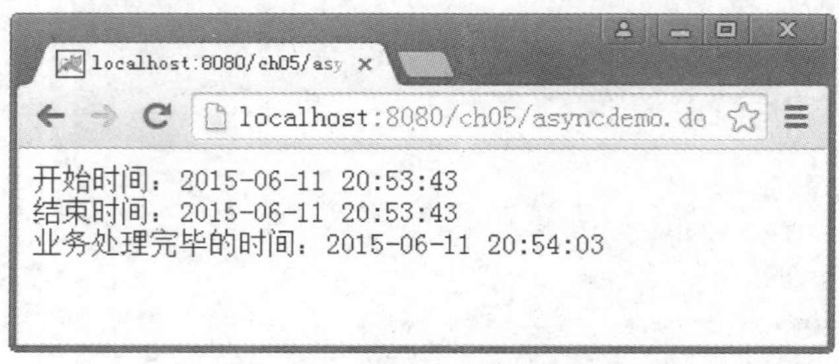

图 5.20　异步处理效果图

运行时如果报 is not surpported 错误，可能是没有将所有的过滤器或者经过的 Servlet 都设置成支持异步处理的。

5.9.2　模拟服务器推送

模拟服务器推送是指模拟由服务器端向客户端推送消息。在 HTTP 协议中，服务器时无法直接对客户端传送消息的，必须得有一个请求服务器端才能够响应。可以利用 Servlet 3.0 的异步处理技术，做到类似服务器主动推送消息到客户端。下面以一个例子说明，这种技术的实现过程。

【例 5.10】模拟服务器推送

首先，编写一个负责存储消息的队列类 ClientService，该类的作用是用 Queue 添加异步所有的 AsyncContext 对象，用 BlockingQueue 阻塞队列存储页面请求的消息队列，当 Queue 队列中有数据时，启动一线程，将 BlockingQueue 阻塞的内容输出到页面中。ClientService 的源代码如下：

```
---------------------- ClientService.java--------------------------
public class ClientService {

    private final Queue<AsyncContext> ASYNC_QUEUE =
        new ConcurrentLinkedQueue<AsyncContext>();//异步 Servlet 上下文队列

    private final BlockingQueue<String> INFO_QUEUE =
        new LinkedBlockingQueue<String>();      //消息队列.

    private static ClientService instance  = new ClientService();

    private ClientService() {                                  //构造方法，启动线程
```

```
            new Thread(this.notifyRunnable).start();
    }

    public static ClientService getInstance() { //获得 ClientService 单例
        return instance;
    }

    public void addAsyncContext(final AsyncContext asyncContext) {
        ASYNC_QUEUE.add(asyncContext);//添加异步 Servlet 上下文
    }

    public void removeAsyncContext(final AsyncContext asyncContext) {
        ASYNC_QUEUE.remove(asyncContext);//删除异步 Servlet 上下文
    }

    /**
     *
     * 发送消息到异步线程，最终输出到 http response 流 .<br>
     * @param str 发送给客户端的消息.<br>
     */
    public void callClient(final String str) {
        try {
            INFO_QUEUE.put(str);
        } catch (Exception ex) {
            throw new RuntimeException(ex);
        }
    }
    /**
     *
     * 将数据发送到 response 流上
     */
    private Runnable notifyRunnable = new Runnable() {
        public void run() {
            boolean done = false;
            while (!done) {
                try {
        final String script = INFO_QUEUE.take();//当消息队列中有数据，则调
                                                  用 take()方法，
        for (AsyncContext ac : ASYNC_QUEUE) {
            try {
                //调用响应中的 getWriter 方法
                PrintWriter writer = ac.getResponse().getWriter();
                writer.println(escapeHTML(script));
                writer.flush();
            } catch (IOException ex) {
```

```
56                      ASYNC_QUEUE.remove(ac);
57                              throw new RuntimeException(ex);
58                              }
59                          }
60                  }catch(InterruptedException e) {
61                      //抛异常，开关为 true
62                      done = true;
63                      e.printStackTrace();
64                  }
65              }
66          }
67      };
68
69          /**
70           * 删除多余的空格
71           */
72          private String escapeHTML(String str) {
73              return "<script type='text/javascript'>\n"
74              + str.replaceAll("\n", "").replaceAll("\r", "")
                //替换空格或者换行符
75          + "</script>\n";
76      }
77  }
```

上述代码中，代码第 12 行表示启动一个线程，该线程的作用是输出内容到页面中。代码第 49~66 行遍历所有的异步队列，将内容输出到页面中，代码第 72~76 行是过滤回车符和换行符。

其次，编写异步的 Servlet 类，该类的作用是将客户端注册到发送消息的监听队列中，当产生超时、错误等事件时，将异步上下文对象从队列中移除。同时当访问该 Servlet 的客户端时，在 ASYNC_QUEUE 中注册一个 AsyncContext 对象，这样当服务端需要调用客户端时，就会输出 AsyncContext 内容到客户端。其源代码如下：

```
----------------------- AsyncContextServlet.java--------------------------
@WebServlet(urlPatterns = { "/AsyncContextServlet" }, asyncSupported =
 true)
public class AsyncContextServlet extends HttpServlet {

    private static final long serialVersionUID = 1L;

    @Override
    protected void doGet(HttpServletRequest request, HttpServletResponse
    response)
    throws ServletException, IOException {

```

```
10            request.setCharacterEncoding("UTF-8");                    //设置字符集
11            response.setContentType("text/html;charset=UTF-8");
              //设置响应类型和字符集
12            final AsyncContext asyncContext = request.startAsync();
13            //注册 AsyncContext 对象超时时间
14            asyncContext.setTimeout(1000000);
15            asyncContext.addListener(new AsyncListener() { //添加异步监听器
16                public void onComplete(AsyncEvent event) throws IOException {
17                ClientService.getInstance().removeAsyncContext(asyncContext);
18                }
19                public void onTimeout(AsyncEvent event) throws IOException {
20                ClientService.getInstance().removeAsyncContext(asyncContext);
21                }
22                public void onError(AsyncEvent event) throws IOException {
23                ClientService.getInstance().removeAsyncContext(asyncContext);
24                }
25                public void onStartAsync(AsyncEvent event) throws IOException {
26                }
27            });
28            //添加异步 asyncContext 对象
29            ClientService.getInstance().addAsyncContext(asyncContext);
30        }
31    }
```

上述代码中，第 14 行是注册 AsyncContext 对象超时时间，第 15~27 行为异步对象添加监听器。

为了显示，通过一个隐藏的 frame 去读取这个异步 Servlet 发出的信息，其源代码如下：

```
<html>
    <head>
        <script type="text/javascript"
        src="${pageContext.request.contextPath}/js/jquery-1.8.1.js"></script>
        <style>
. textareaStyle {
    width: 100%;
    height: 100%;
    border: 0;
}
</style>
<script type="text/javascript">
    function update(data) {
        var result = $('#result')[0];
            result.value = result.value + data + '\n';
    }
</script>
    </head>
```

```
19      <body style="margin: 0; overflow: hidden">
20          <table width="100%" height="100%" border="0" >
21              <tr>
22                  <td colspan="2">
23                      <textarea name="result" id="result" readonly="true"
                        wrap="off"
24                          style="padding: 10; overflow: auto"
25                    class=" textareaStyle"></textarea>
26                  </td>
27              </tr>
28          </table>
29          <iframe id="autoFrame" style="display:
30            none;" src="/ch07/AsyncContextServlet"></iframe>
31      </body>
32  </html>
```

上述代码中，第 30 行 iframe 的 src 写死为支持异步的 Servlet 路径。

然后写个测试的 Servlet，每隔 2 秒钟调用一次客户端方法。

```
---------------------- TestServlet.java-------------------------
01  @WebServlet("/test.action")
02  public class TestServlet extends HttpServlet {
03
04      public void doGet(HttpServletRequest request,
05              HttpServletResponse response)throws ServletException, IOException {
06          try {
07              //隔2秒钟调用一次客户端方法
08              for (int i = 0; i < 20; i++) {
09                  final String str = "window.parent.update(\""
10                          + String.valueOf(i) + "\");";
11                  ClientService.getInstance().callClient(str);
12                  Thread.sleep(2 * 1000); //线程暂停2秒钟，模拟业务执行方法
13                  if (i == 10) {          //执行到第10次时，跳出循环
14                      break;
15                  }
16              }
17          } catch (InterruptedException e) {
18              e.printStackTrace();
19          }
20      }
21  }
```

上述代码中，第 8~16 行就是循环调用客户端方法，客户端推送信息。其运行结果如图 5.21 所示。

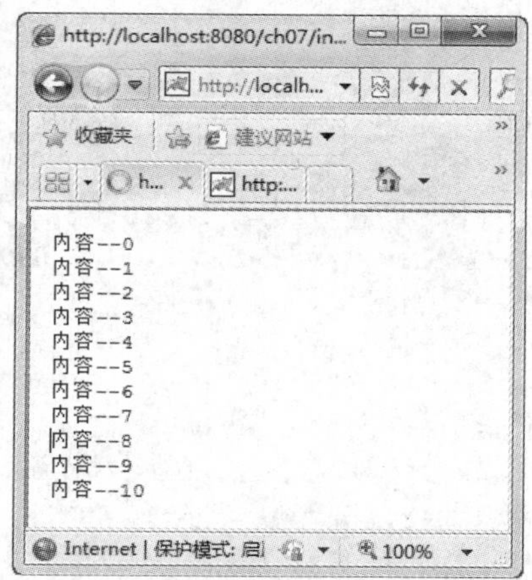

图 5.21　模拟服务器推送效果图

5.10 上机实践

1. 请编制一个 Servlet 程序，输出一个九九乘法表在页面中显示。
2. 编写一个监听器，实现记录在线用户数的功能。
3. 编写一个过滤器，过滤 IP 地址，使得某些 IP 地址不能访问。

第 6 章

EL标签的应用

在 JSP 页面中，经常用 JSP 表达式<%=变量或者表达式%>来输出声明的变量以及页面传递的参数，当变量很多的时候，书写这样的表达式会显得累赘，那么 EL 标签就解决了这个问题，它简化了表达式。本章主要内容就是介绍 EL 标签的使用方法，包括 EL 标签的语法、功能、操作符的运算和隐含变量。

6.1 认识 EL 标签

EL 标签在 JSP 开发中可以使业务逻辑与代码分工更明确，便于代码维护功能发挥，在开发中占据着极其重要的位置。

6.1.1 EL 标签的语法

EL 标签的语法形式如下：

```
${参数名}
```

例如：${param.name}获得参数 name 值，它等同于<%=request.getParameter('name')%>。

从形式和用法上看，这种 EL 表达式简化了 JSP 原有的表达式，也是目前开发中经常使用的方式。

【例 6.1】EL 标签用法示例

ex6_1.jsp 输出基本运算值，其源代码如下：

```
----------------- ex6_1.jsp -----------------
<%@ page language="java" import="java.util.*" pageEncoding="UTF-8"%>
<!DOCTYPE HTML PUBLIC "-//W3C//DTD HTML 4.01 Transitional//EN">
<html>
 <head>
  <title>EL 标签示例</title>
 </head>
 <body>
   <center>
```

```
            <p>2+3=${2+3}   </p>          //输出2+3的值
        </center>
    </body>
</html>
```

ex6_1.jsp 的运行结果如图 6.1 所示。

图 6.1 ex6_1.jsp 运行结果

从结果可以看出，页面得到了正常的结果，相当于<%=2+3%>这样的表达式输出。

6.1.2 EL 标签的功能

上一节介绍了 EL 标签的语法，让读者初步了解其使用规范，本节讲述其主要的功能点。总体而言，EL 标签提供了更为简洁方便的形式访问变量，它不仅可以简化 JSP 页面代码，而且使得开发者的逻辑更加清晰。

EL 标签具有以下功能：

- 可以访问 JSP 中不同域的对象；
- 可以访问 Java Bean 中的属性；
- 可以访问集合元素；
- 支持简单的运算符操作。

下面将分别介绍这 4 个功能。

1. 访问 JSP 中不同域的对象

在 Web 中有 4 个作用域分别是：PageContext、Request、Session、和 application。EL 标签都可以对这 4 个作用域的参数进行访问。

【例 6.2】EL 标签访问示例

ex6_2.jsp 演示 EL 标签的访问方法，其源代码如下：

```
----------------- ex6_2.jsp ----------------
<%@ page language="java" import="java.util.*" pageEncoding="UTF-8"%>
<!DOCTYPE HTML PUBLIC "-//W3C//DTD HTML 4.01 Transitional//EN">
<html>
 <head>
  <title>EL 标签访问示例</title>
 </head>
 <body>
```

```
08      <%
09        pageContext.setAttribute("name","Smith");
10        request.setAttribute("age",20);
11        session.setAttribute("address","china");
12        application.setAttribute("sex","male");
13       %>
14    <center><h3>访问演示</h3></center>
15    <table border="1" width="100%">
16     <tr>
17      <td align="center">姓名</td>
18      <td align="center">年龄</td>
19      <td align="center">性别</td>
20      <td align="center">地址</td>
21     </tr>
22     <!--范围.参数名称 -->
23     <tr>
24      <td align="center">${pageScope.name}</td>
25      <td align="center">${requestScope.age}</td>
26      <td align="center">${sessionScope.address}</td>
27      <td align="center">${applicationScope.sex}</td>
28     </tr>
29     <!-- 直接写参数名称 -->
30     <tr>
31      <td align="center">${name}</td>
32      <td align="center">${age}</td>
33      <td align="center">${address}</td>
34      <td align="center">${sex}</td>
35     </tr>
36     </table>
37    </body>
38   </html>
```

上述代码中，有两种方法取得参数值，代码第 23~28 行是用 EL 隐含变量取得，第 30~35 行直接写明参数名，由此可知 EL 标签会依序从 Page、Request、Session、Application 范围内查找，若找到 name 参数名，就直接回传参数值，不再继续查找，若没有参数则返回 null。其运行结果如图 6.2 所示。

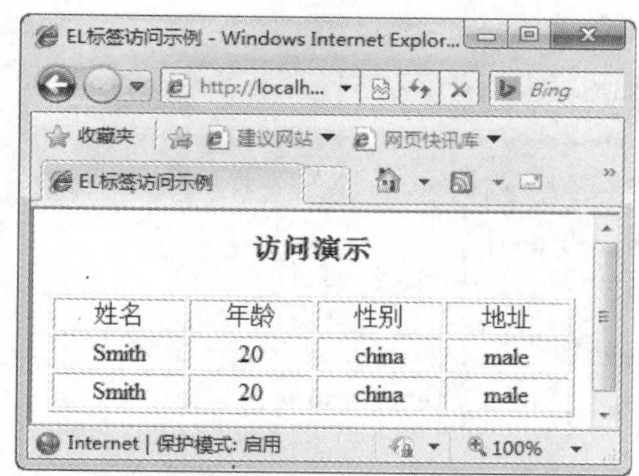

图 6.2　ex6_2.jsp 页面运行结果

2. 访问 Java Bean 中的属性

在 Java Bean 中，经常是用 Java Bean 做内部变量（数据成员）。在 JSP 表达式中访问这些 Java Bean 中的变量比较麻烦，但是应用 EL 标签却极其简单方便。

【例 6.3】EL 标签访问 Java Bean 中的属性示例

ex6_3.jsp 演示 EL 标签的访问 Java Bean 中的属性，其源代码如下：

```
----------------- ex6_3.jsp ----------------
01  <%@ page language="java" import="java.util.*" pageEncoding="UTF-8"%>
02  <%@page import="com.eshore.pojo.Users" %>
03  <!DOCTYPE HTML PUBLIC "-//W3C//DTD HTML 4.01 Transitional//EN">
04  <html>
05   <head>
06    <title>EL 标签访问示例</title>
07   </head>
08   <body>
09     <%
10      Users user = new Users();
11      user.setAddress("中国");
12      user.setAge(20);
13      user.setName("王五");
14      request.setAttribute("user",user);
15     %>
16    用户信息：${user }; <br/>
17    用户年龄：${user.age } ,用户姓名:${user.name }
18   </body>
19  </html>
```

上述代码中第 14 行向页面传递参数 user 对象，第 16 行取得 user 对象，第 17 行确定 user 对象中的 age 和 name 属性值。页面运行结果如图 6.3 所示。

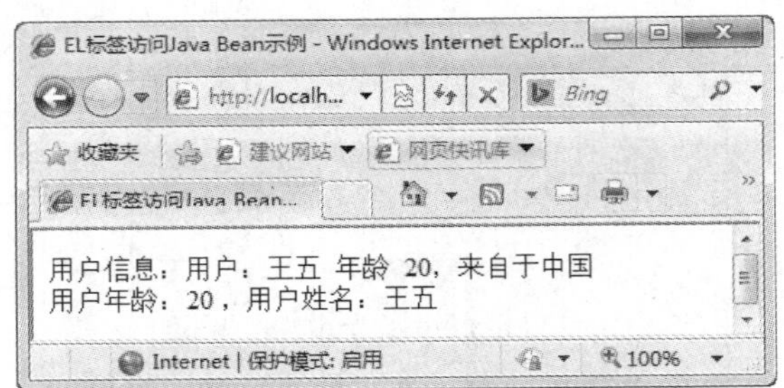

图 6.3　ex6_3.jsp 页面运行结果

User 对象的 POJO 如下：

```
public class User {
  private String name;        //姓名
  private int age;            //年龄
  private String address;         //地址
 public String getName() {
     return name;
 }
 public void setName(String name) {
     this.name = name;
 }
 public int getAge() {
     return age;
 }
 public void setAge(int age) {
     this.age = age;
 }
 public String getAddress() {
     return address;
 }
 public void setAddress(String address) {
     this.address = address;
 }
 @Override
 public String toString() {
     return "用户: "+name+" 年龄 "+age+", 来自于"+address;
 }

}
```

3. 可以访问集合元素

EL 标签可以很方便地访问集合中的元素。

4. 支持简单的运算符操作

在 EL 标签中，可以使用一些简单的运算符进行运算操作。

6.1.3　EL 标签的操作符

上一节介绍 EL 标签的两大功能点，也介绍了些示例，使读者了解其基本的访问方法。本节主要介绍 EL 标签的操作符，它可以帮助解决各种所需的功能。操作符大致可以分为算术运算符、逻辑运算符、关系运算符、条件运算符等，如表 6-1 所示。

表 6.1　EL 标签的操作符

运算符类别	运算符
算术运算符	+、-、*、/或 div、%或 mod
逻辑运算符	&&或 and、||或 or、!或 not
关系运算符	==或 eq、!=或 ne、<或 lt、>或 gt、<=或 le、>=或 ge
其他运算符	Empty 运算符、条件运算符、()运算符、[]运算符

下面通过例子说明运算符的使用方法。

【例 6.4】EL 标签运算符示例

ex6_4.jsp 演示说明其运算符的使用方法，其源代码如下：

```
----------------- ex6_4.jsp ----------------
<%@ page language="java" import="java.util.*" pageEncoding="UTF-8"%>
<%@page import="com.eshore.pojo.Users" %>
<!DOCTYPE HTML PUBLIC "-//W3C//DTD HTML 4.01 Transitional//EN">
<html>
 <head>
  <title>EL 标签运算符示例</title>
 </head>
 <body>
 <%
     Users user = new Users();
     user.setAddress("中国");
     request.setAttribute("user",user);
     request.setAttribute("str",null);
     String[] arr = new String[]{"第1个","第2个"};
     request.setAttribute("arr",arr);
    %>
 <table border="1">
   <tr>
     <td>算术运算符：</td>
     <td>逻辑运算符：</td>
     <td>关系运算符：</td>
     <td>其他运算符：</td>
```

```
    </tr>
    <tr>
      <td>
   加：3+3 = ${3+3}<br/>
   减：4-3 = ${4-3}<br/>
   乘：2*3 = ${2*3}<br/>
   除：3/6 = ${3/6}<br/>
   求模：10%3 = ${10%3}</td>
      <td>
   逻辑与：${2<15 && 15<20}<br/>
   逻辑与： ${2<15 and 15 <20}<br/>
   逻辑或： ${2<15||15<20}<br/>
  逻辑或：${2<15 or 15<20}<br/>
  逻辑否： ${!(2<15)}<br/>
  逻辑否： ${not(2<15)}<br/></td>
      <td>
   符号左右两端是否相等：2==15 : ${2 == 15 }或${2 eq 15 }<br/>
   符号左右两端是否不相等：2!=15：${2 != 15 }或${2 ne 15 }<br/>
  符号左边是否小于右边： 2<15：${2 < 15 }或${2 lt 15 }<br/>
  符号左边是否大于右边 ：2>15：${2 > 15 }或${2 gt 15 }<br/>
  符号左边是否小于或者等于右边： 2<=15：${2 <= 15 }或${2 le 15 }<br/>
  符号左边是否大于或者等于右边： 2>=15：${2 >= 15 }或${2 ge 15 }<br/></td>
      <td>
   str 是否为空：${empty str }<br/>
   user 对象是否为空：${empty user }<br/>
   2小于15输出 yes 否则输出 no：${2<15?'yes':'no' }<br/>
     输出 user 对象的 address 属性：${user.address}<br/>
     输出 arr 数组的第一个值：${arr[0]}<br/></td>
    </tr>
  </table>
 </body>
</html>
```

“.”与“[]”运算符的区别在于[]是访问数组或者集合，而“.”是访问 Java Bean 属性或者 Map Entry。变量必须用“[]”运算符，“.”不可以使用。

页面 ex6_4.jsp 运行结果如图 6.4 所示。

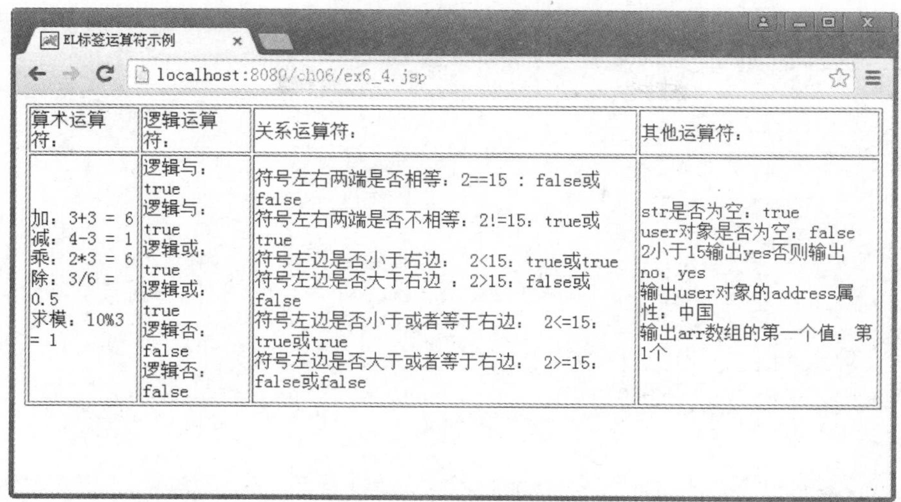

图 6.4　ex6_4.jsp 页面运行结果

6.2 EL 标签的隐含变量

上一节介绍 EL 标签操作符使用方法，让读者了解使用 EL 标签可以给 JSP 开发带来巨大的便利，而且是逻辑更加的清晰。本节介绍 EL 标签的隐含变量，这些隐含变量跟 JSP 中的隐藏对象很相似。下面逐一地进行介绍。

6.2.1 隐含变量 pageScope、requestScope、sessionScope、applicationScope

隐含变量 pageScope、requestScope、sessionScope、applicationScope 分别对应于 JSP 隐含变量 pageContext、request、session、application，用 JSP 中对应的作用域发送请求的参数变量，可以用相应的 EL 标签变量取参数值，例如【例 6.2】所示。

6.2.2 隐含变量 param、paramValues

隐含变量 param、paramValues 包含请求参数集合变量，param 是取得某一个参数，paramValues 是取得参数集合中的变量值，它们相当于 JSP 中的：

```
    request.getParameter(String name)
request.getParameterValues(String name)
```

【例 6.5】参数 param、paramValues 示例

ex6_5.jsp 演示隐含变量 param、paramValues 的使用方法，其源代码如下：

```
    ---------------- ex6_5.jsp ----------------
```

```
<%@ page language="java" import="java.util.*" pageEncoding="UTF-8"%>
<!DOCTYPE HTML PUBLIC "-//W3C//DTD HTML 4.01 Transitional//EN">
<html>
 <head>
  <title>EL 标签隐含对象 param、paramValues 使用示例</title>
 </head>
 <body>
  <form action="test.jsp">
  <input type="text" name="sampleVal" value="1">
  <input type="text" name="sampleVal" value="17">
  <input type="text" name="sampleVal" value="16">
  <input type="text" name="sampleSingleVal" value="single">
  <input type="submit" value="Submit">
  </form>
 </body>
</html>
```

跳转页面 test.jsp 源码如下：

```
-----------------------test.jsp--------------------------
<%@ page language="java" import="java.util.*" pageEncoding="UTF-8"%>
<!DOCTYPE HTML PUBLIC "-//W3C//DTD HTML 4.01 Transitional//EN">
<html>
 <head>
  <title>跳转页面</title>
 </head>
 <body>
 请求参数值：${param.sampleSingleVal} ,
 ${paramValues.sampleVal[1]}</br>
 </body>
</html>
```

运行结果如图 6.5 所示。

图 6.5　ex6_5.jsp 运行结果

6.2.3 其他变量

1. header、headerValues 变量

header、headerValues 这两变量是获取请求 HTTP 的表头信息，header 表示取得 HTTP 某表头信息，headerValues 表示取得表头数组信息，它们分别对应于：request.getHeader()和 request.getHeaders()。

2. cookie 变量

cookie 变量取得所有请求的 cookie 参数，参数中的每个对象对应 javax.servlet.http.Cookie。

例如：要获取 cookie 中名称为 username 的值，可以直接使用${cookie.username}来获取。

3. initParam 变量

initParam 取得应用程序初始化参数，相当于 application.getInitParameter()方法。但通常很少使用。

例如：一般页面获取初始化的方法 String url= application.getInitParameter("url")，那么可以用${initParam.url}来代替。

4. pageContext 变量

pageContext 变量时取得其他相关用户请求或页面的详细信息，其等同于 JSP 中的 PageContext 对象。下面列出几个常用的方法：

- ${pageContext.request.queryString}：取得请求的参数名；
- ${pageContext.request.requestURL}：取得请求的 URL；
- ${pageContext.request.contextPath}：取得服务应用的名称；
- ${pageContext.request.method}：取得 HTTP 的提交方法；
- ${pageContext.request.protocol}：取得使用的协议(HTTP/1.1、HTTP/1.0)；
- ${pageContext.request.remoteUser}：获取登录用户名称；
- ${pageContext.request.remoteAddr }：获取登录用户 IP 地址；
- ${pageContext.session.new}：判断 session 是否为新的会话；
- ${pageContext.session.id}：取得 session 的 ID 号；
- ${pageContext.servletContext.serverInfo}：取得主机端的服务信息。

【例 6.6】其他变量示例

ex6_6.jsp 演示隐含变量的使用方法，其源代码如下：

```
----------------- ex6_6.jsp -----------------
<%@ page language="java" import="java.util.*" pageEncoding="UTF-8"%>
<!DOCTYPE HTML PUBLIC "-//W3C//DTD HTML 4.01 Transitional//EN">
<html>
```

```
  <head>
   <title>EL 标签隐含对象使用示例</title>
  </head>
  <body>
  会话 ID：${pageContext.session.id}</br>
  Accept-Encoding 报头：${header["Accept-Encoding"] }</br>
    connection 报头:${header["connection"] }
  </body>
</html>
```

ex6_6.jsp 页面运行结果如图 6.6 所示。

图 6.6　ex6_6.jsp 页面运行结果

从以上叙述的 EL 隐含变量可以看出，它大多和 JSP 中的隐含对象的处理方法对应，所以基本可以应用 EL 隐含变量来取代 JSP 中隐含对象的大多数方法，即方便又简单。

6.3 禁用 EL 标签

上一节中介绍了 EL 标签中的隐含变量，使读者了解隐含变量的基本用法。虽然 EL 标签给编码带来了遍历，但有时候并不想使用 EL 标签，例如：如何让某页面禁用 EL 标签或者在页面中禁用表达式。本节将介绍如何禁用 EL 标签。

6.3.1 在整个 Web 应用中禁用

EL 标签是在 JSP 2.0 之后才有的新特性，因此需要指定 JSP 版本并设置为较低版本（Servlet2.3 或者更早），那么 JSP 页面将不再支持 EL 标签。可以通过修改 web.xml 文件实现。

修改后的 web.xml 配置为：

```
<?xml version="1.0" encoding="UTF-8"?>
<!DOCTYPE web-app
  PUBLIC "-//Sun Microsystems, Inc.//DTD Web Application 2.3//EN"
  "http://java.sun.com/dtd/web-app_2_3.dtd">
```

```
<web-app>
 <welcome-file-list>
  <welcome-file>index.jsp</welcome-file>
 </welcome-file-list>
</web-app>
```

与能支持 EL 标签的 web.xml 对比，不用 DTD 对 XML 文档进行约束，而是用 XSD，且 dtd 版本是 2.3 版本。

能支持 EL 标签的 web.xml 如下：

```
<?xml version="1.0" encoding="UTF-8"?>
<web-app version="2.5"
    xmlns="http://java.sun.com/xml/ns/javaee"
    xmlns:xsi="http://www.w3.org/2001/XMLSchema-instance"
    xsi:schemaLocation="http://java.sun.com/xml/ns/javaee
    http://java.sun.com/xml/ns/javaee/web-app_2_5.xsd">
 <welcome-file-list>
  <welcome-file>index.jsp</welcome-file>
 </welcome-file-list>
</web-app>
```

6.3.2　在单个页面中禁用

如果想在某个页面中禁用 EL 标签 那么可以直接通过 page 指令的 isELIgnored 属性实现。命令如下：

```
<%@page isELIgnored="true" %>
```

6.3.3　在页面中禁用个别表达式

通常在一个应用中要么全部允许使用 EL 标签，要么都不允许。因此很少会出现部分页面使用，而另一部分不用的情形。但 EL 标签允许在一页面中部分支持标签，部分不支持，要实现此功能，只需在不需要解析的 EL 表达式前的“$”符号加入一反斜杠就行，变成“\$”。

6.4　上机实践

1. 请使用 EL 标签编写四则混合运算，并分析运算符的优先级。
2. 编写一个 JSP 页面，应用 EL 标签输出用户信息。
3. 请编写一个 JSP 页面，应用 EL 标签隐含对象输出表头等信息。

第 7 章

网页的请求、响应与会话管理

在 JSP 开发中，请求、响应、会话管理是最基本的 3 个内置对象。一个 Web 应用系统，必须得有请求、响应、会话才能构建一个完整的程序。本章会依次介绍 JSP 中这 3 个内置对象的应用方法和技巧，包括如何获取网页内容、如何输出页面上的字符、如何重写 URL 等等。最后还会通过两个比较完整的案例复习本章的这 3 个对象。

7.1 从容器到 HttpServlet

在第 4、5 章中，介绍了 JSP 的内置对象和 Servlet 相关知识，以及如何部署和开发一个 Servlet。但是，并没有详细介绍如何将 Servlet 与 JSP 结合起来使用。Web 容器是 JSP 唯一可以识别的 HTTP 服务器，所以必须了解 Web 容器如何生成请求和响应对象。本节将介绍容器与 HTTPServlet 之间的来龙去脉。

7.1.1 Web 容器做了什么

Web 容器的作用就是创建一个 Servlet 实例，并完成 Servlet 注册以及根据 web.xml 中的 URL 响应。当请求来到容器时，Web 容器会转发给对应的 Servlet 来处理请求。

当客户端请求 HTTP 服务器时，会使用 HTTP 来传递请求和标头、参数等信息。HTTP 协议是无意识的协议，通过文本信息传递消息，而 Servlet 其实是 Java 对象，运行在 Web 容器中。当 HTTP 服务器将请求转给 Web 容器时，Web 容器会创建一个 HttpServletRequest 和 HttpServletResponse 对象，将请求中的信息传递给 HttpServletRequest 对象，而 HttpServletResponse 对象则作为对客户端响应的 Java 对象。这一过程如图 7.1 所示。

Web 容器会根据配置信息例如 web.xml 或者@WebServlet，查找相对应的 Servlet 并调用它的 service()方法。service()方法会根据 HTTP 请求的方式，决定是调用 doPost()还是 doGest()方法。例如，HTTP 请求的方式为 post，则调用 doPost()方法。

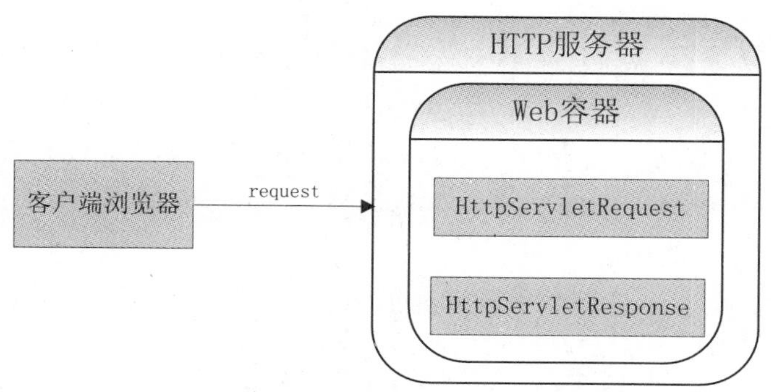

图 7.1　Web 容器收集请求信息并创建请求和响应对象

在 doPost()方法中，可以使用 HttpServletRequest 对象和 HttpServletResponse 对象。例如，使用 getParameter()取得请求参数值，使用 getWriter()取得输出流对象 PrintWriter，并进行响应处理。最后由 Web 容器转换为 HTTP 响应，由 HTTP 服务器对浏览器做出响应。随之，Web 容器将 HttpServletRequest 对象、HttpServletResponse 对象销毁回收，请求响应结束。过程如图 7.2 所示。

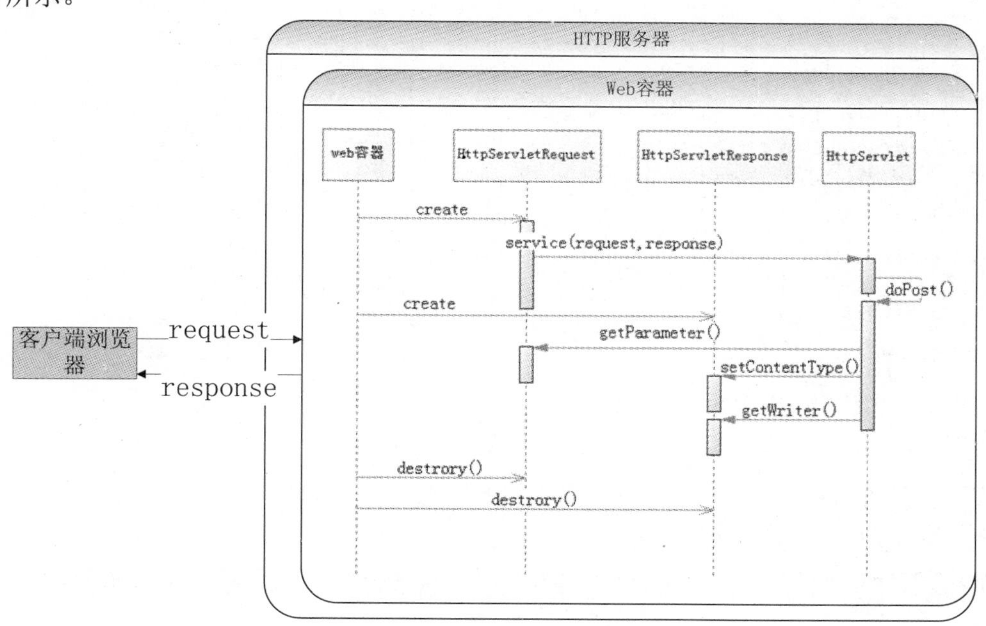

图 7.2　从请求到响应流程示意图

前面提到过 HTTP 协议是一种无意识的协议，每一次的请求、响应后，浏览器不会记得客户端的信息。Web 容器根据每次请求都会创建新的 request、response 对象，做出响应后就销毁该次的 request、response 对象。下次的请求、响应就与上一次的请求、响应对象无关了。所以，对于请求、响应的设置，是不能保存到下一次请求的。

像这种 request、response 对象的创建、服务、销毁，就是 Web 容器提供的请求、响应的生命周期管理。

7.1.2 doXXX()方法有什么用

Servlet 中的 service()方法有多种方法：doGet()、doPost()、doHead()等。当 Method 是 POST 时，请求会调用 doPost()方法；当 Method 是 GET 时，请求会调用 doGet()方法。在定义 Servlet 时，一般是继承 HttpServlet 类，然后定义 doPost()或者 doGet()方法。在 HttpServlet 中 doGet()和 doPost()方法实现过程如下：

```
protected void doGet(HttpServletRequest req, HttpServletResponse resp)
        throws ServletException, IOException
    {
        String protocol = req.getProtocol();
        String msg = lStrings.getString("http.method_get_not_supported");
        if (protocol.endsWith("1.1")) {
            resp.sendError(HttpServletResponse.SC_METHOD_NOT_ALLOWED, msg);
        } else {
            resp.sendError(HttpServletResponse.SC_BAD_REQUEST, msg);
        }
}
protected void doPost(HttpServletRequest req, HttpServletResponse resp)
        throws ServletException, IOException {
        String protocol = req.getProtocol();
        String msg = lStrings.getString("http.method_post_not_supported");
        if (protocol.endsWith("1.1")) {
            resp.sendError(HttpServletResponse.SC_METHOD_NOT_ALLOWED, msg);
        } else {
            resp.sendError(HttpServletResponse.SC_BAD_REQUEST, msg);
        }
}
```

假如，在自定义的 Servlet 不实现 doGet()或者 doPost()方法，那么程序就会调用以上方法。当客户端发出 POST 请求时，就会收到错误消息，如图 7.3 所示。

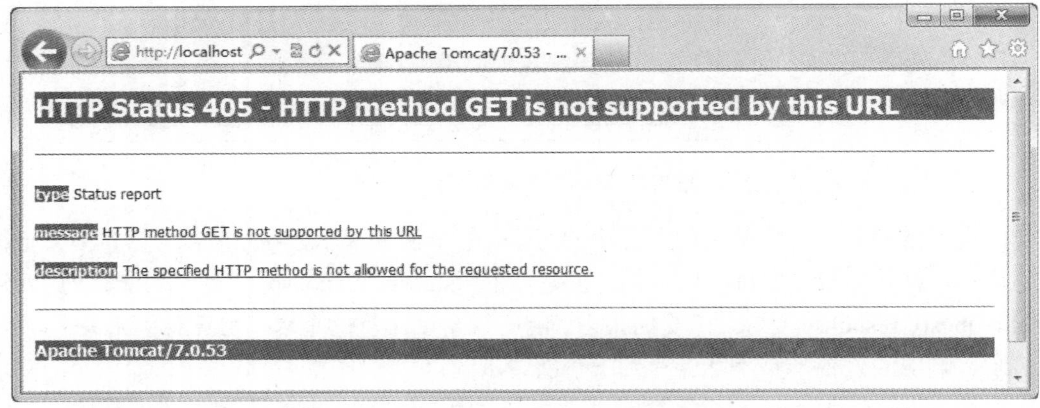

图 7.3　不实现 doGet()或者 doPost()方法显示的错误信息

因此，在定义完 Servlet 后，要实现 doGet()或者 doPost()方法。

7.2 HttpServletRequest 对象的应用

HttpServletRequest 对象是请求封装对象，由 Web 容器生成。使用该对象可以取得 HTTP 请求中的信息。在 Servlet 中，也是使用该对象进行请求的处理。如果要共享 request 中的属性值，就可以将请求对象设置到该对象中，那么 Servlet 在同一请求中就可以共享其对象。在第 4 章中，曾介绍过 JSP 的内置对象 request 实质上它就是 HttpServletRequest。有关参数的请求和设置标头等都可以参考该章节。本节要讨论的是如何运用 HttpServletRequest 对象读取 Body 内容、getParts()取得上传文件等内容。

7.2.1 使用 getReader()、getInputStream()读取 Body 内容

在 HttpServletRequest 对象中，可以运用 getReader()方法获取 Body 内容，而使用 getInputStream()方法获得上传文件的内容。下面用一个示例来说明如何获取 Body 内容。

【例 7.1】利用 getReader()方法获取 Body 内容

首先，编写 GetReaderBody 的 Servlet 类，用于获取 BufferedReader 对象，并且逐行遍历其内容，随后输出到浏览器中。其源代码如下。

```
-----------------------GetReaderBody.java--------------------------
package com.eshore;

import java.io.BufferedReader;
import java.io.IOException;
import java.io.PrintWriter;

import javax.servlet.ServletException;
import javax.servlet.http.HttpServlet;
import javax.servlet.http.HttpServletRequest;
import javax.servlet.http.HttpServletResponse;
public class GetReaderBody extends HttpServlet {

    public void doPost(HttpServletRequest request, HttpServletResponse response)
            throws ServletException, IOException {
        BufferedReader br = request.getReader();        //获取BufferedReader 对象
        String input ="";
        String body = "";
        while((input = br.readLine())!=null){           //遍历 body 内容
```

```
19              body+=input+"<br/>";
20          }
21          response.setContentType("text/html;charset=UTF-8");
            //设置响应的类型和编码
22          PrintWriter out = response.getWriter(); //取得 PrintWriter()对象
23          out.
24      println("<!DOCTYPE HTML PUBLIC \"-//W3C//DTD HTML 4.01
        Transitional//EN\">");
25          out.println("<HTML>");
26          out.println("  <HEAD><TITLE>A Servlet</TITLE></HEAD>");
27          out.println("  <BODY>");
28          out.print(body);
29          out.println("  </BODY>");
30          out.println("</HTML>");
31          out.flush();
32          out.close();
33      }
34  }
```

如上代码，第 15 行获得 BufferedReader 对象，第 18~20 行循环遍历 body 内容，代码第 21~32 行向浏览器输出内容。

其次，编写 getReaderBody.jsp 页面用于提交到该 Servlet 中，其源代码如下：

```
-----------------------getReaderBody.jsp--------------------------
01  <%@ page language="java" import="java.util.*" pageEncoding="UTF-8"%>
02  <!DOCTYPE HTML PUBLIC "-//W3C//DTD HTML 4.01 Transitional//EN">
03  <html>
04    <head>
05      <title>获取 body 内容</title>
06    </head>
07
08    <body>
09        <form action="<%=request.getContextPath()%>/servlet/GetReaderBody"
10              method="POST">
11              用户名：<input name="username"/><br/>
12              密  码：<input name="password" type="password"/><br/>
13              <input type="submit" name="user_submit" value="提交" />
14        </form>
15    </body>
16  </html>
```

如上述代码，代码第 9~14 行用 form 表单提交 Servlet，并在 input 中分别定义 name 值。部署项目，在浏览器中输入该页面地址并提交，程序的运行效果如图 7.4 所示。

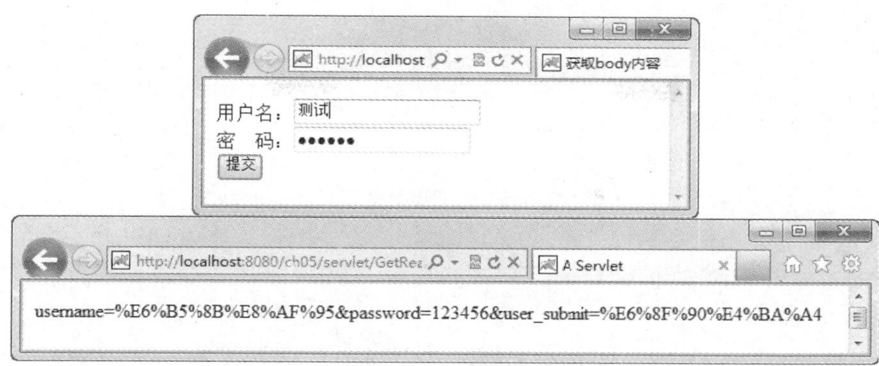

图 7.4　浏览器发出的请求 Body 内容

从图中可以看出，页面用的是 UTF-8 编码，所以在 Servlet 中都统一被转化为 UTF-8 编码。并且参数值都是一一对应的。以上是用 getReader()方法获取 Body 内容。如果在 Form 中设置 enctype 的值为 multipart/form-data 表示是客户端是上传文件，获取 Body 的内容就会获得一堆很奇怪的字符，如例子 7.2 所示。

【例 7.2】获取上传文件 Body 内容

更改下页面的 form 内容，其源代码如下。

```
01  <%@ page language="java" import="java.util.*" pageEncoding="UTF-8"%>
02  <!DOCTYPE HTML PUBLIC "-//W3C//DTD HTML 4.01 Transitional//EN">
03  <html>
04    <head>
05      <title>获取上传文件 body 内容</title>
06    </head>
07
08    <body>
09        <form action="<%=request.getContextPath()%>/servlet/GetReaderBody"
10              method="POST" enctype="multipart/form-data">
11              选择文件：<input  type="file" name="filename"/>
12              <input type="submit" name="file_submit" value="提交" />
13        </form>
14    </body>
15  </html>
```

上述代码中，第 10 行设置了 enctype 类别为 multipart/form-data，代码第 11 行设置 input 类型为 file。运行后的效果如图 7.5 所示。

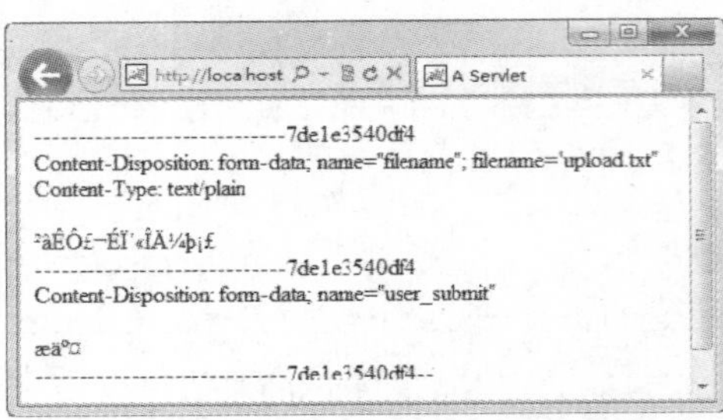

图 7.5　上传文件输出 body 内容示意图

从图中可以看出，一堆的乱码就是上传文件的内容。显然这不是想要的结果，可以使用 getInputStream()方法来获取内容，如例子 7.3 所示。

【例 7.3】获取上传文件 Body 内容

```
----------------------- GetReaderBody2.java--------------------------
package com.eshore;

import java.io.FileNotFoundException;
import java.io.FileOutputStream;
import java.io.IOException;
import java.io.InputStream;
import java.io.PrintWriter;

import javax.servlet.ServletException;
import javax.servlet.http.HttpServlet;
import javax.servlet.http.HttpServletRequest;
import javax.servlet.http.HttpServletResponse;

public class GetReaderBody2 extends HttpServlet {

    public void doPost(HttpServletRequest request, HttpServletResponse response)
            throws ServletException, IOException {
        //读取请求 Body
        byte[] body = readBody(request);
        //将内容进行统一编码
        String textBody = new String(body,"UTF-8");
        //获得文件名字
        String filename = getFilename(textBody);
        //写到指定位置
        writeToFile(filename, body);
        //返回页面消息
        response.setContentType("text/html;charset=UTF-8") ;
        //设置响应的类型和编码
        PrintWriter out = response.getWriter(); //取得 PrintWriter()对象
        out.println("<!DOCTYPE HTML PUBLIC \"-  //W3C//DTD HTML 4.01
            Transitional//EN\">");
```

```
31          out.println("<HTML>");
32          out.println("  <HEAD><TITLE>A Servlet</TITLE></HEAD>");
33          out.println("  <script>alert(\"上传成功\")</script><BODY>");
34          out.println("  </BODY>");
35          out.println("</HTML>");
36          out.flush();
37          out.close();
38      }
39      //读取请求的输入流
40      private byte[] readBody(HttpServletRequest request)
41                  throws IOException{
42          //获取请求内容长度
43          int len = request.getContentLength();
44          //获取请求的输入流
45          InputStream is = request.getInputStream();
46          //新建一字节数组
47          byte b[] = new byte[len];
48          int total = 0;
49          while(total<len){                                          //读取字节流
50              int bytes = is.read(b, total, len);
51              total += bytes;
52          }
53          return b;
54      }
55      //获取文件名称
56      private String getFilename(String bodyText){
57          String filename = bodyText.substring(bodyText.indexOf
            ("filename=\"")+10);
58          filename = filename.substring(filename.lastIndexOf("\\")+1,
            filename.indexOf("\""));
59          return filename;
60      }
61      //写入到指定文件
62      private void writeToFile(String filename,byte[] body)
63              throws FileNotFoundException,IOException{
64          //写到指定的文件地址
65          FileOutputStream fileOutputStream =
66              new FileOutputStream("d:/file/"+filename);
67          fileOutputStream.write(body);
68          fileOutputStream.flush();
69          fileOutputStream.close();                        //注意关闭输出流
70      }
71  }
```

代码第 40~54 行读取请求中的字节流，并用字节数组保存；代码第 56~60 行获取文件名称；代码第 62~69 行将字节流输出到指定的位置；代码第 27~37 行输出页面的响应。页面代码使用例 7.2 中的代码，将 Servlet 更改为 GetReaderBody2，并在 web.xml 中配置请求路径。程序运行结束会在“d:/file/”文件夹下新建一上传的文件，打开文件发现多了很多页面中的信息，这是因为没有进行位置的过滤。其实在 Servlet 3.0 中，可以利用 getPart()、getParts()去获取上传文件。

7.2.2　使用 getPart()、getParts()取得上传文件

上一节中，介绍了用 Servlet 上传文件的内容，从例 7.3 可以看出代码比较长，而且得到的结果也不是很理想。在 Servlet 3.0 中新增 getPart()和 getParts()方法来处理上传文件的功能。getPart()处理单文件内容，getParts()处理多文件内容。在使用 getPart()和 getParts()方法时，必须使用 MultipartConfig 注解，这样 Servlet 才能获得到 Part 对象。MultipartConfig 属性见表 7.1。

表 7.1　@MultipartConfig属性介绍

属性	说明
fileSizeThreshold	数值类型，当上传文件大小大于该值时，内容将先写入缓存文件，默认值为 0
location	字符串类型，设置存放生成的文件目录地址，它既是保存路径(在写入的时候，可以忽略路径设定)，又是上传过程中临时文件的保存路径，执行 Part.write()方法之后，临时文件将被清除
maxFileSize	数值类型，设置允许文件上传的最大值。默认值为 -1L，表示不限制大小
maxRequestSize	数值类型，限制 multipart/form-data 请求的最大数，默认值为-1L，表示没有限制大小

通过改写 GetReaderBody2 来说明 getPart()的用法，见例 7.4 所示。

【例 7.4】用 getPart()方法获取上传文件 Body 内容

```
---------------------- GetPartBodyContent.java----------------------------
import java.io.PrintWriter;

import javax.servlet.ServletException;
import javax.servlet.annotation.MultipartConfig;
import javax.servlet.annotation.WebServlet;
import javax.servlet.http.HttpServlet;
import javax.servlet.http.HttpServletRequest;
import javax.servlet.http.HttpServletResponse;
import javax.servlet.http.Part;

import org.apache.commons.lang.StringUtils;

@MultipartConfig(location = "D:/tmp/", maxFileSize = 1024 * 1024 * 10)
@WebServlet(name = "getPartBodyContentServlet", urlPatterns =
{ "/upload.do" },
loadOnStartup = 0)
```

```
13  public class GetPartBodyContent extends HttpServlet {

14  public void doPost(HttpServletRequest request, HttpServletResponse response)
15          throws ServletException, IOException {
16          //设置处理编码
17          request.setCharacterEncoding("UTF-8");
18          Part part = request.getPart("filename");//取得 Part 对象
19          // 获得文件名字
20          String filename = getFilename(part);
21          part.write(filename);
22          // 返回页面消息
23          response.setContentType("text/html;charset=UTF-8");
            // 设置响应的类型和编码
24          PrintWriter out = response.getWriter();// 取得 PrintWriter()对象
25          out.println("<!DOCTYPE HTML PUBLIC \"-//W3C//DTD HTML
26              4.01 Transitional//EN\">");
27          out.println("<HTML>");
28          out.println("  <HEAD><TITLE>A Servlet</TITLE></HEAD>");
29          out.println("  <script>alert(\"上传成功\")</script><BODY>");
30          out.println("  </BODY>");
31          out.println("</HTML>");
32          out.flush();
33          out.close();
34      }
35      // 获取文件名称
36      private String getFilename(Part part) {
37          if (part == null)
38              return null;
39          String fileName = part.getHeader("content-disposition");
40          if (StringUtils.isBlank(fileName)) {
41              return null;
42          }
43          return StringUtils.substringBetween(fileName, "filename=\"", "\"");
44      }
45  }
```

从上述代码可以看出，使用 Part 对象来处理上传文件程序变得简单很多。代码第 10 行注入 MultipartConfig，并设置上传路径和最大的上传文件大小；代码第 11 行用注入的方式配置 Servlet 路径；代码第 17 行设置文件编码；代码第 18 行取得 Part 对象；代码第 21 行直接调用 Part 对象的 write()方法输出文件。页面代码 getReaderBody3.jsp 的代码如下：

```
---------------------- getReaderBody3.jsp--------------------------
<%@ page language="java" import="java.util.*" pageEncoding="UTF-8"%>
<!DOCTYPE HTML PUBLIC "-//W3C//DTD HTML 4.01 Transitional//EN">
<html>
```

```
04    <head>
05      <title>getPart()获取上传文件 body 内容</title>
06    </head>
07
08    <body>
09          <form action="<%=request.getContextPath()%>/upload.do"
10                  method="POST" enctype="multipart/form-data">
11                  选择文件：<input type="file" name="filename"/>
12                  <input type="submit" name="file_submit" value="提交" />
13          </form>
14    </body>
15  </html>
```

上述代码中，与 getReaderBody2.jsp 的区别在于第 9 行，其余都一样。代码运行之后，在"D:/tmp/"文件夹下新建了上传文件，打开文件内容，发现与原文件一样。

代码中运用了 StringUtils 类，这个是 commons.lang 包中的 String 工具类。Part 对象必须使用@MultipartConfig 才能取得 Part 对象，否则 getPart()对象会得到 null。

上述演示单个文件的处理过程，如果要处理多个文件则使用 getParts()方法，getParts()方法返回 Collection 集合，见例 7.5 所示。

【例 7.5】用 getParts()方法获取批量上传文件

```
------------------------ GetPartsBodyContent.java----------------------------
package com.eshore;

import java.io.FileNotFoundException;
import java.io.FileOutputStream;
import java.io.IOException;
import java.io.InputStream;
import java.io.PrintWriter;

import javax.servlet.ServletException;
import javax.servlet.annotation.MultipartConfig;
import javax.servlet.annotation.WebServlet;
import javax.servlet.http.HttpServlet;
import javax.servlet.http.HttpServletRequest;
import javax.servlet.http.HttpServletResponse;
import javax.servlet.http.Part;

import org.apache.commons.lang.StringUtils;

@MultipartConfig(location = "D:/tmp/", maxFileSize = 1024 * 1024 * 10)
@WebServlet(name = "getPartsBodyContentServlet",
```

```
21  urlPatterns = { "/uploads.do" }, loadOnStartup = 0)
22  public class GetPartsBodyContent extends HttpServlet {
23
24      public void doPost(HttpServletRequest request, HttpServletResponse
response)
25              throws ServletException, IOException {
26          //设置处理编码
27          request.setCharacterEncoding("UTF-8");
28          request.getParts();
29          for(Part part:request.getParts()){
30              //只处理上传文件，因为提交按钮也被当作是一个 Part 对象
31              if(part.getName().startsWith("filenam")){
32                  // 获得文件名字
33                  String filename = getFilename(part);
34                  part.write(filename);
35              }
36          }
37          // 返回页面消息
38          response.setContentType("text/html;charset=UTF-8");
            // 设置响应的类型和编码
39          PrintWriter out = response.getWriter(); // 取得 PrintWriter()对象
40          out
41                  .println("<!DOCTYPE HTML PUBLIC \"-//W3C//DTD HTML
42          4.01 Transitional//EN\">");
43          out.println("<HTML>");
44          out.println("  <HEAD><TITLE>A Servlet</TITLE></HEAD>");
45          out.println("  <script>alert(\"上传成功\")</script><BODY>");
46          out.println("  </BODY>");
47          out.println("</HTML>");
48          out.flush();
49          out.close();
50      }
51      // 获取文件名称
52      private String getFilename(Part part) {
53          if (part == null)
54              return null;
55          String fileName = part.getHeader("content-disposition");
56          if (StringUtils.isBlank(fileName)) {
57              return null;
58          }
59          return StringUtils.substringBetween(fileName, "filename=\"",
"\"");
60      }
61  }
```

如上程序，代码第 29~36 行取得 Part 集合对象并遍历，接着逐个输出单个文件，在这里

需要注意的是提交按钮也会被当成是一个 Part 对象，所以要判断只输出上传文件的内容。其余代码与例 7.4 的一样。页面请求代码如下。

```
01  <%@ page language="java" import="java.util.*" pageEncoding="UTF-8"%>
02  <!DOCTYPE HTML PUBLIC "-//W3C//DTD HTML 4.01 Transitional//EN">
03  <html>
04    <head>
05      <title>getPart()获取上传文件 body 内容</title>
06    </head>
07    <body>
08          <form action="<%=request.getContextPath()%>/uploads.do"
09                method="POST" enctype="multipart/form-data">
10                选择文件1：<input type="file" name="filename1"/><br/>
11                选择文件2：<input type="file" name="filename2"/><br/>
12                选择文件3：<input type="file" name="filename3"/><br/>
13                <input type="submit" name="file_submit" value="提交" />
14          </form>
15    </body>
16  </html>
```

上述代码中，代码第 10~12 行分别定义 3 个 type 为 file 的 input，action 更改为新定义的 Servlet 路径 uploads.do。运行程序，在“D:/tmp/”路径下新增 3 个上传文件。

7.2.3 使用 RequestDispatcher 调派请求

Web 程序中，经常是由多个 Servlet 来完成请求。RequestDispatcher 接口就是为了多个 Servlet 之间的调整而实现的。该接口可以由 HttpServletRequest 的 getRequestDispatcher()方法取得。调用时指定跳转的 URL 地址能完成跳转动作。RequestDispatcher 接口有两种方法跳转 Servlet：include()和 forward()。

1. include()方法

include()方法的含义是，可以在当前的 Servlet 中显示另外一个 Servlet 中的内容，见例 7.6 所示。

【例 7.6】演示 include()方法

```
---------------------GetRequestDispatcherDemo.java------------------------
package com.eshore;

import java.io.IOException;
import java.io.PrintWriter;

import javax.servlet.RequestDispatcher;
import javax.servlet.ServletException;
import javax.servlet.annotation.WebServlet;
```

```
09  import javax.servlet.http.HttpServlet;
10  import javax.servlet.http.HttpServletRequest;
11  import javax.servlet.http.HttpServletResponse;
12
13  @WebServlet(
14          name = "getDispatcherDemo",
15          urlPatterns = { "/include.do" },
16          loadOnStartup = 0)
17  public class GetRequestDispatcherDemo extends HttpServlet {
18
19      private static final long serialVersionUID = 1L;
20
21      public void doGet(HttpServletRequest request, HttpServletResponse
        response)
22          throws ServletException, IOException{
23          doPost(request,response);
24      }
25      public void doPost(HttpServletRequest request, HttpServletResponse
        response)
26              throws ServletException, IOException {
27          // 返回页面消息
28          response.setContentType("text/html;charset=UTF-8");
            // 设置响应的类型和编码
29          PrintWriter out = response.getWriter(); // 取得 PrintWriter()对象
30          out.println("<!DOCTYPE HTML PUBLIC \"-//W3C//DTD HTML
31              4.01 Transitional//EN\">");
32          out.println("<HTML>");                        //输出页面信息
33          out.println("  <HEAD><TITLE>A Servlet</TITLE></HEAD>");
34          out.println("  <BODY>");
35          out.println("  The First Servlet<br/>");
36          //获取 RequestDispatcher 对象
37          RequestDispatcher dispatcher =
38              request.getRequestDispatcher("/includeSeconde.do");
39          dispatcher.include(request, response);
40          out.println("  Including Servlet<br/>");
41          out.println("  </BODY>");
42          out.println("</HTML>");
43          out.close();
44      }
45  }
```

上述代码中，第 13~16 行用注入的方式声明一个 Servlet；代码第 29 行取得 PrintWriter 对象，向浏览器输出内容；代码第 37~38 行设置包含的 Servlet 路径。包含的 Servlet 类源代码如下：

```
-----------------------
```

```
GetRequestDispatcherSecondeDemo.java---------------------------
package com.eshore;

import java.io.IOException;
import java.io.PrintWriter;

import javax.servlet.ServletException;
import javax.servlet.annotation.WebServlet;
import javax.servlet.http.HttpServlet;
import javax.servlet.http.HttpServletRequest;
import javax.servlet.http.HttpServletResponse;

@WebServlet(name = "getDispatcherSecondeDemo",
        urlPatterns = { "/includeSeconde.do" },
        loadOnStartup = 0)
public class GetRequestDispatcherSecondeDemo extends HttpServlet {

    private static final long serialVersionUID = 1L;

    public void doGet(HttpServletRequest request, HttpServletResponse response)
            throws ServletException, IOException {
        doPost(request, response);
    }
    public void doPost(HttpServletRequest request, HttpServletResponse response)
            throws ServletException, IOException {
        // 返回页面消息
        response.setContentType("text/html;charset=UTF-8");
       // 设置响应的类型和编码
        PrintWriter out = response.getWriter();// 取得 PrintWriter()对象
        out.println("  The seconcde Servlet<br/>");
    }
}
```

上述代码相对简单，首先声明一个 Servlet，然后获得 PrintWriter 对象并输出页面内容。程序运行效果见图 7.6 所示。

图 7.6　例 7.6 效果图

2. forward()方法

forward()方法是 RequestDispatcher 中的另一跳转 Servlet 方法，只是它的含义是跳转到别的 Servlet，而不会再返回跳转前的 Servlet。forward()方法在 Servlet 中经常用到，因为很多的业务逻辑都是跳转到别的 Servlet 中进行处理。下面用一个例子说明其用法并且建立简单的模型架构。

【例 7.7】演示 forward ()方法

```
----------------------- HelloServlet.java---------------------------
package com.eshore;

import java.io.IOException;

import javax.servlet.ServletException;
import javax.servlet.annotation.WebServlet;
import javax.servlet.http.HttpServlet;
import javax.servlet.http.HttpServletRequest;
import javax.servlet.http.HttpServletResponse;

import com.eshore.pojo.HelloUser;

@WebServlet(
        name = "helloServlet",
        urlPatterns = { "/hello.do" })
public class HelloServlet extends HttpServlet{
    private static final long serialVersionUID = 1L;

    private HelloUser user = new HelloUser();
    public void doGet(HttpServletRequest request, HttpServletResponse response)
        throws ServletException, IOException{
        doPost(request,response);//调用 doPost()方法
    }
    public void doPost(HttpServletRequest request, HttpServletResponse
response)
            throws ServletException, IOException {

        String userName = request.getParameter("username");
        //获取参数 username 的值
        String message = user.sayHello(userName);  //获取消息值
        request.setAttribute("message", message);  //设置参数值
        request.getRequestDispatcher("/hello.htm").//跳转到 hello.htm
                forward(request, response);
    }
}
```

上述代码中，doPost()方法非常简单，代码第 27 行获取参数 username 的值，通过调用类 HelloUser 中的 sayHello 方法获取欢迎信息，代码第 30 行跳转到以 hello.htm 为 URL 的 Servlet 类中。类 HelloUser 是普通的 pojo 类，作用是模拟业务上的操作，其源代码如下：

```
------------------------ HelloUser.java---------------------------
01  package com.eshore.pojo;
02
03  import java.util.HashMap;
04  import java.util.Map;
05
06  public class HelloUser {
07
08      private Map<String,String> helloMessage = new HashMap<String,String>();
09      //构建用户值，模拟从数据库中取出的值
10      public HelloUser(){
11          helloMessage.put("John", "Hello,John");
12          helloMessage.put("Smith", "Welcome,Smith!");
13          helloMessage.put("Rose", "Hi,Rose");
14      }
15      //根据用户，返回 Map 中的消息
16      public String sayHello(String userName){
17          return helloMessage.get(userName);
18      }
19  }
```

上述代码中，第 10~13 行用构造方法在 Map 中存放用户数据，根据用户 ID 获取欢迎信息。路径为 hello.htm 的 Servlet 类，主要作用是显示并打印出欢迎消息，其源代码如下。

```
------------------------ HelloHtml.java---------------------------
package com.eshore;

import java.io.IOException;
import java.io.PrintWriter;

import javax.servlet.ServletException;
import javax.servlet.annotation.WebServlet;
import javax.servlet.http.HttpServlet;
import javax.servlet.http.HttpServletRequest;
import javax.servlet.http.HttpServletResponse;

@WebServlet(name = "helloHtml", urlPatterns ={"/hello.htm"}, loadOnStartup = 0)
public class HelloHtml extends HttpServlet {
    private static final long serialVersionUID = 1L;

    public void doGet(HttpServletRequest request, HttpServletResponse response)
            throws ServletException, IOException {
```

```
18          doPost(request, response);
19      }
20
21      public void doPost(HttpServletRequest request, HttpServletResponse response)
22              throws ServletException, IOException {
23          String message = (String)request.getAttribute("message");
24          // 返回页面消息
25          response.setContentType("text/html;charset=UTF-8");
            // 设置响应的类型和编码
26          PrintWriter out = response.getWriter(); // 取得PrintWriter()对象
27          out.println("<!DOCTYPE HTML PUBLIC \"-//W3C//DTD HTML 4.01
28              Transitional//EN\">");
29          out.println("<HTML>");
30          out.println("  <HEAD><TITLE>A Servlet</TITLE></HEAD>");
31          out.println("  <BODY>");
32          out.println(message);                           //打印出消息到页面
33          out.println("  </BODY>");
34          out.println("</HTML>");
35          out.close();
36      }
37  }
```

上述代码中，第 23 行获取上一个 Servlet 中传递的参数 message，第 26 行获取 PrintWriter() 对象输出页面信息。在浏览器中输入带参数的地址：http://localhost:8080/ch05/hello.do?username=Smith。运行效果见图 7.7 所示。

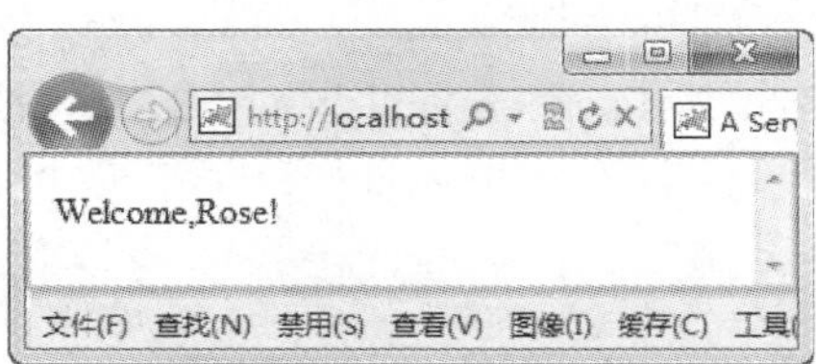

图 7.7　例 7.7 效果图

在上述代码中，除演示 include()方法外也演示了简单的 MVC 模型。HelloServlet 类可以看作是 Controller，Controller 中不会有 HTML 代码，它只负责进行业务中的处理。HelloUser 是 Model，Model 是负责对数据中的处理，例如本例中是获取欢迎信息。HelloHtml 是 View，用于展示页面结果。当然这只是对 MVC 一个简单的描述，主要是让读者对 MVC 有个初步的概念，随后章节中的例子会继续介绍。

7.3 HttpServletResponse 对象的应用

HttpServletResponse 是用于对浏览器的做出响应的操作对象。用它可以设置响应类型，可以直接输出 HTML 内容。

通常，使用 setContentType()设置 JSP 响应类型，使用 getWriter()取得 PrintWriter 对象或者 getOutputStream()取得 ServletOutputStream 流对象。使用 setHeader()、addHeader()设置标头，使用 sendRedirect()、sendError()对页面进行重定向或者是发送错误消息。

7.3.1 使用 getWriter()输出字符

在前面的例子中，经常可以看到使用 getWriter()方法获取 PrintWriter 对象，指定字符串对浏览器输出 HTML 代码。例如：

```
    PrintWriter out = response.getWriter();// 取得 PrintWriter()对象
    out.println("<!DOCTYPE HTML PUBLIC \"-//W3C//DTD HTML 4.01
Transitional//EN\">");
    out.println("<HTML>");
    out.println("  <HEAD><TITLE>A Servlet</TITLE></HEAD>");
    out.println("  <BODY>");
    out.println("  </BODY>");
    out.println("</HTML>");
```

通常，在对浏览器做出响应的同时会设置字符编码，因为默认的编码是 GBK 或者是 ISO-8859-1，那么浏览器输出中文时，会显示乱码。所以，需要设置支持中文的字符编码格式，这里有两种方式设置编码格式：使用 setContentType()或者 setCharacterEncoding()方法

设置编码语句如下：

```
    response.setContentType("text/html;charset=UTF-8");
```

或者

```
    response.setCharacterEncoding("UTF-8");
```

设置编码的时候需要在获得 PrintWriter 对象或者 ServletOutputStream 流对象之前设置，否则无效果。如果请求的参数中有中文，则还需要设置请求的支持中文字符编码，而且要注意与响应的编码保持一致。

在 Servlet 开发中，必须告诉浏览器是以何种方式处理响应即告诉内容类型，所以要设置响应标头。在设置 context-type 中，指定 MIME 类型浏览器就能知道标头类型。MIME 类型有 text/html、application/pdf、application/jar、application/x-zip、image/jpeg 等。在应用程序中，可以在 web.xml 中设置 MIME 类型。例如：

```
    <mime-mapping>
```

```
    <extension>jar</extension>
    <mime-type>application/jar</mime-type>
</mime-mapping>
```

<extension>设置文件的后缀，<mime-type>设置对应的 MIME 类型名称。在前面的例子中，曾介绍过请求参数是中文的例子。这里再通过例子说明请求参数是中文和响应中也有中文该如何发送请求的例子，见例 7.8 所示。

【例 7.8】演示发送中文字符请求

getResponse.jsp 页面用一个 form 表单描述喜欢的明星，下拉列表中的值特意设置为中文，然后提交到 Servlet 中进行处理。getResponse.jsp 源代码如下：

```
------------------------Index.jsp---------------------------
01  <%@ page language="java" import="java.util.*" pageEncoding="UTF-8"%>
02  <!DOCTYPE HTML PUBLIC "-//W3C//DTD HTML 4.01 Transitional//EN">
03  <html>
04    <head>
05      <title>明星调查</title>
06    </head>
07
08    <body>
09        <center>
10                调查问卷
11            <form action="<%=request.getContextPath()%>/diaocha.do"
              ethod="POST">
12                姓名：<input name="username"/><br/>
13                邮箱：<input name="email"/><br/>
14                你喜欢的明星：<br/>
15                <select name="starname" multiple="multiple">
16                    <option value="成龙">成龙</option>
17                    <option value="李连杰">李连杰</option>
18                    <option value="邓超">邓超</option>
19                    <option value="元彪">元彪</option>
20                    <option value="洪金宝">洪金宝</option>
21                    <option value="周润发">周润发</option>
22                </select>
23                <input type="submit" value="提交"/>
24            </form>
25        </center>
26    </body>
27  </html>
```

如上代码，代码第 15~22 行中用下拉列表提供可选项，并且可以多选。diaocha.do 的 Servlet 代码如下：

```
------------------------StarSurvey.java---------------------------
```

```
package com.eshore;

import java.io.IOException;
import java.io.PrintWriter;

import javax.servlet.ServletException;
import javax.servlet.annotation.WebServlet;
import javax.servlet.http.HttpServlet;
import javax.servlet.http.HttpServletRequest;
import javax.servlet.http.HttpServletResponse;
@WebServlet(
        name = "starSurvey",
        urlPatterns = { "/diaocha.do" })
public class StarSurvey extends HttpServlet{

    private static final long serialVersionUID = 1L;

    public void doGet(HttpServletRequest request, HttpServletResponse response)
        throws ServletException, IOException{
        doPost(request,response);
    }
    public void doPost(HttpServletRequest request, HttpServletResponse response)
            throws ServletException, IOException {
        // 设置请求的编码类型
        request.setCharacterEncoding("UTF-8");
        String username = request.getParameter("username");
        String email = request.getParameter("email");
        String[] starname = request.getParameterValues("starname");
        // 设置响应的类型和编码
        response.setContentType("text/html;charset=UTF-8");
        // 取得 PrintWriter()对象
        PrintWriter out = response.getWriter();
        out.println("<!DOCTYPE HTML PUBLIC \"-//W3C//DTD HTML 4.01
        Transitional//EN\">");
        out.println("<HTML>");
        out.println("  <HEAD><TITLE>感谢您的调查</TITLE></HEAD>");
        out.println("  <BODY>");
        out.println("联系人：<a href='"+email+"'>"+username+"</a>");
        out.println("<br/>喜欢的明星：");
        String str = "";
        for(int i=0;i<starname.length;i++){
            str +=starname[i]+", ";
        }
        str = str.substring(0, str.length()-1);
        out.println(str);
```

```
46            out.println("   </BODY>");
47            out.println("</HTML>");
48            out.close();
49        }
50    }
```

如上代码中，代码第 25 行设置请求的编码类型，这一行代码必须在获取参数前设置；代码第 26~28 行取得请求中的参数值；代码第 30 行设置响应的类型和编码；代码第 32~47 行对浏览器的响应；在范例中示范了如何设置超链接以及添加邮箱。请求的字符编码和响应的字符编码必须设置一致，这样显示的时候才不会乱码。程序的运行效果见图 7.8 所示。

图 7.8　例 7.8 显示效果图

7.3.2　使用 getOutputStream()输出二进制字符

getOutputStream()是为了取得输出流而设置的，在绝大数情况下，PrintWriter 对象就能解决问题。但是对于上传文件和下载文件则需要用到字节输出流。通过 getOutputStream()方法可以取得 ServletOutputStream 对象，它是 OutputStream 的子类。

在例 7.2 中，介绍了如何上传文件到指定目录。本小节将介绍下载文件来说明 getOutputStream()方法的使用，具体见例 7.9 所示。

【例 7.9】演示下载文件

DownloadServlet 类是简单的下载文件类，设置响应的内容类型和标头。其原代码如下：

```
-----------------------DownloadServlet.java---------------------------
package com.eshore;

import java.io.IOException;
import java.io.InputStream;

import javax.servlet.ServletException;
import javax.servlet.ServletOutputStream;
import javax.servlet.annotation.WebServlet;
import javax.servlet.http.HttpServlet;
import javax.servlet.http.HttpServletRequest;
import javax.servlet.http.HttpServletResponse;

@WebServlet(name = "downloadServlet", urlPatterns = { "/download.do" })
```

```
14  public class DownloadServlet extends HttpServlet {
15
16      private static final long serialVersionUID = 1L;
17
18      public void doGet(HttpServletRequest request, HttpServletResponse response)
19              throws ServletException, IOException {
20          doPost(request,response);
21      }
22
23      public void doPost(HttpServletRequest request, HttpServletResponse response)
24              throws ServletException, IOException {
25          //设置响应的内容类型
26          response.setContentType("application/msword");
27          //设置响应的标头内容
28          response.addHeader("Content-disposition","attachment;
            filename=test.doc");
29          //获取资源文件
30          InputStream in =
getServletContext().getResourceAsStream("/doc/test.doc");
31          //输出到浏览器中
32          ServletOutputStream os = response.getOutputStream();
33          byte[] bytes = new byte[1024];
34          int len = -1;
35          while((len=in.read(bytes))!=-1){
36              os.write(bytes, 0, len);
37          }
38          //关闭输出输入流
39          in.close();
40          os.close();
41      }
42  }
```

如上代码，代码第 13 行用注入的方式声明 Servlet；代码第 26 行设置响应的内容类型为 doc 文档，所以设置内容类型为 application/msword，这样若系统装了 Office 软件，就会直接打开 DOC 文档，还可以进行另存为操作；代码第 30 行使用 HttpServlet 的 getServletContext()取得 ServletContext 对象，并利用 getResourceAsStream()方法以流形式获取资源文件，指定的路径时相对于 Web 应用程序的根目录；代码第 32~37 行利用 Java IO 流的特性从 DOC 文件中读取字节数据，并利用 ServletOutputStream 输出文件到浏览器中响应。在浏览器中输入 Servlet 地址，即可弹出下载对话框，如图 7.9 所示。

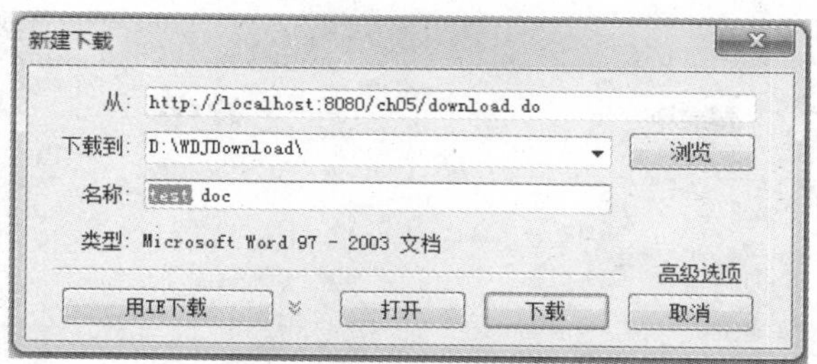

图 7.9　下载文件效果图

7.3.3　使用 sendRedirect()、sendError()方法

1. sendRedirect()方法

在 7.2.3 小节中，介绍了 getRequestDispatcher()方法进行 Servlet 的跳转，forward()方法会将请求转发到指定的 URL，这动作是不会被浏览器所知的，因此地址栏也不会发生变化。在转发过程中，处于同一 Request 周期中，因此可以用 setAttribute()方法设置属性对象，用 getAttribute()获取属性对象。本小节介绍 sendRedirect()方法，它重定向到另外一个 URL。例如：

```
response.sendRedirect("http://localhost://ch05/showview.do");
```

sendRedirect()方法会在响应中设置 HTTP 状态码和 Location 标头，当客户端接收到这个标头时，会重新请求指定的 URL，所以地址栏上的地址会发生改变。

下面用例子演示 sendRedirect()方法，见例 7.10 所示。

【例 7.10】演示 sendRedirect()方法

ResponseRedirectDemo 类是跳转前的 Servlet 类，重定向到 redirSeconde.do 方法中，观察浏览器地址栏是否会发生变化。ResponseRedirectDemo 类的源代码如下：

```
----------------------ResponseRedirectDemo.java---------------------------
package com.eshore;

import java.io.IOException;
import java.io.PrintWriter;

import javax.servlet.RequestDispatcher;
import javax.servlet.ServletException;
import javax.servlet.annotation.WebServlet;
import javax.servlet.http.HttpServlet;
import javax.servlet.http.HttpServletRequest;
import javax.servlet.http.HttpServletResponse;

@WebServlet(name = "responseRedirectDemo",
        urlPatterns = { "/redirect.do" })
public class ResponseRedirectDemo extends HttpServlet {

```

```
17      private static final long serialVersionUID = 1L;
18      public void doGet(HttpServletRequest request, HttpServletResponse response)
19          throws ServletException, IOException{
20          doPost(request,response);
21      }
22      public void doPost(HttpServletRequest request, HttpServletResponse response)
23              throws ServletException, IOException {
24          response.setContentType("text/html;charset=UTF-8");
            // 设置响应的类型和编码
25          PrintWriter out = response.getWriter();// 取得 PrintWriter()对象
26          out.println(" Redirect 跳转页面的第一个页面<br/>");
27          response.sendRedirect(request.getContextPath()+"/redirSeconde.do");
28          out.println(" Redirect Servlet<br/>");
29          out.close();
30      }
31  }
```

如上代码，代码第 24 行设置响应的类型和编码；代码第 25 行取得响应的 PrintWriter()对象；代码第 27 行重定向到 redirSeconde.do 链接中。链接为 redirSeconde.do 的 Servlet 源代码如下：

```
-------------------- ResponseRedirectSecondeDemo.java--------------------
package com.eshore;
import java.io.IOException;
import java.io.PrintWriter;
import javax.servlet.RequestDispatcher;
import javax.servlet.ServletException;
import javax.servlet.annotation.WebServlet;
import javax.servlet.http.HttpServlet;
import javax.servlet.http.HttpServletRequest;
import javax.servlet.http.HttpServletResponse;

@WebServlet(name = "responseRedirectSecodDemo",
        urlPatterns = { "/redirSeconde.do" })
public class ResponseRedirectSecondeDemo extends HttpServlet {

    private static final long serialVersionUID = 1L;
    public void doGet(HttpServletRequest request, HttpServletResponse response)
        throws ServletException, IOException{
        doPost(request,response);
    }
    public void doPost(HttpServletRequest request, HttpServletResponse response)
            throws ServletException, IOException {
        // 返回页面消息
        response.setContentType("text/html;charset=UTF-8");
```

```
            // 设置响应的类型和编码
            PrintWriter out = response.getWriter(); // 取得 PrintWriter()对象
            out.println(" Redirect 跳转页面的第二个页面<br/>");
        }
    }
```

上述代码中，程序逻辑简单就是输出页面内容。程序运行效果如图 7.10 所示。

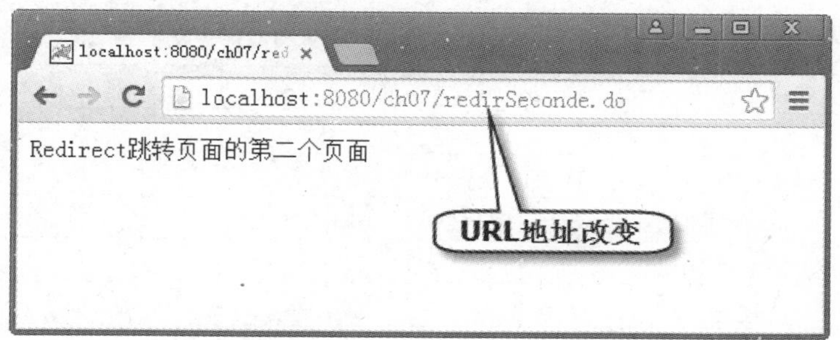

图 7.10　sendRedirect()示意图

从图中可以看出，跳转到 ResponseRedirectSecondeDemo 后，只显示其内容，而不显示跳转前的 Servlet 内容。

2. sendError()方法

如果在处理请求的时候发生错误，这时可以用 sendError()方法传递服务器的状态和错误消息。例如，请求的页面地址不存在，则可以发送如下错误信息：

```
response.sendError(HttpServletResponse.SC_NOT_FOUND);
```

SC_NOT_FOUND 表示资源文件不存在，服务器会响应 404 错误代码，错误代码统一定义在 HttpServletResponse 接口上。还可以自定义错误信息，程序如下：

```
response.sendError(HttpServletResponse.SC_NOT_FOUND, "页面错误");
```

在 HttpServlet 中 doPost()方法就实现了 sendError()方法：

```
protected void doPost(HttpServletRequest req, HttpServletResponse resp)
        throws ServletException, IOException {

        String protocol = req.getProtocol();
        String msg = Strings.getString("http.method_post_not_supported");
        if (protocol.endsWith("1.1")) {
            resp.sendError(HttpServletResponse.SC_METHOD_NOT_ALLOWED, msg);
        } else {
            resp.sendError(HttpServletResponse.SC_BAD_REQUEST, msg);
        }
}
```

上述代码中，判断是否有该方法，如果没有则抛出 SC_METHOD_NOT_ALLOWED 错误即 405。常见的错误见表 7.2 所示。

表 7.2　常见的 HTTP 错误列表

错误代码	说明
401	访问被拒绝
401.2	服务器配置导致登录失败
403	禁止访问
403.6	IP 地址被拒绝
403.9	用户数过多
404	没有找到文件或目录
405	用来访问本页面的 HTTP 不被允许（方法不被允许）
406	客户端浏览器不接受所请求页面的 MIME 类型
500.	内部服务器错误
504	网关超时
505	HTTP 版本不受支持

7.4 会话管理基本原理

实现会话管理，可以通过以下 3 种解决方案：使用隐藏域、使用 Cookie、URL 重写。以下分别介绍各方案。

7.4.1 使用隐藏域

隐藏域是指将表单中的内容，在显示页面时隐藏不显示数据。在 JSP 中将 input 标签的 type 属性值设定为 hidden 即生成一个隐藏表单域。再将会话的惟一标识记录到隐藏域中的 value 属性中，并设定 name 属性值。当表单提交时，会话标识也被提交到服务端，服务端则根据它找到对应的会话对象。

使用隐藏域则需要在每个页面都必须包含有会话标识表单，这样才能在许多页面跳转之间保存会话，并对它进行管理。此方法实现起来比较繁琐，在目前的开发中较少使用。

7.4.2 使用 Cookie

Cookie 实现会话管理是最容易的实现方式也是用的较多的方法。其原理是在服务端保存的会话对象中设定会话的惟一标识，客户端将会话标识保存在 Cookie 中，当浏览器发起请求时，从 Cookie 中取得会话标识一起发送给服务端，服务端在收到请求后，根据发送过来的会话标识查找到对应的会话对象，这样服务端就清楚当前是哪个客户端在连接，并且可以从会话中获得信息。

利用 Cookie 实现会话管理是目前开发中采用的主流方法，大多数的动态页面开发技术都实现了这一功能，并且其管理流程是自动完成的，在实现上并不需要多大的开发工作量。其实现会话管理的过程如图 7.11 所示。

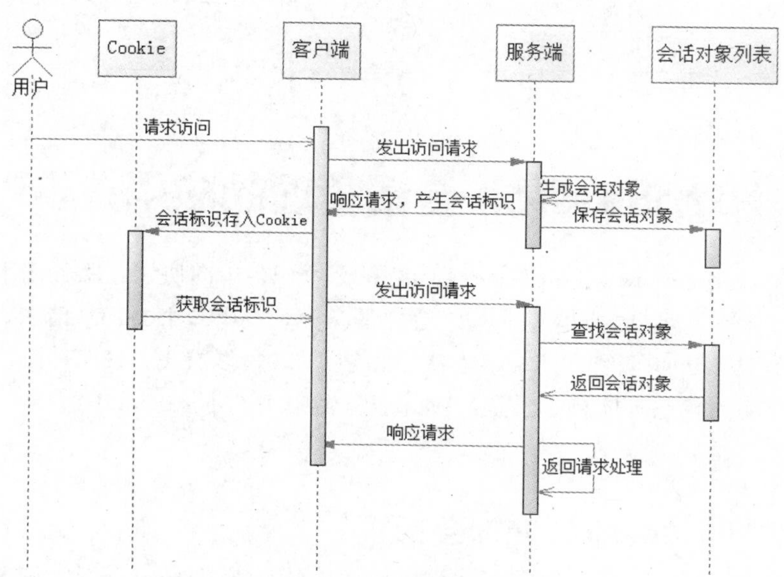

图 7.11　Cookie 实现会话管理的过程

7.4.3　使用 URL 重写

顾名思义，URL 重写是指在 URL 地址的末尾添加会话标识，改写了原先的 URL 地址。本质是用于惟一标识会话的信息以参数的形式添加到了 URL 中，服务端接收到请求时，解析出会话标识，然后利用会话标识查找出与当前请求对应的会话对象。该方法是用在浏览器的 Cookie 被禁用的情况，利用该方法可以很好实现地会话跟踪而不会受到浏览器参数设定的影响。

利用 URL 重写的方法，其特点是在 URL 后面添加会话标识，因此整个 Web 应用中的超链接或者脚本中用到的 URL 都需要添加会话标识，这样 Web 应用中的每一个页面都需要动态生成，页面中的每一个链接或者由客户端生成的跳转指令都必须加上会话标识，这样才能确保是当前会话。当客户端访问静态页面时，会话标识将会丢失，当重回动态页面时将不能继续此前的会话。以上是它的特点也是它的缺陷。

利用该方法还有一个问题是当用户在浏览网页时，可能会将 URL 复制下来分享给朋友，因为 URL 中会包含有会话标识，那么其他人可能与当前浏览者使用同一会话对象，这样当前浏览者的个人隐私就被暴露了，信息不安全。

无论采用哪种方式实现会话管理，在服务端都需要完成相关代码才能完整地实现会话管理的任务。这些任务主要是：生成惟一的会话标识、存储会话对象、从 Web 容器中取得当前请求的会话、回收空闲会话。

会话管理会产生会话对象，它存储在服务器的内存中，会占有内存。所以需要注意空闲会话的回收，以提高服务器的性能。如果 Web 应用不需要会话的存在，那么就禁用会话，这样就能提高服务器的运行速率。

7.5 HttpSession 会话管理的应用

HTTP 协议（http://www.w3.org/Protocols/）是无意识、单向协议。服务端不能主动连接客户端，只能等待并答复客户端请求。客户端连接服务端，发出一个 HTTP 请求，服务端处理请求，并返回一个 HTTP 响应给客户端，本次会话结束。 可以看出，HTTP 协议本身并不支持服务端保存客户端的状态等信息。于是，Web 服务器中引入了 session 的概念，用来保存客户端的信息。

7.5.1 使用 HttpSession 管理会话

在 Java 中，使用 javax.servlet.http.HttpSession 类来实现 session 会话。每个请求者对应一个 Session 对象，客户的所有状态信息都保存在该对象里。当用户第一次请求服务器器时，就创建了 Session 对象。它以 key-value 的形式进行保存，通过相对应的读写方法来保存客户的状态信息，例如 getAttribute(String key)获得参数和 setAttribute(String key,Object value)设定参数和参数值。Servlet 中可以通过 request.getSession()方法获取客户的 Session 对象，例如：

```
HttpSession session = request.getSessioin();        //获取 Sessioin 对象
Session.setAttribute(“username”,”John”);            //设置 Session 中的属性
```

在 JSP 中内置了 Sessioin 对象，可以直接使用，Servlet 中使用上述方法获取 Session 对象。若 JSP 页面中设定了<%@page session=”false”%>，则该 JSP 页面的 Session 对象是不可用的。下面通过例子说明利用 Session 来保存登录信息。

【例 7.11】利用 Session 来保存登录信息

首先，新建一个登录页面 login.jsp，源代码如下：

```
----------------------login1.jsp---------------------------
<%@ page language="java" import="java.util.*" pageEncoding="UTF-8"%>
<%
String path = request.getContextPath();
%>
<!DOCTYPE HTML PUBLIC "-//W3C//DTD HTML 4.01 Transitional//EN">
<html>
  <head>
    <title>用户登录</title>
  </head>
```

```
10
11    <body>
12        <p>用户登录</p>
13        <form action="<%=path%>/servlet/CheckUser" method="post">
14            <table border="1" width="250px;">
15              <tr>
16                 <td width="75px;">用户名：</td>
17                 <td ><input name="userId"/></td>
18              </tr>
19              <tr>
20                 <td width="75px;">密  码：</td>
21                 <td ><input name="passwd" type="password"/></td>
22              </tr>
23              <tr>
24                 <td colspan="2">
25                    <input type="submit" value="提交"/>  
26                    <input type="reset" value="重置"/>
27                 </td>
28              </tr>
29            </table>
30        </form>
31    </body>
32  </html>
```

上述代码中，表单内容很简单，有用户名和密码两个输入框，代码第 13 行填写提交表单时跳转的 action。

本例是用 Servlet 实现的，所以需要编写跳转的 Servlet 类 CheckUser，其源代码如下：

```
------------------------CheckUser.java----------------------------
public void doPost(HttpServletRequest request, HttpServletResponse response)
            throws ServletException, IOException {

        request.setCharacterEncoding("UTF-8");
        String userId = request.getParameter("userId");
        String passwd = request.getParameter("passwd");

        //判断是否是 linl 用户且密码相符
        if(userId!=null&&"linl".equals(userId)
                &&passwd!=null&&"123456".equals(passwd)){
            //获得 sessioin 对象
            HttpSession session = request.getSession();
            //设置 user 参数
            session.setAttribute("user", userId);
            //跳转页面
            RequestDispatcher dispatcher = request.
            getRequestDispatcher("/welcome.jsp");
            dispatcher.forward(request, response);
        }else{
            RequestDispatcher dispatcher = request.
```

```
21                getRequestDispatcher("/login1.jsp");
22                dispatcher.forward(request, response);
23            }
24    }
```

上述代码中，第 05~06 行是获取页面请求的用户名跟密码参数，第 09~18 行是判断用户是否存在，如果存在则存放到 session 中并跳转到欢迎页面，如果用户不存在，则跳转到登录页面。

欢迎页面是显示当前用户的用户名。其实现比较简单，使用 JSP 页面中的 session 对象就可以取出 user 的对象，其源代码如下：

```
-----------------------welcome.jsp--------------------------
01  <%@ page language="java" import="java.util.*" pageEncoding="UTF-8"%>
02  <!DOCTYPE HTML PUBLIC "-//W3C//DTD HTML 4.01 Transitional//EN">
03  <html>
04    <head>
05      <title>欢迎页面</title>
06    </head>
07    <%
08       String user = (String)session.getAttribute("user");
09       if(user==null){
10     %>
11     <jsp:forward page="login1.jsp"/>
12     <%} %>
13     <body>
14        欢迎您：<%=user%>。
15    </body>
16  </html>
```

上述代码中，第 07~10 行是利用 JSP 的 session 获取 user 对象，如果用户为空，返回登录页面。上述例子，部署运行，在浏览器中输入地址，运行结果如图 7.12 所示。

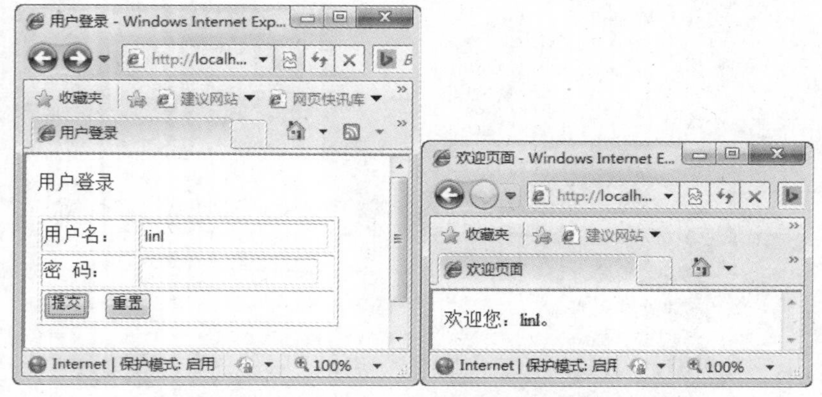

图 7.12　Session 保存登录信息效果图

7.5.2　HttpSession 管理会话的原理

HttpSession 会话管理是利用服务器来管理会话的机制，当程序为某个客户端的请求创建了一个 session 的时候，服务器会检查客户端的请求是否已经包含了一个 session 标识，如果已经有了 session 标识，服务器就把该 session 检索出来使用；如果请求不包含 session 标识，则为客户端创建一个该请求的惟一 session 标识。其会话管理流程图如图 7.13 所示。

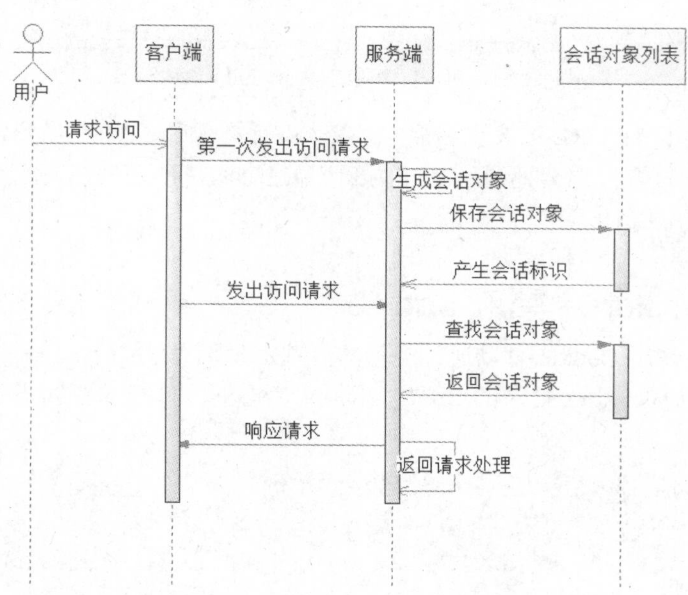

图 7.13　HttpSession 会话管理原理示意图

7.5.3　HttpSession 与 URL 重写

在 7.4.3 节中，提到 URL 重写是用在客户端不支持 Cookie 的情况下使用。它的实现方式是将 Session 的标识 id 添加进 URL 中。服务器解析重写后的 URL 获取 Session 的标识 id。这样即使客户端不支持 Cookie，也可以用 Session 来记录状态信息。重写的方法如下：

```
response.encodeURL("login.jsp");
```

或者

```
response.sendRedirect(response.encodeRedirectURL("login.jsp"));
```

这两种的效果是一样的，如果客户端支持 Cookie，则生成的 URL 不变，如果不支持，生成的 URL 中会带有 jsessionid 字符串的地址，如图 7.14 所示中的超链接。

```
文件(F) 编辑(E) 格式(O) 查看(V) 帮助(H)

<!DOCTYPE HTML PUBLIC "-//W3C//DTD HTML 4.01 Transitional//EN">
<html>
  <head>
    <title>欢迎页面</title>
  </head>

   <body>
      <a href="login.jsp;jsessionid=F5CABAE353CCF6DAF57F570705158B19?username=john"></a>
      欢迎您：linl。
  </body>
</html>
```

图 7.14　URL 中带有 jsessionid 示意图

从上图可以看出，在 URL 的文件名后面，参数的前面添加了字符串“;jsessionid=标识 id”。用户单击这个链接时，客户端会把 Session 的 id 值通过 URL 提交到服务器上，服务器解析 URL 地址获得 Session 的 id 值。

7.5.4　HttpSession 中禁用 Cookie

在 Web 项目中禁用 Cookie 可以通过配置的方式，禁用方法如下：

在 Web 项目的 WebRoot 目录的 META-INF 文件夹下，打开或者创建 context.xml 文件，编辑内容：

```
<?xml version="1.0" encoding="UTF-8"?>
<Context cookies="false" path="/ch06">
</Context>
```

或者

打开 Tomcat 的配置文件 context.xml，编辑内容：

```
<?xml version="1.0" encoding="UTF-8"?>
<Context cookies="false" >
……
</Context>
```

都是在 Context 元素中添加属性 cookies="false"，二者的区别在于前者是对单个项目的 Cookie 禁止，后者是禁止部署在 Tomcat 服务器里的 Web 项目使用 Cookie。

7.5.5　HttpSession 的生命周期

Servlet 有它的生命周期，同样 HttpSession 也有其生命周期：创建-->使用-->消亡。这就是它的生命周期过程。

1. 创建 HttpSession 对象

当客户端第一次访问服务器时，服务器为每个浏览器创建不同 Session 的 ID 值。在服务器端使用 request.getSession()或者 request.getSession(true)方法来获得 HttpSession 对象。

2. 使用 HttpSession 对象

创建 HttpSession 对象后，使用 Session 对象进行数据的存取和传输。具体的过程如下：

将产生的 sessionID 存入到 Cookie 中；

当客户端再次发送请求时，会将 sessionID 与 request 一起传送给服务端；

服务器根据请求过来的 sessionID 与保村在服务器端的 session 对应起来判断是否是同一 session。

3. HttpSession 对象的消亡

有 3 种方式可以结束 session 对象：

- 将浏览器关闭
- 调用 HttpSession 的 invalidate()方法
- session 超时

在 session 结束时，服务器会清空当前浏览器相关的数据信息。

以上就是 HttpSession 的生命周期过程，在请求处理时不断地循环第 2 步直到 session 对象消亡。

Session 是保存在服务器内存中的，每个用户都有一个独立的 Session。Session 中的内容应该尽量少，这样当有大量客户访问服务器时才不会内存溢出。

只有访问 JSP、Servlet 时才会创建 Session，访问静态页面是不会创建 Session 对象的。

7.5.6　HttpSession 的有效期

通常都给 Session 设定一个有效期，当某用户访问的 session 超过这个有效期即不是活跃的 Session，那么它的 session 就失效了，服务器会将它从内存中清除。

设定 Session 的有效期有以下两种方法：

- 调用 Session 的 setMaxInactiveInterval(long interval)设定；
- 在 web.xml 中修改，例如：

```
<session-config>
    <!-- 会话超时时长为30分钟 -->
    <session-timeout>30</session-timeout>
  </session-config>
```

调用 Session 的 invalidate()可以销毁 Session。

7.6 实例：用 Servlet 实现网站的注册和登录

本节用例子来说明请求与响应之间是如何进行传输的，网站的注册与登录也是普遍网站都要实现的功能，只是实现的方法、美工略有不同，但是其原理都是一样的。为了说明请求与响应中的方法，这两个示例都是用 Servlet 书写的，包括页面部分。当然读者也可以使用 JSP 来实现。网站的实现逻辑如图 7.15 所示。

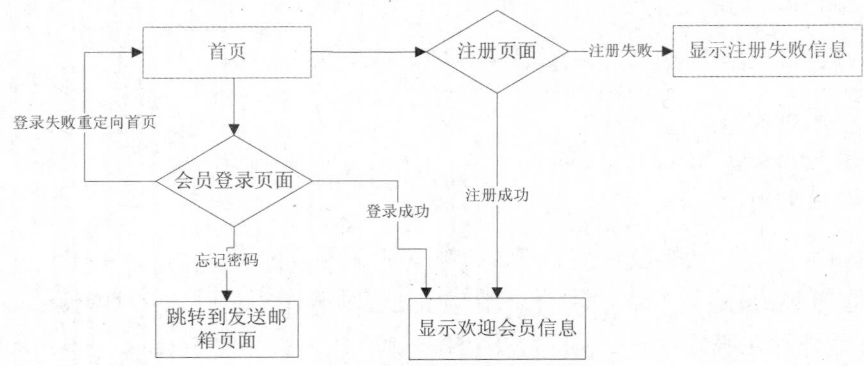

图 7.15　注册与登录的实现过程

7.6.1　实现网站注册功能

注册页面 register.jsp 提供给未注册用户进行注册的功能，由简单的 form 组成包含有用户名、密码和邮箱。其源代码如下：

```
-----------------------register.jsp---------------------------
<%@ page language="java" import="java.util.*" pageEncoding="UTF-8"%>
<!DOCTYPE HTML PUBLIC "-//W3C//DTD HTML 4.01 Transitional//EN">
<html>
  <head>
    <title>会员注册页面</title>
  </head>

  <body>
       <center>注册会员信息<br/>
              <form action="<%=request.getContextPath()%>/register.htm"
                  method="post">
                 <table border="1">
                     <tr>
                         <td>登录名：</td>
                         <td><input name="username"/></td>
                     </tr>
                     <tr>
```

```
18                      <td>密码：</td>
19                      <td><input name="passwd" type="password"/></td>
20                  </tr>
21                  <tr>
22                      <td>密码确认：</td>
23                      <td><input name="confirdPasswd" type=
                        "password"/></td>
24                  </tr>
25                  <tr>
26                      <td>邮箱地址：</td>
27                      <td><input name="email"/></td>
28                  </tr>
29                  <tr align="center">
30                      <td colspan="2"><input type="submit" value=
                        "提交"/></td>
31                  </tr>
32              </table>
33          </form>
34      </center>
35  </body>
36  </html>
```

如上代码中，form 中的 action 提供注册的 Servlet 地址；代码第 12~29 行提供简单的输入框和提交按钮。

注册的 Servlet 类 RegisterServlet，提供接收从页面中的参数并验证用户名、密码和邮箱是否有效。其源代码如下：

```
-----------------------RegisterServlet.java--------------------------
package com.eshore;
……省略
import com.eshore.pojo.User;

@WebServlet(name = "registeServlet", urlPatterns = { "/register.htm" })
public class RegisterServlet extends HttpServlet {

    private static final long serialVersionUID = 1L;

    public void doPost(HttpServletRequest request, HttpServletResponse response)
            throws ServletException, IOException {

        String userName = request.getParameter("username");
        //获取参数 username 的值
        String passwd = request.getParameter("passwd");
        //获取参数 passwd 的值
        String confirdPasswd = request.getParameter("confirdPasswd");
```

```
                                         //获取参数 confirdPasswd 的值
        String email = request.getParameter("email");//获取参数 email 的值
        List<String> errors = new ArrayList<String>(); //装载错误信息
        if(!isValidEmail(email)){                      //验证邮箱
            errors.add("无效的邮箱号码！");
        }
        if(isValidUsername(userName)){             //验证用户名
            errors.add("用户为空或者已经存在！");
        }
        if(isValidPassword(passwd,confirdPasswd)){ //验证密码
            errors.add("密码为空或者密码不一致！");
        }
        if(!errors.isEmpty()){             //如果 List 不为空，则跳转到错误页面
            request.setAttribute("errors", errors);
            request.getRequestDispatcher("/error.htm").//跳转到错误的页面
            forward(request, response);
        }else{                        //验证通过，获取 User 中的 Map 集合添加用户
            User user = User.getInstance();;
            Map<String,String> map = user.getUserMap();
            map.put(userName, passwd+"##"+email);
            request.getRequestDispatcher("/success.htm").//跳转到成功的页面
            forward(request, response);
        }
    }
    //使用正则表达式验证邮箱
    public boolean isValidEmail(String email){
        boolean flag = false;
        if(email==null||"".equals(email)) flag = false;
        if(email!=null&&!"".equals(email)){
            flag = email.matches("^[\\w-]+(\\.[\\w-]+)*@[\\w-]+
            (\\.[\\w-]+)+$");
        }
        return flag;
    }
    //验证用户名是否存在，获取 User 类中的 Map 集合
    //根据 key 判断 Map 中是否存在用户
    public boolean isValidUsername(String userName){
        User user = User.getInstance();;
        Map<String,String> map = user.getUserMap();
        if(userName!=null&&!userName.equals("")){
            if(map.get(userName)!=null&&!map.get(userName).equals("")){
                return true;                              //存在用户
            }else return false;
        }
        if(userName==null&&userName.equals("")){
```

```
60                  return true;
61              }
62              return true;
63          }
64          //验证密码，如果密码为空且长度小于6并且跟确认密码不统一
65          //返回 true
66          public boolean isValidPassword(String passwd,String confirdPasswd){
67              return passwd==null||confirdPasswd==null
68              ||passwd.length()<6||confirdPasswd.length()<6
69              ||!passwd.equals(confirdPasswd);
70          }
71      }
```

如上代码中，代码第 13~17 行获取参数值；第 19~38 行分别验证用户名、密码、邮箱的有效性，如果无效的将错误写进 List 中，通过验证则跳转到欢迎界面；代码第 40~48 行用正则表达式验证邮箱的合法性；第 51~62 行验证用户的合法性，如果新注册的用户名存在则提示用户名已经存在，否则新增一个用户；代码第 66~70 行验证密码的合法性，如果密码为空、确认密码为空、密码的长度小于 6、确认的密码长度小于 6、密码与确认的密码不一致就不合法。在实际开发中这些验证应该放在 JavaScript 中，但是这里先在后台进行验证。在用户验证中，用户类本来是要保存到数据库中的，但是这里用单例模式保存新增用户。User 类的源代码如下：

```
----------------------- User.java--------------------------
package com.eshore.pojo;

import java.util.HashMap;
import java.util.Map;

public class User {

    private Map<String,String> userMap = new HashMap<String,String>();
    private static User user = null;
    private User(){
        userMap.put("zhangsan", "111111##zhangsan@sian.com");
        userMap.put("lisi", "222222##lisi@sian.com");
        userMap.put("wangwu", "333333##wangwu@sian.com");
        userMap.put("zhaoliu", "444444##zhaoliu@sian.com");
    }

    public static User getInstance(){
        if(user==null){
            user = new User();
        }
        return user;
```

```
22      }
23      public Map<String, String> getUserMap() {
24          return userMap;
25      }
26  }
```

如上代码，代码 10~15 行用私有的构造方法，初始化 Map 中的值；代码第 17~21 行建立一个静态的方法来获取 User 对象。

如果用户注册失败了，会转发到 ErrorServlet 这个类中。其源代码如下：

```
---------------------- ErrorServlet.java---------------------------
package com.eshore;

import java.io.IOException;
…
import javax.servlet.http.HttpServletResponse;

@WebServlet(name = "errorServlet", urlPatterns = { "/error.htm" })
public class ErrorServlet extends HttpServlet {

    private static final long serialVersionUID = 1L;

    public void doGet(HttpServletRequest request, HttpServletResponse response)
            throws ServletException, IOException {
        doPost(request, response);                // 调用 doPost()方法
    }

    public void doPost(HttpServletRequest request, HttpServletResponse response)
            throws ServletException, IOException {

        response.setContentType("text/html;charset=UTF-8");
        //设置响应的类型和编码
        PrintWriter out = response.getWriter();//取得 PrintWriter()对象
        out.println("<!DOCTYPE HTML PUBLIC \"-//W3C//DTD HTML
                4.01 Transitional//EN\">");
        out.println("<HTML>");
        out.println("  <HEAD><TITLE>新增会员失败</TITLE></HEAD>");
        out.println("  <BODY>");
        out.print("<h2>新增会员失败</h2>");
        //遍历错误信息
        List<String> list = (List<String>)request.getAttribute("errors");
        for(String str:list){
            out.println(str+"<br>");
        }
        out.print("<a href=\""+request.getContextPath()+"/register.jsp\">
                    返回注册首页</a>");
```

```
35          out.println("  </BODY>");
36          out.println("</HTML>");
37          out.flush();
38          out.close();
39      }
40  }
```

ErrorServlet 这个类主要就是显示错误信息，它也可以用 JSP 来实现，只是在这里为了进一步演示响应的方法，用 Servlet 来叙写 HTML 代码。如上代码，第 20 行设置响应类型和字符编码；代码第 29~32 行遍历错误信息；代码第 33 行创建超链接返回注册页面。

如果用户注册成功，则转发到成功页面 SuccessServlet。其源代码如下：

```
---------------------- SuccessServlet.java---------------------------
package com.eshore;

import java.io.IOException;
......省略
@WebServlet(name = "successServlet", urlPatterns = { "/success.htm" })
public class SuccessServlet extends HttpServlet {

    private static final long serialVersionUID = 1L;

    public void doGet(HttpServletRequest request, HttpServletResponse response)
            throws ServletException, IOException {
        doPost(request, response);// 调用 doPost()方法
    }

    public void doPost(HttpServletRequest request, HttpServletResponse response)
            throws ServletException, IOException {

        response.setContentType("text/html;charset=UTF-8");
        // 设置响应的类型和编码
        PrintWriter out = response.getWriter();// 取得 PrintWriter()对象
        out.println("<!DOCTYPE HTML PUBLIC \"-//W3C//DTD HTML
                4.01 Transitional//EN\">");
        out.println("<HTML>");
        out.println("  <HEAD><TITLE>新增会员成功</TITLE></HEAD>");
        out.println("  <BODY>");
        out.print("<h2>会员,"+request.getParameter("username")+
        "注册成功</h2>");
        out.print("<a href=\"" + request.getContextPath()
                + "/login.jsp\">返回首页登录</a>");
        out.println("  </BODY>");
        out.println("</HTML>");
        out.flush();
```

```
31          out.close();
32      }
33  }
```

上述代码中，主要的作用是为了显示注册用户成功，如代码第 25 行；代码第 26~27 行创建超链接可以返回到登录页面。程序运行效果图依次如图 7.16、图 7.17、图 7.18 所示。

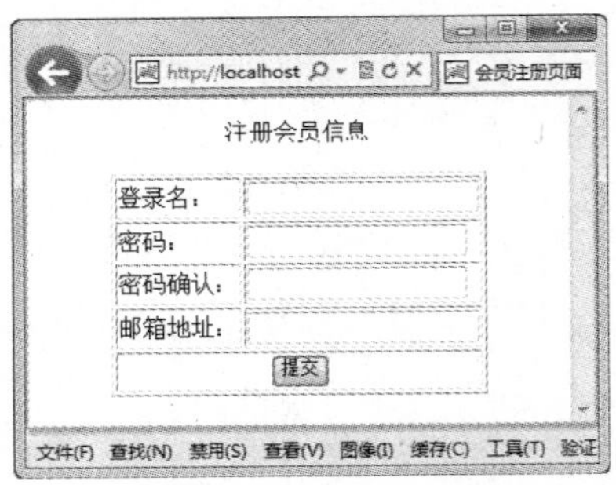

图 7.16　注册页面

图 7.17　错误页面

图 7.18　成功页面

7.6.2　实现网站登录功能

登录页面 login.jsp 提供给已注册用户进行登录的功能，由简单的 form 组成，包含用户名和登录密码。其源代码如下：

```
----------------------- login.jsp---------------------------
<%@ page language="java" import="java.util.*" pageEncoding="UTF-8"%>
<!DOCTYPE HTML PUBLIC "-//W3C//DTD HTML 4.01 Transitional//EN">
<html>
  <head>
    <title>会员登录页面</title>
  </head>

  <body>
      <center>会员登录<br/>
            <form action="<%=request.getContextPath()%>/login.htm"
            method="post">
                <table border="1">
                    <tr>
```

```
13                      <td>登录名：</td>
14                      <td><input name="username"/></td>
15                  </tr>
16                  <tr>
17                      <td>密码：</td>
18                      <td><input name="passwd" type="password"/></td>
19                  </tr>
20                  <tr align="center">
21                      <td colspan="2"><input type="submit" value=
                        "提交"/></td>
22                  </tr>
23                  </table>
24              </form>
25          </center>
26      </body>
27  </html>
```

如上代码中，form 中的 action 提供登录的 Servlet 地址；代码第 10~24 行提供简单的输入框和提交按钮。

登录的 Servlet 类 LoginServlet，提供接收从页面中的参数并验证用户名、密码是否有效。其源代码如下：

```
----------------------- LoginServlet.java ---------------------------
package com.eshore;

import java.io.IOException;
…
import com.eshore.pojo.User;

@WebServlet(name = "loginServlet", urlPatterns = { "/login.htm" })
public class LoginServlet extends HttpServlet {

    private static final long serialVersionUID = 1L;
    public void doPost(HttpServletRequest request, HttpServletResponse response)
            throws ServletException, IOException {

        String userName = request.getParameter("username");
        // 获取参数 username 的值
        String passwd = request.getParameter("passwd");
        // 获取参数 passwd 的值
        if(checkLogin(userName,passwd)){    //用户登录成功，跳转到用户页面
            request.getRequestDispatcher("/member.htm").
            forward(request, response);
        }else{
            response.sendRedirect("login.jsp");            //重定向到登录首页
```

```
21            }
22        }
23        //验证登录用户
24        public boolean checkLogin(String username,String passwd){
25            User user = User.getInstance();
26            Map<String,String> map = user.getUserMap();
27            if(username!=null&&!"".equals(username)&&  //用户不为空才判断
28                passwd!=null&&!"".equals(passwd)){
29                String[] arr = map.get(username).split("##");//分割 Map 中的值
30                if(arr[0].equals(passwd)) return true;
31                else return false;
32            }else
33                return false;
34        }
35    }
```

如上代码，接收用户名和密码，验证如果是合法用户，则跳转到会员欢迎页面，否则重定向到登录页面也可以重定向到首页或者读者自定义的页面。上述代码中第 14~15 行接收页面参数；代码第 16~20 判断是否为合法用户选择相应的跳转页面；代码第 24~33 行获取用户实例判断用户是否存在，如果存在验证密码是否正确。

如果用户登录成功，则跳转到会员欢迎页面，其源代码如下：

```
---------------------- MemberServlet .java---------------------------
package com.eshore;

import java.io.IOException;
......省略
import javax.servlet.http.HttpServletResponse;

@WebServlet(name = "memberServlet", urlPatterns = { "/member.htm" })
public class MemberServlet extends HttpServlet {

    private static final long serialVersionUID = 1L;

    public void doGet(HttpServletRequest request, HttpServletResponse response)
            throws ServletException, IOException {
        doPost(request, response);                 // 调用 doPost()方法
    }
    public void doPost(HttpServletRequest request, HttpServletResponse
response)
            throws ServletException, IOException {
        response.setContentType("text/html;charset=UTF-8");
        // 设置响应的类型和编码
        PrintWriter out = response.getWriter();// 取得 PrintWriter()对象
        out.println("<!DOCTYPE HTML PUBLIC \"-//W3C//DTD HTML
```

```
21                4.01 Transitional//EN\">");
22            out.println("<HTML>");
23            out.println("  <HEAD><TITLE>会员登录成功</TITLE></HEAD>");
24            out.println("  <BODY>");
25            out.print("<h2>会员,"+request.getParameter("username")+",您好!</h2>");
26            out.print("<a href=\"" + request.getContextPath()
27                    + "/register.jsp\">返回首页登录</a>");
28            out.println("  </BODY>");
29            out.println("</HTML>");
30            out.flush();
31            out.close();
32        }
33    }
```

MemberServlet 类主要就是显示会员信息的作用。如上代码第 25 行页面打印欢迎信息；代码第 26 行创建超链接返回登录页面。程序运行效果图如图 7.19、图 7.20 所示。

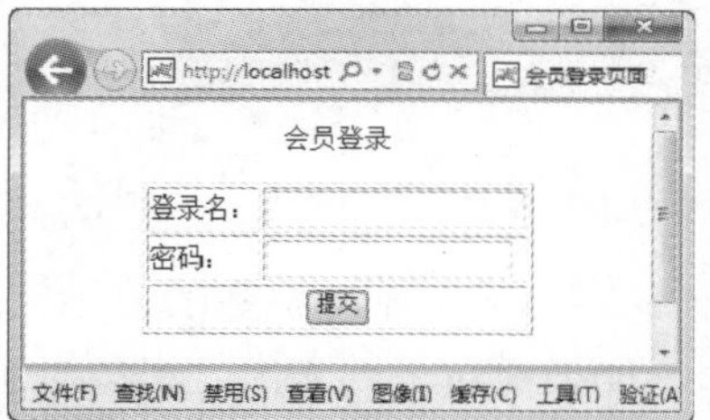

图 7.19　登录界面

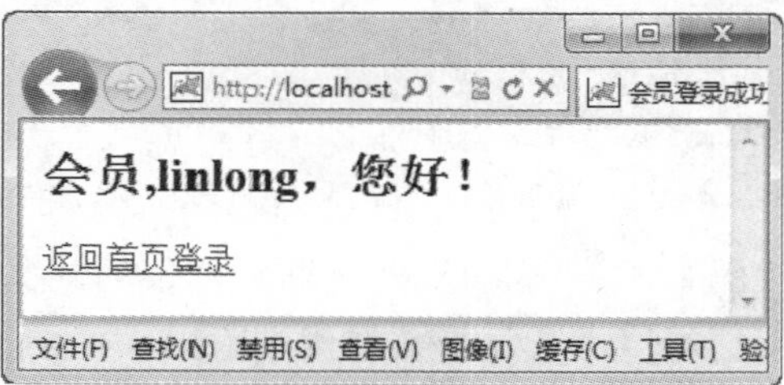

图 7.20　会员欢迎界面

到目前为止，网站的注册和登录功能实现完毕，但是只能检查用户名跟密码是否正确并转发到指定页面，但无法判断用户是否已经登录，读者可以根据后面学习的内容继续完善本案例。

7.7 实例：使用 HttpSession 实现猜字游戏

利用 HttpSession 会话，可以实现 Web 的基本功能与操作，例如，实现简单购物车、在线猜字游戏等。下面通过猜字数字游戏来说明 Session 会话管理。

首先，新建一个猜数字页面 guessNumber.jsp，源代码如下：

```
-----------------------guessNumber.jsp---------------------------
<%@ page language="java" import="java.util.*" pageEncoding="UTF-8"%>
<%
String path = request.getContextPath();
%>
<!DOCTYPE HTML PUBLIC "-//W3C//DTD HTML 4.01 Transitional//EN">
<html>
  <head>
```

```
08      <title>在线猜数字</title>
09    </head>
10    <%
11       String flag = request.getParameter("flag");
12       String message="";
13       if(flag!=null&&"larger".equals(flag)){
14           message="太大了";
15       }else if(flag!=null&&"lessner".equals(flag)){
16           message="太小了";
17       }else if(flag!=null&&"success".equals(flag)){
18           message="您猜对了";
19       }
20     %>
21    <body>
22         <form action="<%=path %>/servlet/Guess" method="post">
23             <span>请输入您猜得数字：</span>
24             <input name="guessNumber" size=""10/>
25             <span style="color: red"><%=message %></span>
26             <input type="submit" value="提交" />
27         </form>
28    </body>
29  </html>
```

上述代码中，第 10~18 行判断返回值的情况，如果猜得值大于随机数的值则输出“太大了”；如果猜得值小于随机数的值则输出“太小了”；如果猜对了则输出“您猜对了”。

其次，编写提交的 Servlet 类 Guess。其源代码如下：

```
-----------------------Guess.java--------------------------
public void doPost(HttpServletRequest request, HttpServletResponse response)
            throws ServletException, IOException {

            //获取页面提交的数字
        String guessNumber = request.getParameter("guessNumber");
        int number = Integer.parseInt(guessNumber);
            //产生一个 Session，并获取存放在 Sessioin 中的 currentNumber
        HttpSession session = request.getSession();
        Integer currentNumber = (Integer)session.getAttribute("currentNumber");
        String context = request.getContextPath();
        if(currentNumber==null){
            //产生1~50的随机数
            currentNumber = 1+(int)(Math.random()*50);
            session.setAttribute("currentNumber", currentNumber);
        }
        //判断猜得数跟 Sessioin 中的 currentNumber 大小
        if(number>currentNumber){
            response.sendRedirect(context+"/guessNumber.jsp?flag=larger");
        }else if(number<currentNumber){
            response.sendRedirect(context+"/guessNumber.jsp?flag=lessner");
        }else {
            currentNumber = 1+(int)(Math.random()*50);
            session.setAttribute("currentNumber", currentNumber);
```

```
24                  response.sendRedirect(context+"/guessNumber.jsp?flag=success");
25              }
26   }
```

上述代码中，第 11~15 行利用 Math.random 方法产生 1~50 之间的随机数，并把它保存在 session 中，代码第 17~25 行判断提交的数和 Sessioin 中的值大小，用重定向的方法跳转到指定页面并携带 flag 标识。代码运行效果如图 7.21 所示。

图 7.21　在线猜数字效果图

本例用的是重定向到 guessNumber.jsp 页面，读者也可以用 RequestDispatcher 中的 forward 方法跳转。

7.8 上机实践

1. 实现一个 Web 程序，同时上传多个文件到本地文件夹中。
2. 实现一个 Web 程序，下载本地的指定目录文件。
3. 实现一个 Web 程序，用 JSP 实现网站的注册和登录功能中显示页面的部分。
4. 编写一个 Web 程序，根据请求者的需要动态地设定会话超时时间。
5. 编写一个简单购物车，实现基本的购物功能。

第 8 章

使用Java Bean读取数据库

在 Web 应用技术中，数据库的操作是必不可少的，包括对数据库表的增加、删除、修改、查询等功能。现如今，数据库可以分为关系型数据库和非关系型数据库，关系型数据库主要有 MySQL、Oracle、DB2、Infomix、SQL Server 等数据库；而非关系型数据库主要有 NoSQL、voltDB 等数据库。

对于 JSP 而言，Java Bean 是最基础的分层技术。Bean 是一种软件组件，在 JSP 开发中常用来封装事务逻辑、数据库操作等。

本章首先介绍 MySQL 的简单使用，然后介绍 Java Bean 的使用，最后会介绍下 DAO 设计模式。

8.1 MySQL 数据库入门

MySQL 数据库是一款小型的关系型数据库，它以其自身的特点（例如：体积小、速度快、成本低等）独树一帜。MySQL 数据库是目前最受欢迎的开源数据库，但目前已经被 Oracle 公司收购。

本节介绍 MySQL 的安装部署和启动。

8.1.1 MySQL 的安装和配置

MySQL 支持不同的操作系统平台，虽然在不同平台下的安装和配置都不相同，但是差别也不是很大。在 Windows 平台下可以使用二进制的安装包或者免安装版的软件包进行安装，安装包提供图形化的安装向导过程，免安装版则直接解压就能用。Linux 平台下使用命令安装 MySQL，但由于 Linux 有很多的版本，因此不同的 Linux 平台需要下载相应的 MySQL 安装包。本小节主要叙述 Windows 平台下的 MySQL 的安装和配置过程。

Windows 平台下提供两种安装方式：MySQL 二进制版和免安装版。一般来说，应该采用二进制版，因为该版本在使用起来比较简单，不用第三方工具来启动就可以运行 MySQL。这里采用二进制的安装方式。

1. 下载 MySQL 安装文件

具体的下载操作步骤如下：

步骤 01　打开常用浏览器，输入网址：http://dev.mysql.com/downloads/mysql，页面自动跳转到 MySQL Community Server 5.6.15 下载页面，选择 Generally Available(GA) Release 选项卡，下载界面如图 8.1 所示。

Generally Available (GA) Releases
Development Releases
MySQL Community Server 5.6.15
Select Platform:
Microsoft Windows
Recommended Download:
MySQL Installer 5.6
for Windows
All MySQL Products. For All Windows Platforms.
In One Package.
Windows (x86, 64-bit), MySQL Installer MSI　Download
Other Downloads:
Windows (x86, 32-bit), ZIP Archive　5.6.15　211.7M　Download
(mysql-5.6.15-win32.zip)　MD5: dc43e6fc9af8f582f1d972f7599a6738 | Signature
Windows (x86, 64-bit), ZIP Archive　5.6.15　216.2M　Download

图 8.1　MySQL 下载界面

步骤 02　在下拉列表框中选择 Microsoft Windows 平台，如图 8.2 所示。

图 8.2　选择 Microsoft Windows 平台

步骤 03　可以选择网络安装二进制或者直接下载二进制文件，在这里选择直接下载二进制文件，单击按钮 **Download** 下载，如图 8.3 所示。

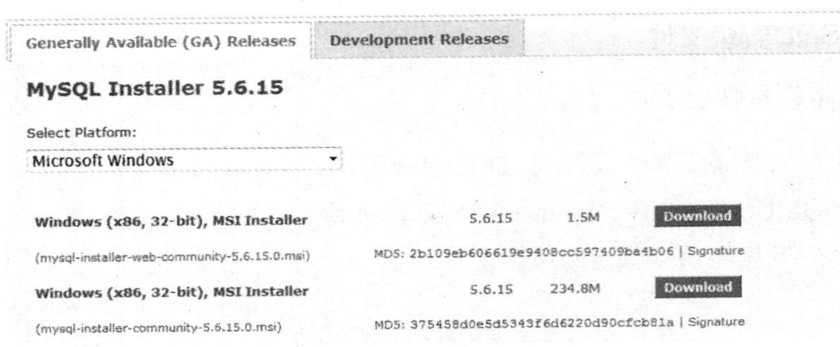

图 8.3　单击下载 MySQL 二进制文件

2. 安装和配置 MySQL 数据库

MySQL 二进制文件下载完成后，找到下载文件（例如 d:\ MySQL-installer-5.6.15.msi），双击进行安装，具体步骤如下。

步骤 01　双击下载的 MySQL-installer-community-5.6.15.0.msi 文件，如图 8.4 所示。

步骤 02　等待 Windows 系统检测 MySQL 安装环境，如图 8.5 所示。

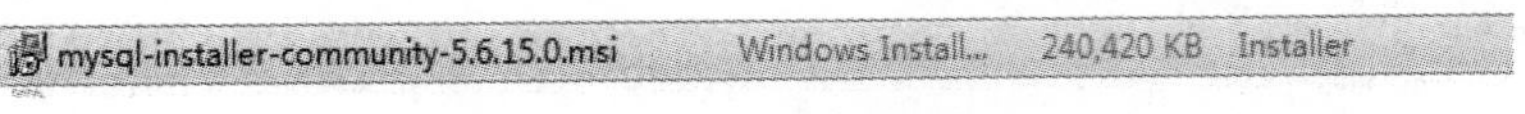

图 8.4　MySQL 安装文件

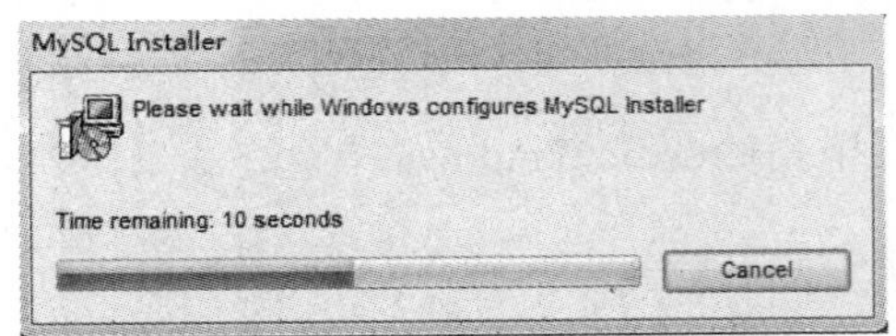

图 8.5　Windows 系统检测 MySQL 安装环境

步骤 03　弹出安装 MySQL 选择对话框，选择“Install MySQL Products”。如图 8.6 所示。

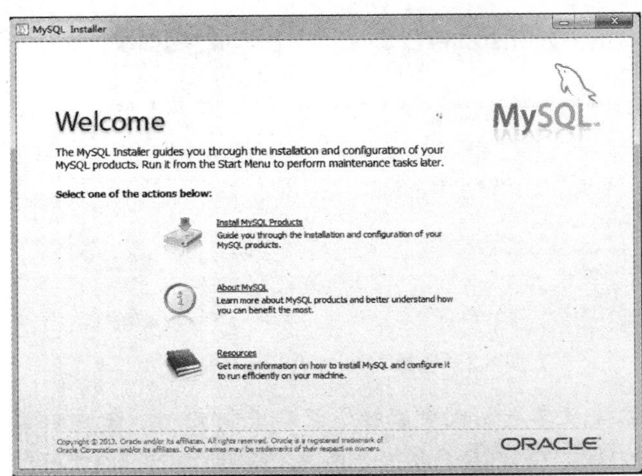

图 8.6　选择 MySQL 产品

步骤 04　弹出安装对话框，选中同意协议，单击 Next 按钮，如图 8.7 所示。

步骤 05　查找最新产品，可以选择跳过此步骤，单击 Execute 按钮，继续执行。如图 8.8 所示。

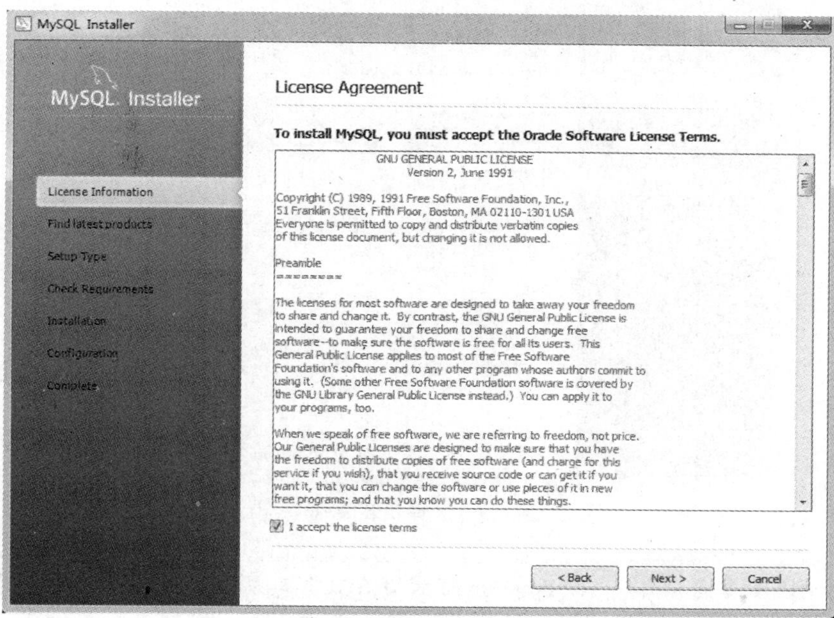

图 8.7　MySQL 协议对话框

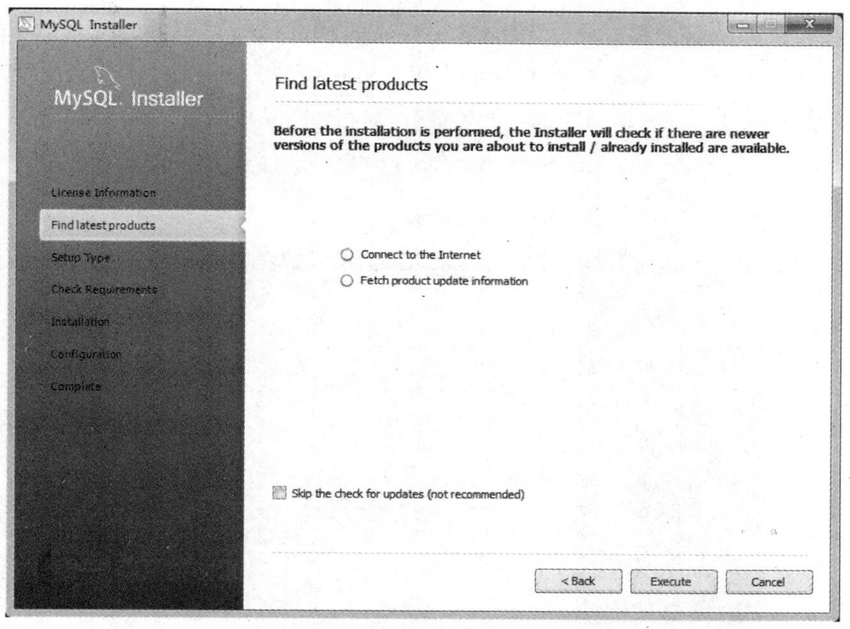

图 8.8　查找最新 MySQL 产品

步骤 06　选择安装类别，有 5 种类别可以选择，分别是开发版（Developer Default）、服务器版（Server Only）、客户端（Client Only）、全部 MySQL（Full）和经典版（Custom）。

在此，本书选择开发版并选择指定的安装目录。如图 8.9 所示。

图 8.9　选择安装类别

MySQL 默认安装路径“C:\Program Files\MySQL”可以单击右侧的“...”更改安装路径，同样 MySQL 数据路径也可以单击右侧的“...”更改。

步骤 07　检查安装的必需产品，如果系统中没有安装相应的必需工具，可以返回上一步选择普通版的 MySQL 类别或者直接单击 Execute 按钮，MySQL 会过滤掉相应的安装程序，等系统有了对应的软件，可以重新运行安装程序进行更新，如图 8.10 所示。

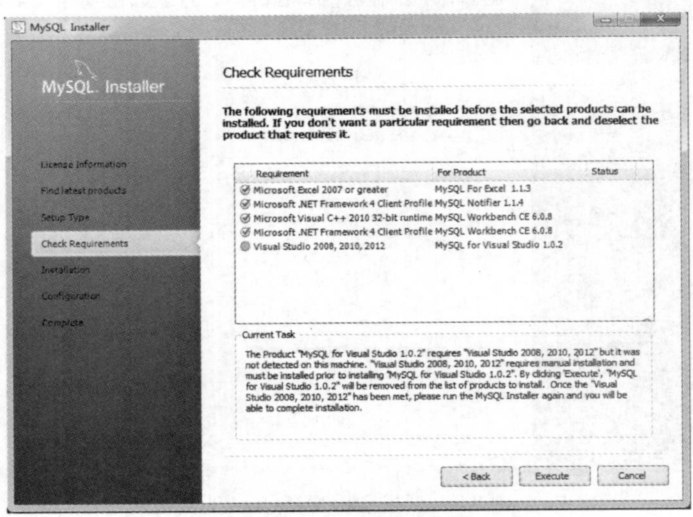

图 8.10　检查必需软件

步骤 08　单击 Execute 按钮，MySQL 开始安装或者更新，如图 8.11 所示。

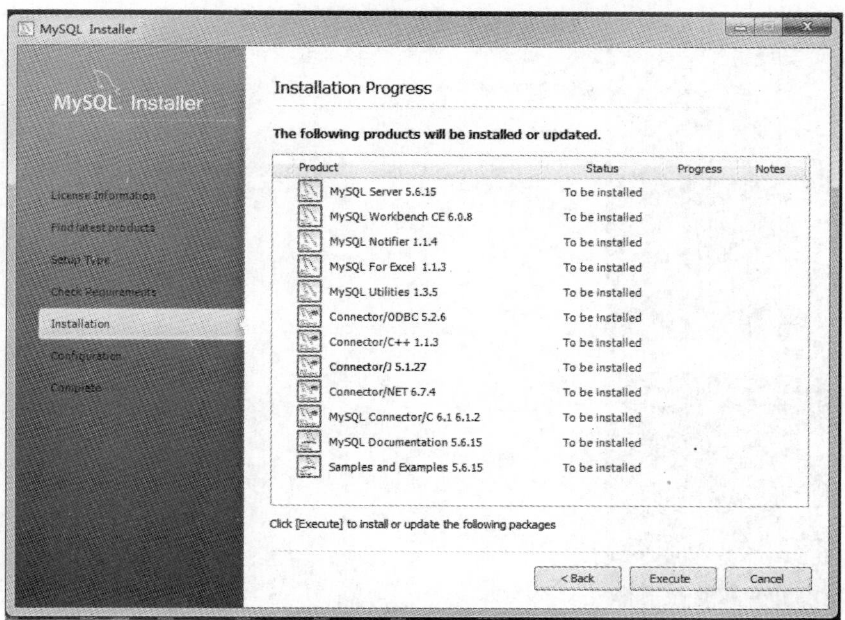

图 8.11　MySQL 安装界面

显示 MySQL 安装进度，如图 8.12 所示。

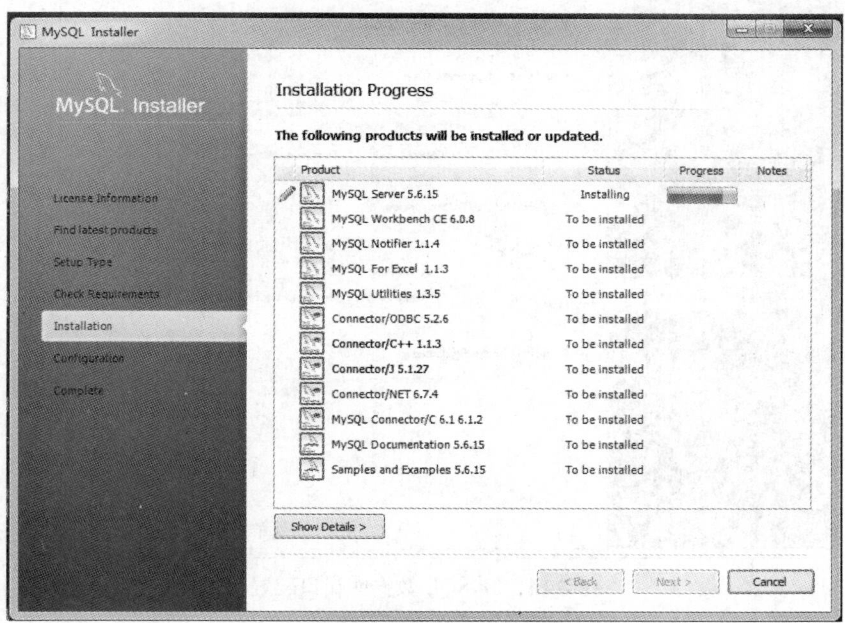

图 8.12　MySQL 安装进度

步骤 09　初始化配置 MySQL，在第 8 步都安装完成后，单击 Next 按钮，跳转到配置页面，单击 Next 按钮，如图 8.13 所示。再单击 Next 按钮，进行 MySQL 端口配置，如图 8.14 所示。

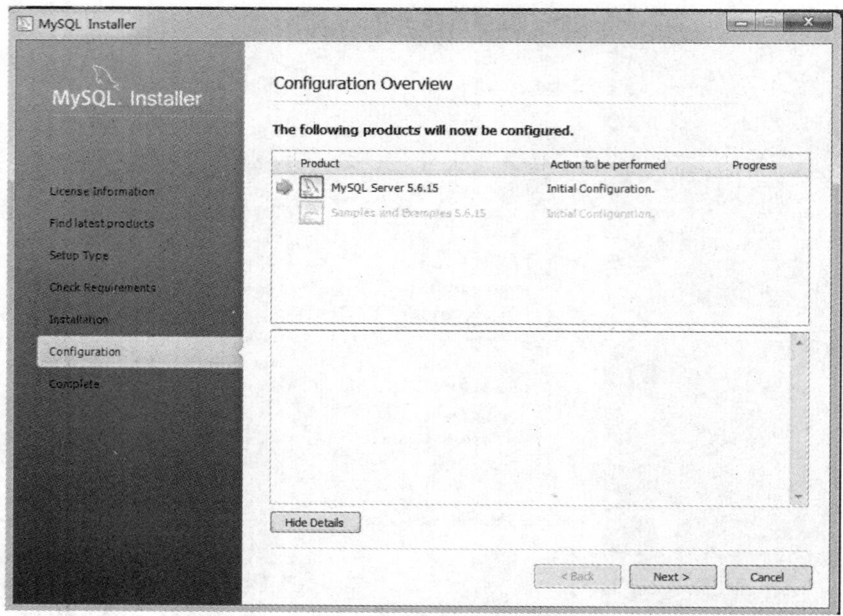

图 8.13　MySQL 配置界面

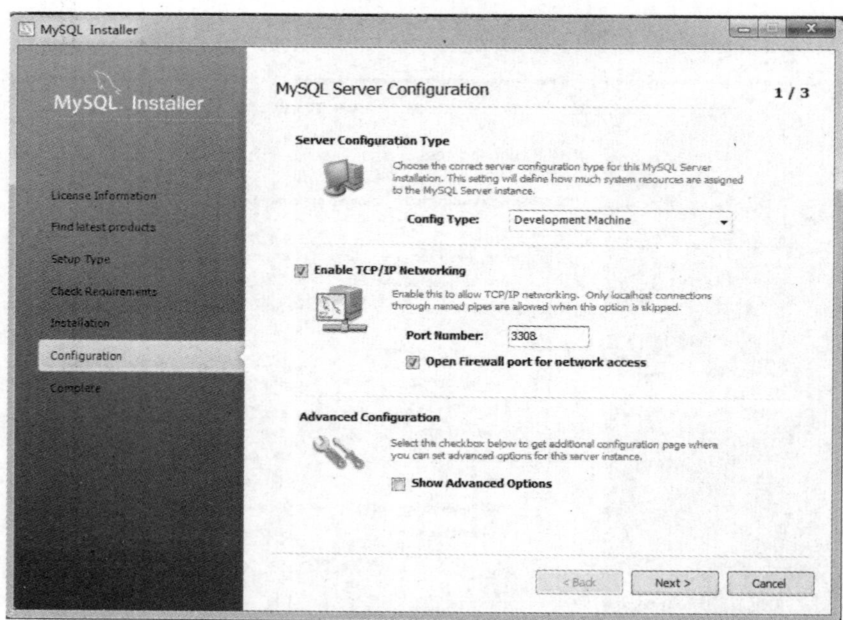

图 8.14　MySQL 数据库端口配置

步骤 10　单击 Next 按钮，进入 MySQL 管理员的密码设置和用户设置，如图 8.15 所示。

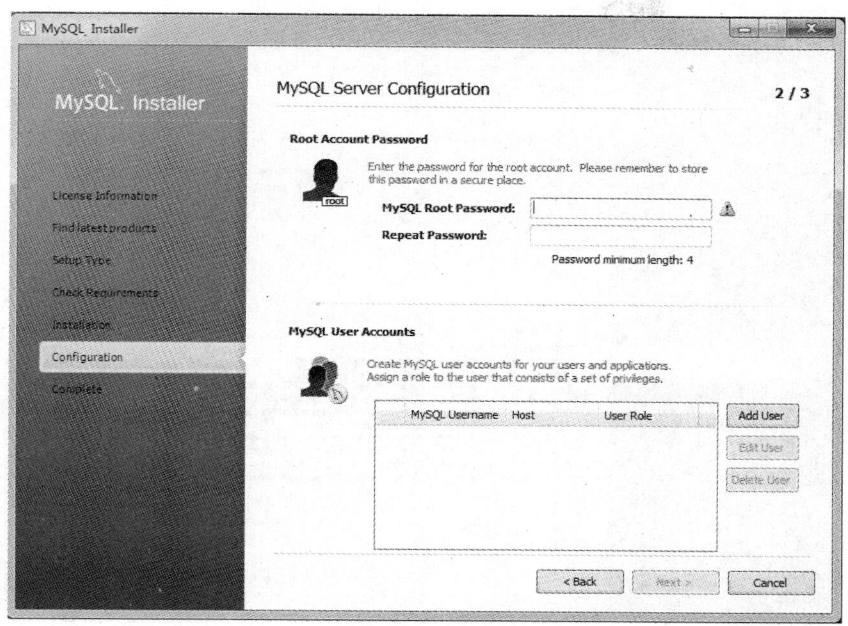

图 8.15　MySQL 用户和密码设置

步骤 11　设置完，单击 Next 按钮，进行 MySQL 服务器名的设置，如图 8.16 所示。

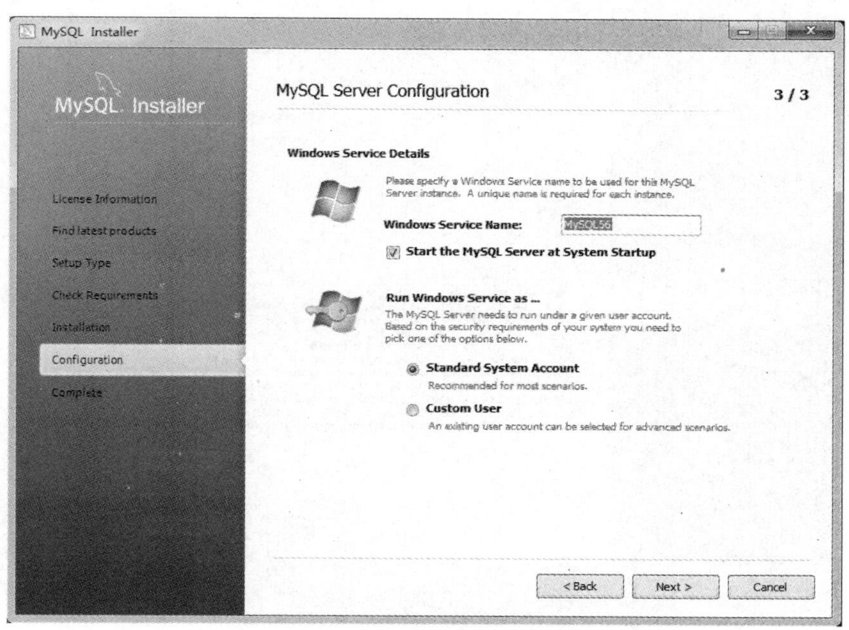

图 8.16　MySQL 服务名设置

步骤 12　配置完如图 8.17 所示。再单击 Next 按钮，进行 MySQL 示例配置，如图 8.18 所示。

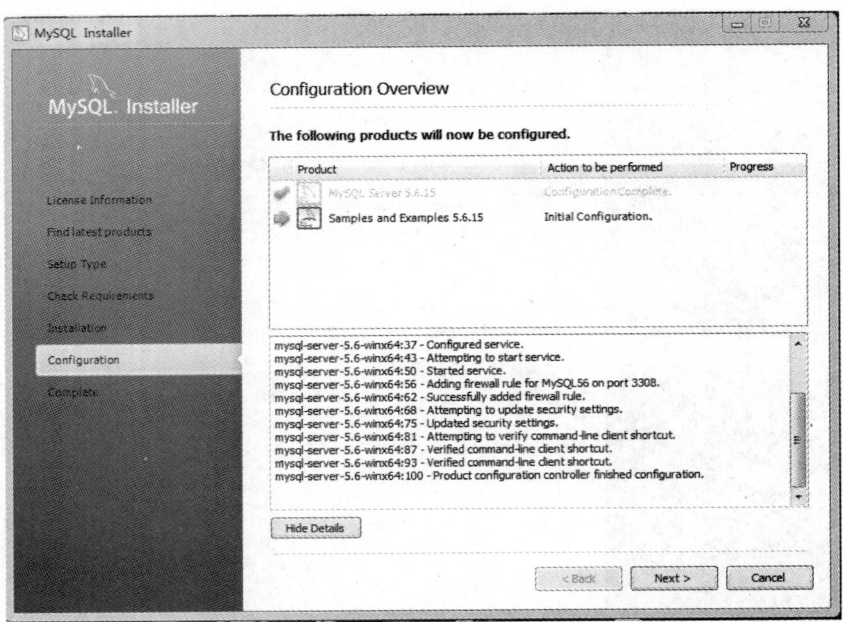

图 8.17　MySQL 配置完初始化

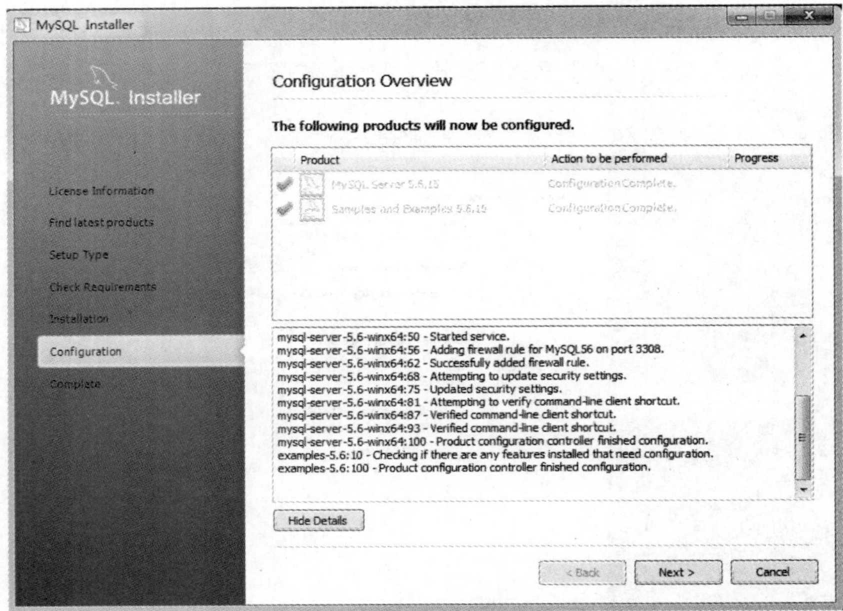

图 8.18　MySQL 示例配置

步骤 13　单击 Next，显示完成，如图 8.19 所示。

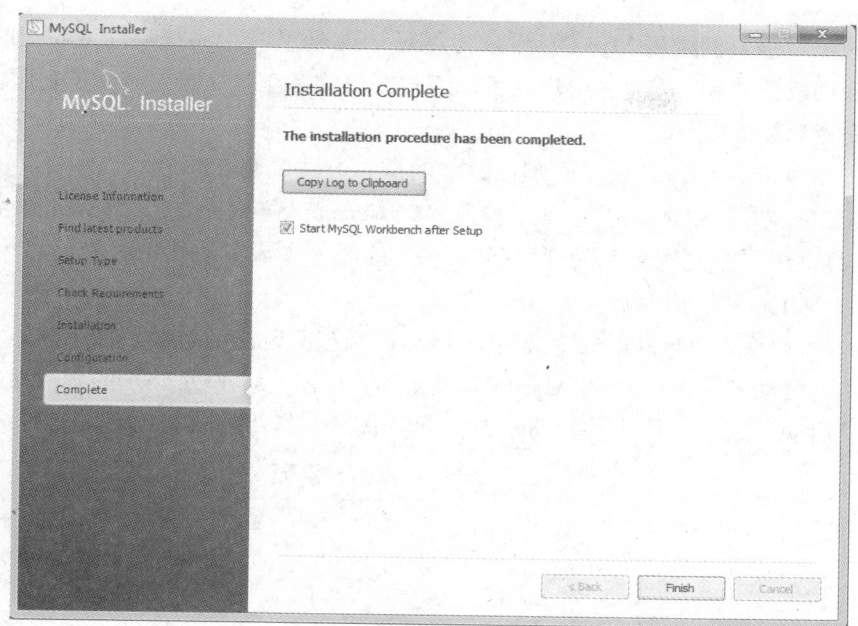

图 8.19　MySQL 安装完成界面

单击 Finish 按钮，系统出现 MySQL Workbench 图形化界面，如图 8.20 所示。

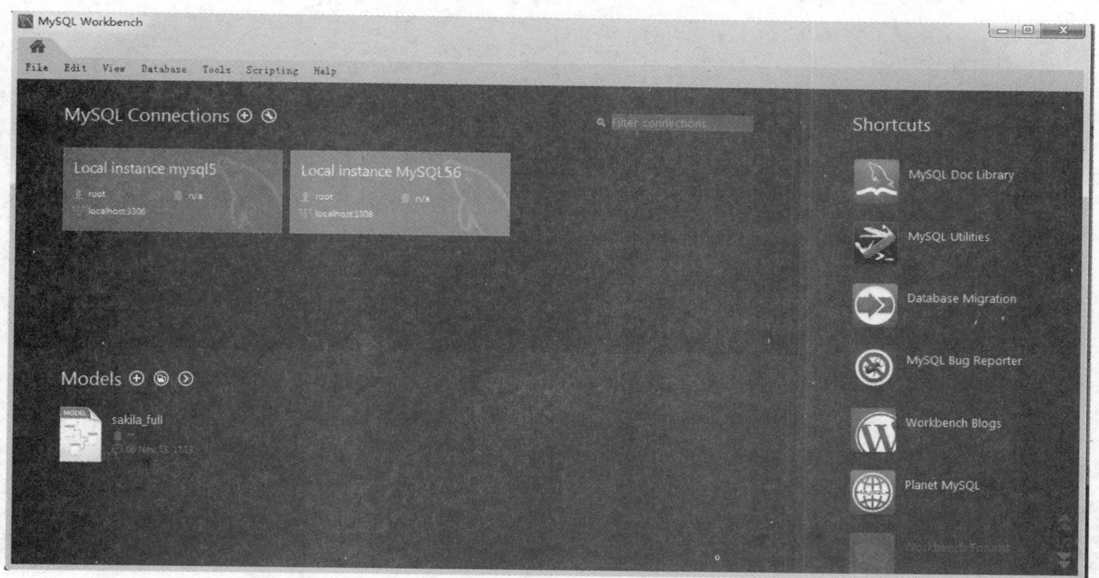

图 8.20　MySQL Workbench 操作界面

8.1.2　启动 MySQL 服务

安装完 MySQL 数据库之后，需要启动服务进程，相应的客户端就可以连接数据库，客户端可以通过命令行或者图形界面工具登录数据库。

在默认的配置中，已经将 MySQL 设置为 Windows 服务，当系统启动或停止时，MySQL 服务会自动启动或者关闭。但是，用户还可以通过图形服务工具来控制 MySQL 服务器或者从命令行使用命令启动。

可以通过 Windows 的服务进行管理，具体的操作方法说明如下。

单击“开始”菜单，在弹出的菜单中输入“services.msc”命令，打开 Windows 的“服务管理器”，在其中可以看到服务名为“MySQL56”的服务项，右边状态为“已启动”，表名该服务已经启动，如图 8.21 所示。

从图 8.21 可以看出 MySQL 启动类型为自动，且该服务已经启动。如果状态为空白，说明服务未启动。启动方法为：双击 MySQL 服务名，打开“MySQL56”的属性对话框，在其中通过单击“启动”或者“停止”按钮来改变服务状态，具体如图 8.22 所示。

图 8.21　服务管理器窗口

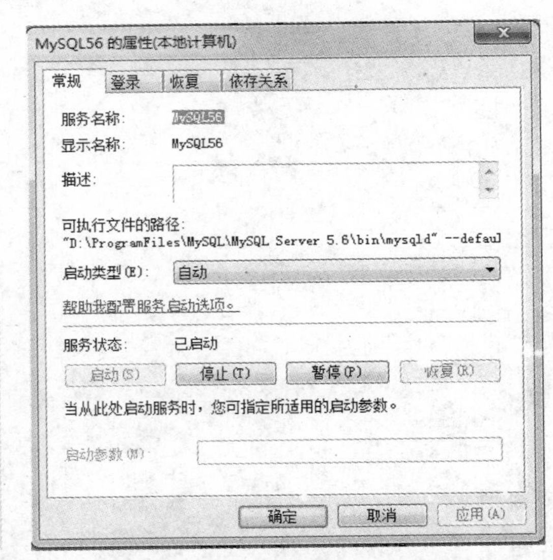

图 8.22　MySQL56 服务属性对话框

也可以通过命令行启动，启动方法为：单击“开始”菜单，在搜索框中输入“cmd”，回车弹出 Windows 命令操作界面，如图 8.23 所示。

然后输入“net start MySQL56”，回车，就可以启动 MySQL56 服务了，停止 MySQL56 服务的命令为“net stop MySQL56”，如图 8.24 所示。

“net start MySQL56”中“MySQL56”是 MySQL 服务的名字。如果 MySQL 服务的名字是其他名字，应该输入“net start XX”。

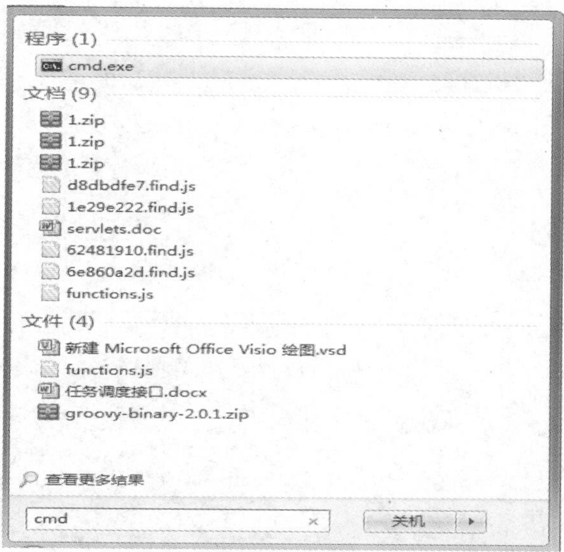

图 8.23 Windows 运行界面

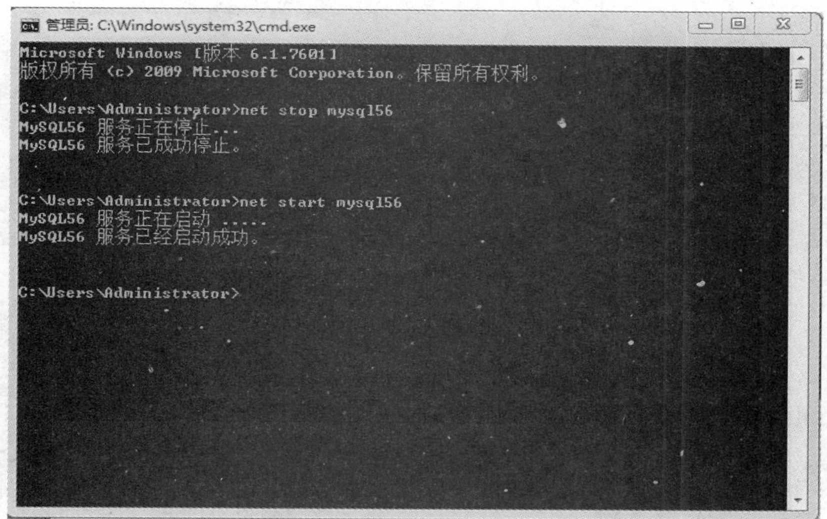

图 8.24 命令行中启动和停止 MySQL

8.1.3 登录 MySQL 数据库

当 MySQL 服务启动后，可以通过客户端来登录 MySQL 数据库。在 Windows 系统中，有两种方式登录 MySQL 数据库。

1. Windows 命令行登录 MySQL

单击“开始”菜单，在弹出的对话框中输入命令“cmd”，如上图 8.23 所示。在 DOS 窗口中通过登录命令连接到 MySQL 数据库，连接 MySQL 的命令为：

```
MySQL -h hostname  -u username -p
```

其中 MySQL 为命令，-h 后面是服务器主机地址，-u 后面是登录数据库的用户名，-p 后面是用户登录密码。在这里由于 MySQL 客户端和服务器是同一台机器，所以输入命令如下：

```
MySQL -h localhost -u root -p
```

回车之后，系统会提示输入密码“Enter password”，如图 8.25 所示，输入前面配置中的密码，验证正确后，即可登录到 MySQL 数据库，如图 8.26 所示。

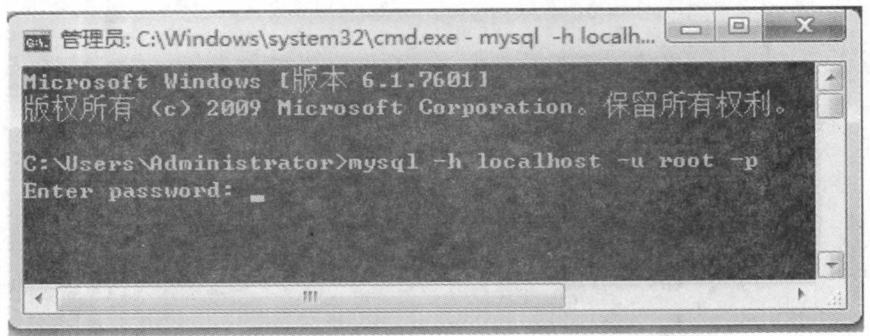

图 8.25　输入命令提示输入密码

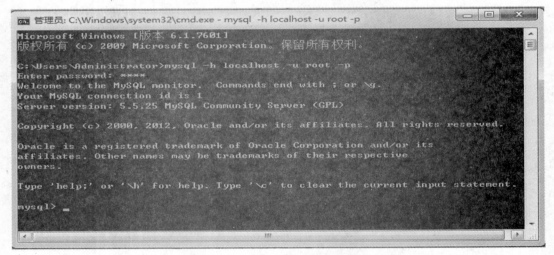

```
管理员: C:\Windows\system32\cmd.exe - mysql  -h localhost -u root -p
Microsoft Windows [版本 6.1.7601]
版权所有 (c) 2009 Microsoft Corporation。保留所有权利。

C:\Users\Administrator>mysql -h localhost -u root -p
Enter password: ****
Welcome to the MySQL monitor.  Commands end with ; or \g.
Your MySQL connection id is 1
Server version: 5.5.25 MySQL Community Server (GPL)

Copyright (c) 2000, 2012, Oracle and/or its affiliates. All rights reserved.

Oracle is a registered trademark of Oracle Corporation and/or its
affiliates. Other names may be trademarks of their respective
owners.

Type 'help;' or '\h' for help. Type '\c' to clear the current input statement.

mysql>
```

图 8.26　Windows 命令行登录窗口

当窗口中出现图 8.26 所示的描述信息，且命令提示符变为“MySQL>”时，表明已成功登录 MySQL 服务器了。

2. 使用 MySQL 命令行登录 MySQL

依次选择“开始”|“所以程序”|“MySQL”|“MySQL Server 5.6”|“MySQL 5.6 Command Line Client”菜单命令，进入密码输入窗口，如图 8.27 所示。

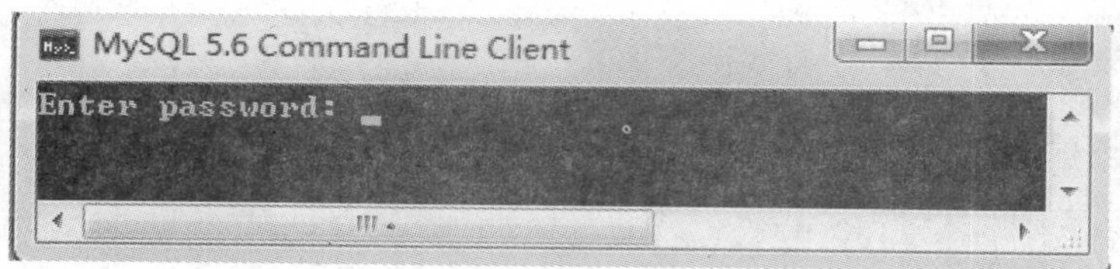

图 8.27　MySQL 命令行登录窗口

输入正确的密码后，就可以登录到 MySQL 数据库中了。显示的结果跟 Windows 命令行登录的结果是一样的。

8.2 MySQL 数据库的基本操作

MySQL 数据库安装完成并启动后，就可以对数据库进行简单的操作，其内容涉及到创建数据库、修改数据库、删除数据库等基本操作。本节将介绍如何在 MySQL 数据库中作简单的操作，在操作数据库之前，创建数据库是实现各种功能的前提。

8.2.1 创建数据库

MySQL 安装完成之后，将会在 data 目录下自动创建几个必需的数据库，可以通过命令

```
show databases;
```

查看当前的所有数据库，如图 8.28 所示。

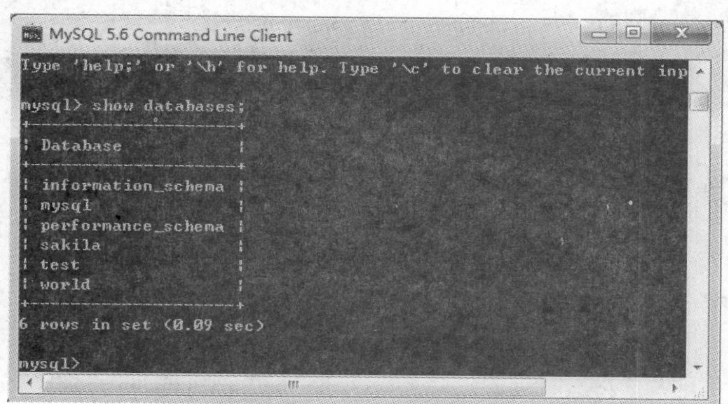

图 8.28　显示必需的数据库界面

可以看出，数据库列表中包含了 4 个数据库，MySQL 是必需的，它描述了用户的访问权限；test 数据库是用户做测试的数据库。

MySQL 创建数据库的基本语法如下：

```
create database databasename;
```

“database_name”是要创建的数据库名称。

【例 8.1】 创建测试数据库 test_database，输入命令如下。

```
create database test_database;
```

如图 8.29 所示。

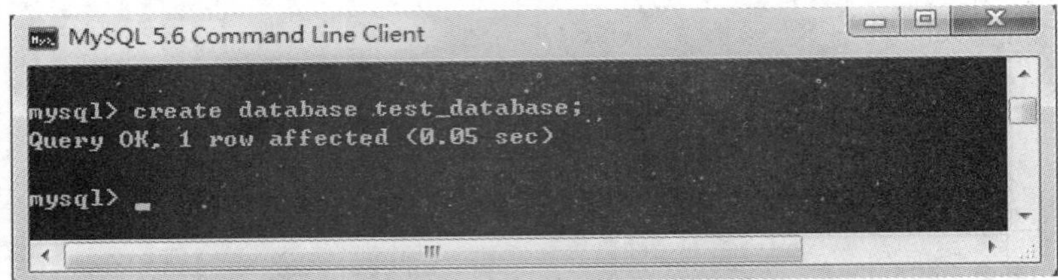

图 8.29　创建数据库

数据库创建好之后，可以使用命令：

```
show create database databasename
```

命令查看数据库的定义，如图 8.30 所示。

```
MySQL 5.6 Command Line Client
mysql> show create database test_database;
+---------------+------------------------------------------------------------------------
| Database      | Create Database
+---------------+------------------------------------------------------------------------
| test_database | CREATE DATABASE `test_database` /*!40100 DEFAULT CHARACTER SET
 utf8 */ |
+---------------+------------------------------------------------------------------------
1 row in set (0.00 sec)

mysql>
```

图 8.30　查看数据库的定义

可以看出，如果数据库创建成功，将显示数据库的创建信息等内容。输入命令

```
show databases
```

再次查看当前的存在的数据库，如图 8.31 所示。

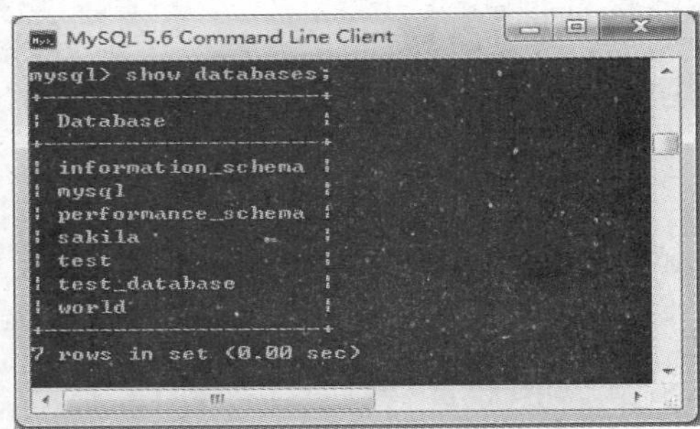

图 8.31　查看当前存在的数据库

可以看到，数据库列表中包含了刚刚创建的数据库 test_database。

8.2.2　删除数据库

删除数据库是将已经创建的数据库和数据库中的数据一并从磁盘空间中删除。删除数据库的基本语法格式如下：

```
drop database database_name;
```

“database_name”是数据库名称，如果数据库不存在，那么删除会出现错误。

【例 8.2】删除测试数据库 test_database，输入命令如下。

```
drop database test_database;
```

如图 8.32 所示。

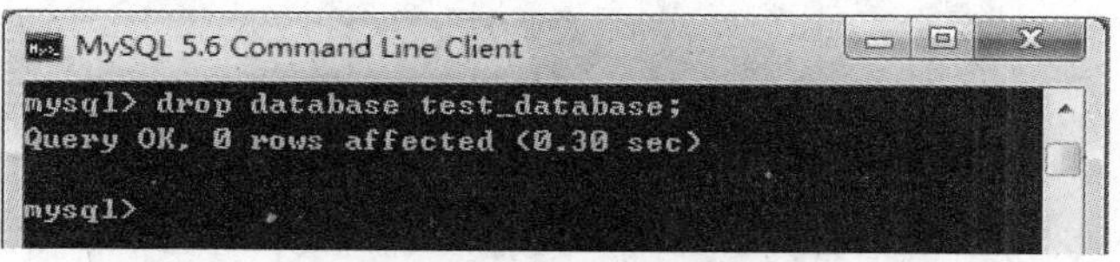

图 8.32　MySQL 删除数据库

执行完删除命令，数据库将被删除，再次执行删除“test_database”数据库时，系统会提示不存在数据库，删除出错，如图 8.33 所示。

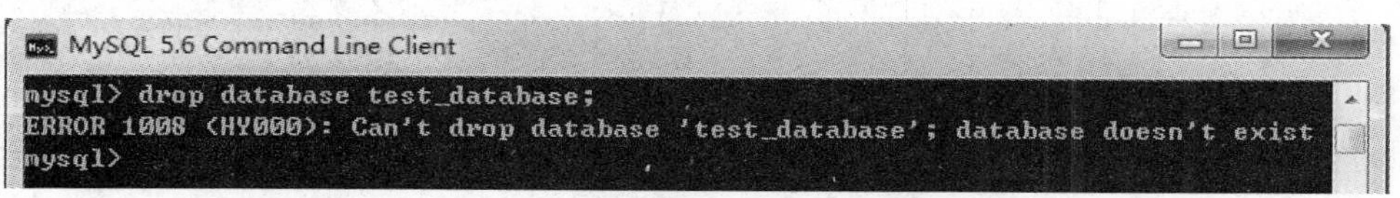

图 8.33　MySQL 再次执行删除数据库

使用删除数据库命令要十分谨慎，因为在执行删除命令时，MySQL 不会给出提醒确认信息。数据库中存储的数据都会被删除掉且不可恢复。

8.2.3　创建数据库表

在创建数据库表之前，要使用语句“use database_name”指定在哪个库中进行新建。创建数据库表的基本语法如下：

```
CREATE TABLE <TABLE_NAME>(字段名，数据类型 [默认值]);
```

【例 8.3】创建数据库表 tb_temp1。

首先选择创建表的数据库，SQL 命令如下：

```
use test_database;
```

创建 tb_temp1 表，SQL 命令如下：

```
CREATE TABLE tb_temp1(
  id        int(10),
  name     varchar(20)
);
```

语句执行后，创建了一个名为 tb_temp1 的表，使用：

```
show tables;
```

查看是否创建成功，如图 8.34 所示。

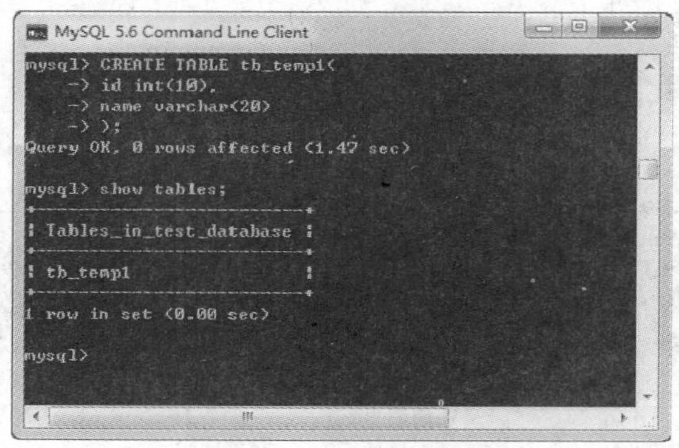

图 8.34　MySQL 创建表 tb_temp1

可以看到，tb_temp1 数据库表已经在 test_database 中创建成功。

8.2.4　修改数据库表

MySQL 修改数据库表的基本语法如下：

```
alter  table  <旧表名>  rename [to] <新表名>
```

在修改表之前，也要选定在哪个库中执行修改操作。

【例 8.4】修改数据库表 tb_temp1 为 tb_temp2。

首先选择创建表的数据库，SQL 命令如下：

```
use test_database;
```

修改 tb_temp1 表，SQL 命令如下：

```
alter table tb_temp1 rename to tb_temp2;
```

语句执行后，修改 tb_temp1 的表后，使用

```
show tables;
```

查看是否修改成功，如图 8.35 所示。

```
mysql> show tables;
+------------------------+
| Tables_in_test_database |
+------------------------+
| tb_temp1               |
+------------------------+
1 row in set (0.00 sec)

mysql>  alter table tb_temp1 rename to tb_temp2;
Query OK, 0 rows affected (0.33 sec)

mysql> show tables;
+------------------------+
| Tables_in_test_database |
+------------------------+
| tb_temp2               |
+------------------------+
1 row in set (0.00 sec)

mysql>
```

图 8.35　MySQL 修改表 tb_temp1

经过比较可以看到，表 tb_temp1 已经修改为 tb_temp2。

可以用命令 desc 查看，修改完的表结构是否与修改前的表结构相同。

8.2.5　修改数据库表字段名

MySQL 中修改表字段名的基本语法规则如下：

```
alter table <表名> change <旧字段名> <新字段名> <新数据类型>;
```

【例 8.5】修改数据库表 tb_temp2 中的 id 字段名称为 temp_id，数据类型修改为 varchar(15)，语句如下：

```
alter table tb_temp2 change id temp_id  varchar(15);
```

使用命令：

```
desc tb_temp2;
```

查看 tb_temp2 表结构，其结果如图 8.36 所示。

```
mysql> alter table tb_temp2 change id temp_id  varchar(15);
Query OK, 0 rows affected (0.26 sec)
Records: 0  Duplicates: 0  Warnings: 0

mysql> desc tb_temp2;
+---------+-------------+------+-----+---------+-------+
| Field   | Type        | Null | Key | Default | Extra |
+---------+-------------+------+-----+---------+-------+
| temp_id | varchar(15) | YES  |     | NULL    |       |
| name    | varchar(20) | YES  |     | NULL    |       |
+---------+-------------+------+-----+---------+-------+
2 rows in set (0.13 sec)

mysql>
```

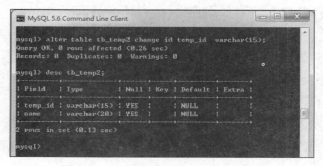

图 8.36　MySQL 修改表 tb_temp2 字段

经过比较发现，字段 id 已经被修改成功，且类型被修改为 varchar。

由于不同类型的数据在 MySQL 中存储的方式并不一定相同，所以修改数据类型可能会影响到表中的数据。因此，当表中存在数据时，最好不要修改数据类型。

8.2.6 删除数据库表

删除表是将库中存在的表删除。MySQL 删除表的基本语法如下：

```
drop table [if exists]table_name1,tablename2,…….;
```

【例 8.6】删除表 tb_temp2 语句如下：

```
drop  table tb_temp2;
```

并使用

```
show tables;
```

查看表 tb_temp2 是否删除成功，其结果如图 8.37 所示。执行结果可以看到，数据库表中已经不存在 tb_temp2，说明删除成功。

```
MySQL 5.6 Command Line Client
mysql> desc tb_temp2;
+---------+-------------+------+-----+---------+-------+
| Field   | Type        | Null | Key | Default | Extra |
+---------+-------------+------+-----+---------+-------+
| temp_id | varchar(15) | YES  |     | NULL    |       |
| name    | varchar(20) | YES  |     | NULL    |       |
+---------+-------------+------+-----+---------+-------+
2 rows in set (0.13 sec)

mysql> drop table tb_temp2;
Query OK, 0 rows affected (0.03 sec)

mysql> show tables;
Empty set (0.00 sec)

mysql>
```

图 8.37　删除表 tb_temp2

8.3 MySQL 数据库的数据管理

上节介绍 MySQL 数据库的基本操作，使读者对数据库的操作有了基本印象。在 MySQL 数据库中，数据的管理是数据库中最重要也是最基本的操作。数据在结构化数据库中是按照行、列的格式进行存储。行代表一条记录，列代表记录中的域。

数据的基本操作，主要涉及：插入数据、修改数据、删除数据等操作。本节将逐一介绍数据的具体操作，通过本节的介绍，读者能够掌握 MySQL 数据库中的数据操作。

8.3.1　插入数据

MySQL 数据库插入数据的语法结构也是遵循标准的 SQL 语法结构，其基本语法如下：

```
insert into table_name(column1,column2,…)  value(value1,value2,…);
```

“table_name”要插入的数据表，column1 是表中的列，value1 为插入的列值，如果不指明列，则 MySQL 按照表中默认的顺序插入值。

【例 8.7】向数据表 tb_temp1 插入数据，输入命令如下。

```
    insert into tb_temp1(id,name,address,phone)
value(1,'TOM','Guangzhou','1890220XXXX');
```

执行命令：

```
select * from tb_temp1;
```

查看是否添加成功，如图 8.38 所示。

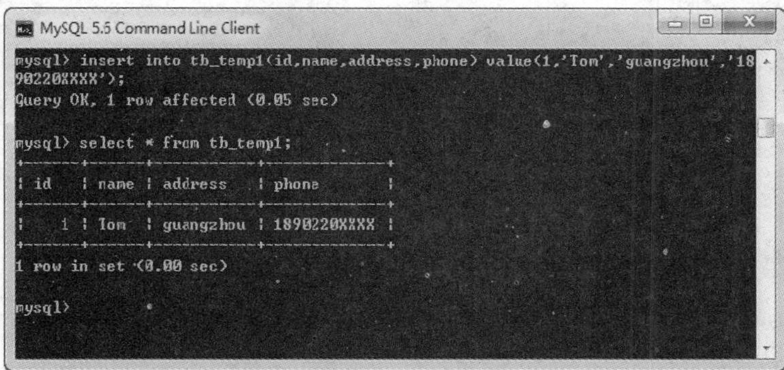

图 8.38　在表 tb_temp1 中插入数据

从结果可以看出，数据添加成功。

8.3.2　修改数据

MySQL 数据库修改数据是修改表中已经存在的数据，其基本语法如下：

```
Update table_name set column1='value1',column2='value2' [condition…..]
```

“table_name”要更新的数据表，column1、column2 是更新的列，value1、value2 为更新的列值，“condition”更新的条件。如果不指明条件，则是将表中的所有的 column1、column2 列值都更新为 value1 和 value2。

【例 8.8】更新数据表 tb_temp1 中的数据，将表中 name 为“Smith”的数据更新为“John”。

```
Update tb_temp1 set name='John' where name='Smith';
```

执行命令：

```
select * from tb_temp1;
```

查看是否修改成功，如图 8.39 所示。

从结果可以看出，数据添加成功。

如果更新的语句为：

```
Update  tb_temp1 set name='John' ;
```

查看结果如图 8.40 所示。

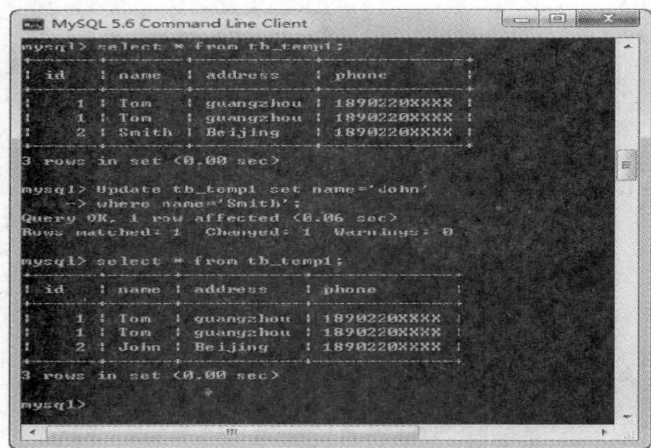

图 8.39　更新表 tb_temp1 中的数据

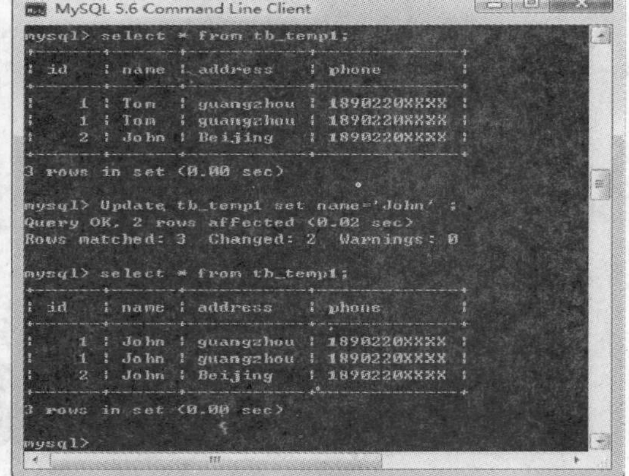

图 8.40　更新表中列的所有数据

8.3.3　删除数据

MySQL 数据库删除数据，可以指定删除行数据或者全部数据，其基本语法如下：

```
delete from table_name  [condition…]
```

“table_name”要删除的数据表，condition 是删除的条件，如果不指明条件，则删除表中的所有数据。

删除表中数据时，要注意备份数据，以免误删的数据有机会可以恢复。

【例 8.9】 删除数据表 tb_temp1 数据，输入命令如下。

```
delete from tb_temp1 where id=1;
```

执行命令：

```
select * from tb_temp1;
```

查看是否删除成功，如图 8.41 所示。

从结果可以看出，数据删除成功。

执行命令：

```
delete from tb_temp1;
```

将表 tb_temp1 中所有的数据都删除，执行结果如图 8.42 所示。

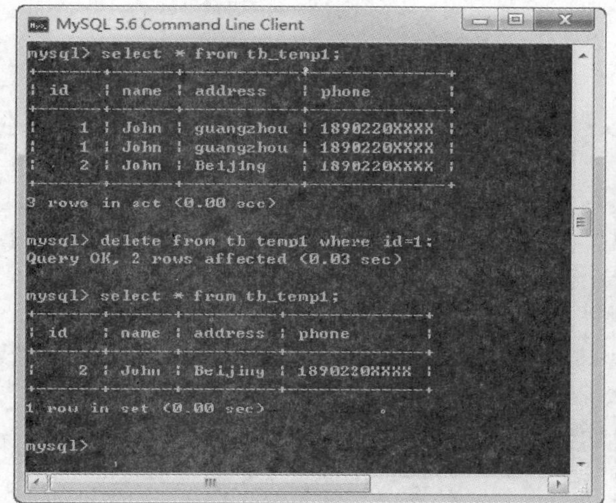

图 8.41　删除表 tb_temp1 中指定数据

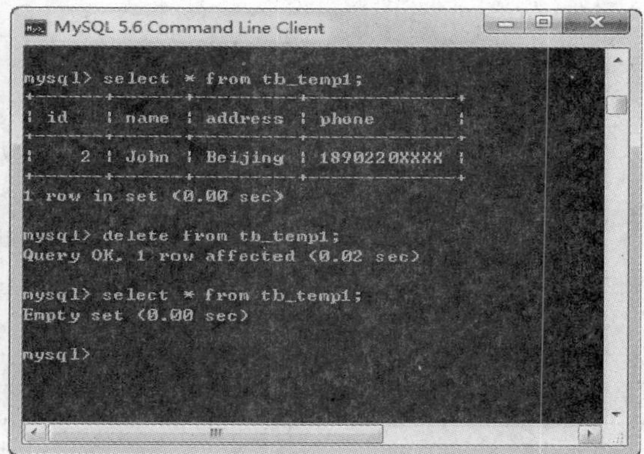

图 8.42　删除表 tb_temp1 中所有数据

8.4 Java Bean 的使用

在开发与数据库有关的软件过程中，要尽量地将业务逻辑和表现层分开，尽量达到完全解耦。这是软件的分层设计的基本理念。在 JSP 中，常利用 Java Bean 实现核心的业务逻辑，而 JSP 页面用于表现层。

8.4.1 认识 Java Bean

Java Bean 它完全符合 Java 语言编码规范的要求和特性，形式上就是纯 Java 代码，它是可以被重复使用的一个组件。在这种设计模式下，JSP 页面只用于接受用户的输入以及显示处理之后的结果，因此不再需要在 JSP 页面中嵌入大量的 Java 代码，这不仅提高了系统的可维护性，而且方便工作的分工。

根据 Java 规范，Java Bean 具有以下特性：

- 支持反射机制：利用反射机制可以分析出 Java Bean 是如何运行的。
- 支持事件：事件是一种简单的通信机制，利用它可以将相应的信息通知给 Java Bean。
- 支持属性：可以自定义属性，利用标准标签与 JSP 页面交互数据。
- 支持持久性：持久性是指可以将 Java Bean 进行保存，在需要的时候又可以重新载入。

下面以一个例子来说明 Java Bean 的创建以及需要遵循的规范。

```
import java.io.Serializable;

```

```
public class User implements Serializable{
    private String username;            //用户名
    private String passwd;          //密码
    private String sex;             //性别
    private String address;             //地址

    public User() {                 //无参的构造方法
        super();

    }

    //以下是属性的 get 和 set 方法，但必须是 public
    public String getUsername() {
        return username;
    }
    public void setUsername(String username) {
        this.username = username;
    }
……
}
```

以上就是新建一个 Java Bean 的基本结构，其要遵循的规范大致如下：

（1）Java Bean 类必须是 public 类。

（2）提供给 JSP 页面调用的方法，必须赋予 public 访问权限。

（3）Java Bean 类中的属性，提供给 JSP 页面调用是必须提供 public 的 get 和 set 方法。

（4）必须拥有不带参数的构造方法。

8.4.2 在 JSP 中使用 Bean

在 JSP 页面中，要正确使用 Bean，要注意以下 3 个问题：

- 按照规范定义 Bean 类，并给出类属性的相应的 get 和 set 方法；
- 在页面中要导入相应的 Bean 类；
- 在 JSP 页面中使用<jsp:useBean>标签使用 Bean 类。

1. 按照规范定义 Bean 类

定义 Bean 类，就跟定义普通的 Java 类一样，如果在 MyEclipse 中，在包中新增一个类就可以，编译会自动编译。如果是用记事本或者其他编辑器，那么编辑完就要手动进行编译。Bean 类文件有两种部署方法：

- 一种是将 Bean 的 Class 文件部署在 Web 服务器的公共目录中；
- 另一种是将 Bean 的 Class 文件部署在 Web 服务器的特定项目的目录中，即将 Bean 类的 Class 文件部署在 Web 项目的 Web-INF 文件夹的 classes 目录中指定文件夹下，例如：ch09\Web-INF\classess\com\eshore\pojo\user.class。

在实际项目开发中，后一种情况比较常见。

2. 在页面中要导入相应的 Bean 类，并用<jsp:useBean>标签获取 Bean 对象

在页面中使用标签 useBean，以便 JSP 页面访问。其语句形式如下：

```
<jsp:useBean
        id="beanInstanceName"
        scope="page | request | session | application"
        {
            class="package.class" |
            type="package.class" |
            class="package.class"  type="package.class" |
            beanName="{package.class | <%= expression %>}"
type="package.class"
        }
  </jsp:useBean>
```

语法中的属性介绍见表 8.1 所示。

表 8.1 <jsp:useBean>标签属性说明

属性值	说明
id	给 Bean 起的一个变量名，可以在指定的范围中使用该变量名
class	Bean 的类路径，必须是严格的 package.class，不能指定其父类
scope	Bean 的有效范围，取值有：page、request、session、application
beanName	实例化的类名称或序列化模板的名称，指定的名称可以为其接口、父类
type	指定其父类或接口的类型，譬如想实例化 HashMap，但是 type 可以填 Map

在使用<jsp:useBean>标签时需要注意以下几点：

- class 或 beanName 不能同时存在。若 JavaBean 对象已存在，class 和 beanName 属性可以不指定，只需指定 type 属性；
- class 可省去 type 独立使用，但 beanName 必须和 type 一起使用；
- class 指定的类必须包含 public、无参的构造方法；
- class 或 beanName 指定的类必须包括包名称，而 type 可以省去包名，通过<%@page import=""%>指定所属包。

class 是通过 new 创建 JavaBean 对象；而 beanName 由 java.beans.Beans.instantiate 初始化 JavaBean 对象。

<jsp:useBean>标签中 Bean 的作用域有 4 个：page、request、session、application。JSP 引擎会根据作用域给用户分配不同的 Bean，运行多个用户拥有相同的 Bean，每个客户的 Bean 是相互独立的。

当指定的范围为 request 时，针对同一用户不同的请求，JSP 引擎都会给用户分配不同的

Bean 对象。当响应结束时，取消分配的 Bean。Bean 的生命周期就是从客户请求开始到响应结束这段时间。·

当指定的范围为 page 时，针对同一用户访问不同的页面，JSP 引擎都会给用户分配不同的 Bean 对象。当客户进入当前页时，JSP 引擎给用户分配一个 Bean 对象，当用户离开当前页时，取消分配的 Bean 对象。因此，该 Bean 的生命周期就是在当前页。

当指定的范围为 session 时，针对同一用户访问同一 Web 项目下不同的页面，JSP 引擎都会给用户分配不同的 Bean 对象。当客户访问 Web 项目中的某一目录下的页面时，JSP 引擎给用户分配一个 Bean 对象，当用户离开 Web 目录时，取消分配的 Bean 对象。因此，该 Bean 的生命周期是 Web 项目的一个 Session 时间。

当指定的范围为 application 时，JSP 引擎为访问用户分配同一个 Bean 对象。它的生命周期就是该 Web 应用的存在时间即 Web 应用在服务器中存在的时间。

8.4.3 访问 Bean 属性

上一节介绍了，在 JSP 中使用 Bean 的方法以及规范。本节介绍如何运用 JSP 标签取得和设置 Bean 的属性值。

1. 设置属性：<jsp:setProperty>

使用标签<jsp:setProperty>可以设置 Bean 的属性值。其语法形式如下：

```
<jsp:setProperty
name="beanInstanceName"
{
property= "*" |
property="propertyName" [ param="parameterName" ] |
property="propertyName" value="{string | <%= expression %>}"
}
/>
```

在使用该标签时，注意必须使用<jsp:useBean>标签创建一个 Bean。<jsp:setProperty>标签的属性说明见表 8.2 所示。

表 8.2　<jsp:setProperty>属性说明

属性	说明
name	是指 Bean 的名字，用来指定被使用的 Bean，它的值必须是<jsp:useBean>标签中的 id 值
property	是指 Bean 的属性名，也就是 Bean 类中的属性，将值赋给 Bean 类的属性
param	是指表单参数名称
value	要设定的属性值

从语法形式中可以看出，使用<jsp:setProperty>标签有 3 种方式：

（1）使用字符串或者表达式给 Bean 属性赋值。其要求是表达式的值与 Bean 属性的值类型要相同。

（2）使用表单参数形式给 Bean 属性赋值。其要求是表单中提供的参数名字与 Bean 属性的名字要相同。这种形式不用具体指定每个 Bean 属性的名字，系统会自动根据表单的参数名字与 Bean 属性名字作一一对应赋值。该形式不用 value 属性。因此其语法形式为：

```
<jsp:setProperty property="" name="" />
```

（3）使用表单参数值给 Bean 属性赋值。其要求是表单中提供的参数名字与<jsp:setProperty>标签中的 param 属性值名字相同。其语法形式为：

```
<jsp:setProperty property="Bean 类中的属性名" param="表单参数名" name="Bean 的变量名 " />
```

下面以例子分别说明这 3 种设置 Bean 属性的方法。

【例 8.10】直接用表达式设置 Bean 属性值

product1.jsp 页面直接用表达式设置 Bean 属性值，其源代码如下：

```
-----------------------product1.jsp--------------------------
<%@ page language="java" import="java.util.*" pageEncoding="UTF-8"%>
<!-- 导入引用的 bean -->
<jsp:useBean id="product" class="com.eshore.pojo.Product" scope="page"/>
<!DOCTYPE HTML PUBLIC "-//W3C//DTD HTML 4.01 Transitional//EN">
<html>
  <head>
    <title>设置 Bean 属性</title>
  </head>

  <body>
      <!-- 设置产品名称 -->
      <jsp:setProperty name="product" property="product_name"
         value="struts 开发教程"/>
      <br/>产品名称是:
      <!-- 获取产品名称 -->
      <%=product.getProduct_name() %>
      <!-- 设置产品编号 -->
      <jsp:setProperty name="product" property="product_id" value="111100123689"/>
      <br/>产品编号是:
      <!-- 获取产品编号 -->
      <%=product.getProduct_id() %>
      <!-- 设置产品价格 -->
      <%
        double price = 68.23;
       %>
      <jsp:setProperty name="product" property="price" value="<%=price+23.67 %>"/>
      <br/>产品价格是:
      <!-- 获取产品价格 -->
      <%=product.getPrice() %>
```

```
30        <!-- 设置产品信息 -->
31        <jsp:setProperty name="product" property="info"
32           value="struts 开发教程是一本介绍，如何使用 Struts 的专业书籍......"/>
33        <br/>产品信息是:
34        <!-- 获取产品信息 -->
35        <%=product.getInfo() %>
36    </body>
37  </html>
```

上述代码中，页面就是简单的设置 Bean 值，代码第 3 行是导入引用的 Bean 类。代码第 11~35 行分别是设定 Bean 属性值和获取 Bean 属性值。Bean 类的源代码如下：

```
---------------------- Product.java--------------------------
package com.eshore.pojo;
 import java.io.Serializable;

 public class Product implements Serializable{

     private static final long serialVersionUID = 1L;
     private String product_id;           //产品号
     private String product_name;         //产品名称
     private double price;                //产品价格
     private String info;                 //产品信息

     public Product() {               //无参的构造方法
         super();
     }
    ……//省略 get 和 set 方法
 }
```

上述代码中，分别是设定 Bean 类中的属性成员以及其 get 方法和 set 方法。成员变量最好用 private，这是编码的良好习惯。页面运行结果如图 8.43 所示。

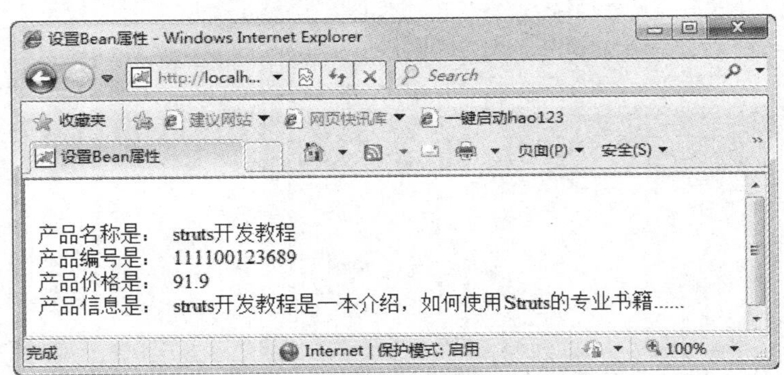

图 8.43　product1.jsp 运行效果图

【例 8.11】通过表单参数名设置 Bean 属性值

通过表单参数设置 Bean 属性值，不用设置 value 值。JSP 引擎会自动根据表单中的参数名与 Bean 属性名对应，并转化为对应的数据类型。

product2.jsp 页面通过表单参数设置 Bean 属性值，其源代码如下：

```
------------------------product2.jsp---------------------------
01  <%@ page language="java"
02  import="java.util.*,com.eshore.pojo.Product" pageEncoding="UTF-8"%>
03  <!-- 设定参数编码 -->
04  <%
05      request.setCharacterEncoding("UTF-8");
06   %>
07  <!-- 导入引用的 bean -->
08  <jsp:useBean id="product" class="com.eshore.pojo.Product" scope="page"/>
09  <!DOCTYPE HTML PUBLIC "-//W3C//DTD HTML 4.01 Transitional//EN">
10  <html>
11    <head>
12      <title>通过表单参数设置 Bean 属性值</title>
13      <meta http-equiv="Content-Type" content="text/html;charset=utf-8">
14    </head>
15  
16    <body>
17        <form action="" method="post">
18            <br>
19            输入产品名称：<input name="product_name"/><br/>
20            输入产品编号：<input name="product_id"/><br/>
21            输入产品价格：<input name="price"/><br/>
22            输入产品信息：<input name="info"/><br/>
23            <input type="submit" value="提交"/>
24        </form>
25        <!-- 设定 product 的属性值 -->
26        <jsp:setProperty property="*" name="product"/>
27        <br/>产品名称是:
28        <!-- 获取产品名称 -->
29        <%=product.getProduct_name() %>
30        <br/>产品编号是:
31        <!-- 获取产品编号 -->
32        <%=product.getProduct_id() %>
33        <br/>产品价格是:
34        <!-- 获取产品价格 -->
35        <%=product.getPrice() %>
36        <br/>产品信息是:
37        <!-- 获取产品信息 -->
38        <%=product.getInfo() %>
39    </body>
40  </html>
```

上述代码中，第 4~6 行设定参数编码，第 17~24 行设定 form 架构。第 26 行，设置 Bean 属性值，但是没有设定具体的值。第 27~38 行获取 form 中设定的 Bean 值。Bean 类不变，继续使用例 8.10 中的 Porduct 类。页面运行效果如图 8.44 所示。

如果在页面输入中文，提交后显示为乱码，则应该设定中文编码，有关参数编码问题参见第 4.1.3 节说明。

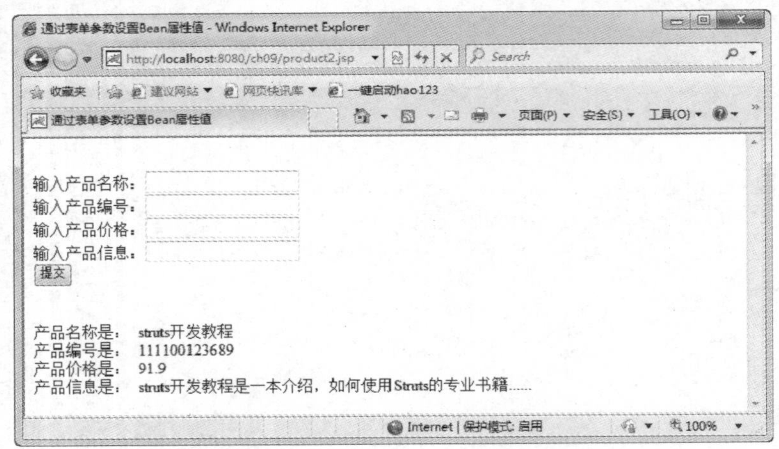

图 8.44　product2.jsp 运行效果图

【例 8.12】通过表单参数值设置 Bean 属性值

通过表单参数值和使用 param 属性设置 Bean 属性值，与第 2 种方式很相似。第 2 种，表单中的 name 名字就是 Bean 名字，第 3 种方式表中的 name 名字与 Bean 名字不同，在标签设置时，用 param 属性引用。

product3.jsp 页面通过表单参数值设置 Bean 属性值，其源代码如下：

```
-----------------------product3.jsp---------------------------
<%@ page language="java"
import="java.util.*,com.eshore.pojo.Product" pageEncoding="UTF-8"%>
<!-- 设定参数编码 -->
<%
   request.setCharacterEncoding("UTF-8");
 %>
<!-- 导入引用的 bean -->
<jsp:useBean id="product" class="com.eshore.pojo.Product" scope="page"/>
<!DOCTYPE HTML PUBLIC "-//W3C//DTD HTML 4.01 Transitional//EN">
<html>
  <head>
    <title>通过表单参数值设置 Bean 属性值</title>
    <meta http-equiv="Content-Type" content="text/html;charset=utf-8">
  </head>

  <body>
     <form action="" method="post">
        <br>
        输入产品名称：<input name="product_name1"/><br/>
        输入产品编号：<input name="product_id1"/><br/>
```

```
21            输入产品价格：<input name="price1"/><br/>
22            输入产品信息：<input name="info1"/><br/>
23            <input type="submit" value="提交"/>
24        </form>
25        <!-- 设置产品名称 -->
26        <jsp:setProperty name="product" property="product_name"
27            param="product_name1"/>
28        <br/>产品名称是：
29        <!-- 获取产品名称 -->
30        <%=product.getProduct_name() %>
31        <!-- 设置产品编号 -->
32        <jsp:setProperty name="product" property="product_id"  param=
         "product_id1"/>
33        <br/>产品编号是：
34        <!-- 获取产品编号 -->
35        <%=product.getProduct_id() %>
36        <!-- 设置产品价格 -->
37        <jsp:setProperty name="product" property="price" param="price1"/>
38        <br/>产品价格是：
39        <!-- 获取产品价格 -->
40        <%=product.getPrice() %>
41        <!-- 设置产品信息 -->
42        <jsp:setProperty name="product" property="info" param="info1"/>
43        <br/>产品信息是：
44        <!-- 获取产品信息 -->
45        <%=product.getInfo() %>
46    </body>
47  </html>
```

上述代码中，与例 8.11 的主要区别在于代码第 25~45 行，应用 param 参数引用表单中的参数名。其代码运行效果如图 8.45 所示。

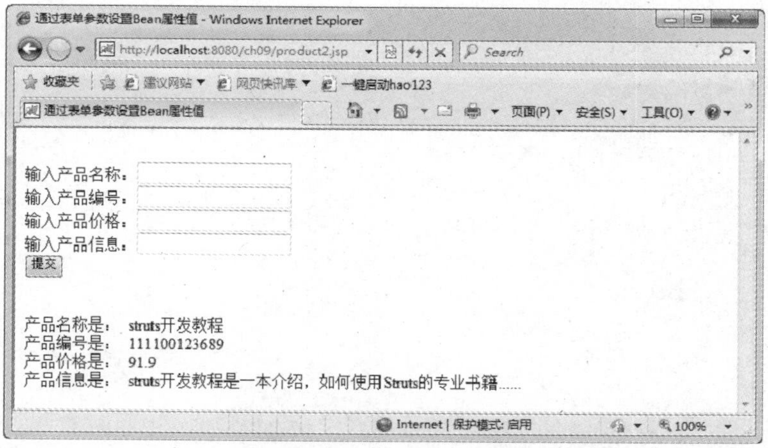

图 8.45　product3.jsp 运行效果图

从图 8.45 可以看出，与图 8.44 的效果是一样的。

2. 取得属性：<jsp:getProperty>

<jsp:getProperty>标签是用来获得 Bean 属性值，并且可以显示在浏览器中。该标签必须和<jsp:useBean>标签一起使用。其语法形式如下：

```
<jsp:getProperty  name="beanInstanceName" property="propertyName"/>
```

上述语法中，name 值是指 Bean 的名字，用来指定被使用的 Bean，它的值必须是<jsp:useBean>标签中的 id 值。property 是指 Bean 的属性名，也就是 Bean 类中的属性，将值赋给 Bean 类的属性。应用<jsp:getProperty>标签获取 Bean 属性值，将例 8.10 修改为如下代码。

```
<%@ page language="java" import="java.util.*" pageEncoding="UTF-8"%>
<!-- 导入引用的 bean -->
<jsp:useBean id="product" class="com.eshore.pojo.Product" scope="page"/>
<!DOCTYPE HTML PUBLIC "-//W3C//DTD HTML 4.01 Transitional//EN">
<html>
  <head>
    <title>设置 Bean 属性</title>
  </head>

  <body>
      <!-- 设置产品名称 -->
      <jsp:setProperty name="product" property="product_name"
          value="struts 开发教程"/>
      <br/>产品名称是:
      <!-- 获取产品名称 -->
      <jsp:getProperty  name="product" property="product_name"/>
      <%=product.getProduct_name() %>
      <!-- 设置产品编号 -->
      <jsp:setProperty name="product" property="product_id"
      value="111100123689"/>
      <br/>产品编号是:
      <!-- 获取产品编号 -->
      <jsp:getProperty  name="product" property="product_id"/>
      <!-- 设置产品价格 -->
      <%
        double price = 68.23;
       %>
      <jsp:setProperty name="product" property="price"  value="<%=
      price+23.67 %>"/>
      <br/>产品价格是:
      <!-- 获取产品价格 -->
      <jsp:getProperty  name="product" property="price"/>
      <!-- 设置产品信息 -->
      <jsp:setProperty name="product" property="info"
        value="struts 开发教程是一本介绍，如何使用 Struts 的专业书籍......"/>
```

```
34          <br/>产品信息是:
35          <!-- 获取产品信息 -->
36          <jsp:getProperty  name="product" property="info"/>
37      </body>
38    </html>
```

在上述代码中，主要的区别在于代码第 16、27、30、36 行，其余一样的。其运行结果如前面的图 8.43 所示。

8.4.4　Bean 的作用域

前面讲过 Bean 的作用域有 4 个，分别是 page、request、session 和 application。Bean 的作用范围是由标签中 scope 属性指定的，默认是 page 范围，即该 Bean 在当前页有效。设置为 request 时，表示该 Bean 对当前用户的当前请求有效。设置为 session 时，表示该 Bean 在当前用户的 session 范围内有效。设置为 application 时，表示该 Bean 对该系统的所有页面有效。scope 属性决定了在使用<jsp:useBean>标签时是否要重新创建新的对象。如果某个 Bean 在其有效的范围，又出现一个 id 和 scope 都相同的 Bean，那么就可以重用已经被实例化的 Bean 而不是重新 new。

【例 8.13】演示 Bean 的 request 生命周期

requestScope.jsp 页面提交圆半径，然后用 reqest.getParameter()获取半径，并计算圆周长和圆面积。默认的圆半径为 1。其源代码为：

```
---------------------- requestScope.jsp-------------------------
<%@ page language="java" import="java.util.*" pageEncoding="UTF-8"%>
<!-- 导入引用的 bean -->
<jsp:useBean id="circle" class="com.eshore.pojo.Circle" scope="request"/>
<!DOCTYPE HTML PUBLIC "-//W3C//DTD HTML 4.01 Transitional//EN">
<html>
  <head>
    <title>测试 scope 为 request</title>
  </head>
  <%
    //获取页面获得的圆半径，如果没有默认是1
     String radius = request.getParameter("radius");
     if(radius==null||radius.equals("")){
         radius = "1";
     }
     double rad = Double.parseDouble(radius);
     //设置圆半径
     circle.setRadius(rad);
   %>
  <body>
      <form action="" method="post">
          请输入圆的半径：<input name="radius"/><br/>
          <input type="submit" value="提交"/><br/>
          该 Bean 类对象为：<%=circle.toString()%><br/>
```

```
24              <br/>圆的半径为：<jsp:getProperty property="radius" name="circle"/>
25              <br/>圆的周长为：<jsp:getProperty property="circumference"
                name="circle"/>
26              <br/>圆的面积为：<jsp:getProperty property="circleArea" name="circle"/>
27          </form>
28      </body>
29  </html>
```

上述代码中，第 9~17 行获取页面获得的半径值，并且转化为 Double 格式。代码第 20~27 行用 form 提交表单数据和获取 Bean 的相关值。第 23 行输出 Circle 类对象值。Circle 类的源代码如下：

```
-----------------------Circle.java---------------------------
package com.eshore.pojo;

import java.io.Serializable;

public class Circle implements Serializable{

    private static final long serialVersionUID = 1L;
    private double radius = 1.0d;                    //半径
    private double circleArea = 0.0d;                //圆面积
    private double circumference=0.0d;               //圆周长
    public Circle() {   //无参数的构造方法
        super();

    }
    //属性的 get 和 set 方法
    public double getRadius() {
        return radius;
    }
    public void setRadius(double radius) {
        this.radius = radius;
    }
    public double getCircleArea() {
        circleArea = Math.PI*radius*radius;          //设置圆面积
        return circleArea;
    }
    public void setCircleArea(double circleArea) {
        this.circleArea = circleArea;
    }
    public double getCircumference() {
        circumference = 2*Math.PI*radius;        //设置圆周长
        return circumference;
    }
    public void setCircumference(double circumference) {
        this.circumference = circumference;
    }
}
```

页面运行效果见图 8.46 所示。

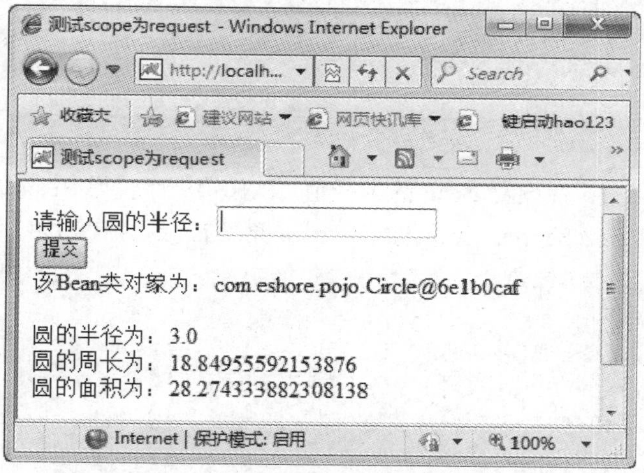

图 8.46　requestScope.jsp 页面运行效果图

从图中可以看出，每次提交显示出的 Circle 类都是不同的。

【例 8.14】演示 Bean 的 page 生命周期

pageScope.jsp 页面先设置圆的半径并计算圆周长和圆面积。然后跳转到 pageScope2.jsp 页面，在 pageScope2.jsp 中设定 Circle 类的范围和 pageScope.jsp 的范围一样。可以发现 Bean 的有效范围就是在当前页有效，在 pageScope2.jsp 中获取的是默认的值。它们的源代码分别为：

```
-----------------------pageScope.jsp--------------------------
<%@ page language="java" import="java.util.*" pageEncoding="UTF-8"%>
<!-- 导入引用的 bean -->
<jsp:useBean id="circle2" class="com.eshore.pojo.Circle" scope="page"/>
<!DOCTYPE HTML PUBLIC "-//W3C//DTD HTML 4.01 Transitional//EN">
<html>
  <head>
    <title>测试 scope 为 page</title>
  </head>
  <body>
        pageScope.jsp 页面信息：<br/>
         该 Bean 类对象为：<%=circle2.toString()%><br/>
         设置该 Bean 的半径为20：
        <%
          circle2.setRadius(20);
        %>
        <!-- 获取 Bean 的属性值 -->
       <br/>圆的半径为：<jsp:getProperty property="radius" name="circle2"/>
       <br/>圆的周长为：<jsp:getProperty property="circumference" name=
       "circle2"/>
       <br/>圆的面积为：<jsp:getProperty property="circleArea" name="circle2"/>
```

```
20          <!-- 跳转 pageScope2.jsp 页面 -->
21          <form action="pageScope2.jsp" method="get">
22               <input type="submit" value="跳转 pageScope2.jsp 页面"/><br/>
23          </form>
24      </body>
25  </html>
```

上述代码中，第 13~15 行设置 Bean 的半径值，第 16~19 行获得 Bean 的各属性值。第 20~23 行，提交 form 跳转到 pageScope2.jsp 页面。pageScope2.jsp 的源代码如下：

```
----------------------- pageScope2.jsp--------------------------
01  <%@ page language="java" import="java.util.*" pageEncoding="UTF-8"%>
02  <!-- 导入引用的 bean -->
03  <jsp:useBean id="circle2" class="com.eshore.pojo.Circle" scope="page"/>
04  <!DOCTYPE HTML PUBLIC "-//W3C//DTD HTML 4.01 Transitional//EN">
05  <html>
06    <head>
07      <title>测试 scope 为 page</title>
08    </head>
09    <body>
10          pageScope2.jsp 页面信息：<br/>
11           该 Bean 类对象为：<%=circle2.toString()%><br/>
12         <br/>圆的半径为：<jsp:getProperty property="radius" name="circle2"/>
13         <br/>圆的周长为:<jsp:getProperty property="circumference" name="circle2"/>
14         <br/>圆的面积为：<jsp:getProperty property="circleArea" name="circle2"/>
15    </body>
16  </html>
```

上述代码中，第 3 行引用 Bean 的 scop 跟 pageScope.jsp 中的一样且 id 值一样。第 12~14 行引用 Bean 的属性值。页面运行效果如图 8.47 所示。

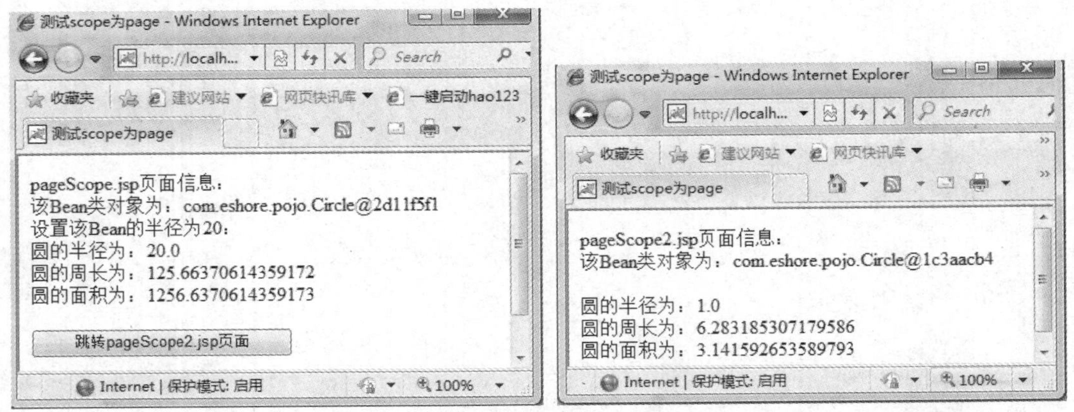

图 8.47　pageScope.jsp 和 pageScope2.jsp 运行效果图

从图 8.47 可以看出，Bean 类的作用域在于同一页面，跳转到别的页面就会失效。

【例 8.15】演示 Bean 的 session 生命周期

将例 8.14 中 pageScope.jsp 的代码更改为如下：

```
----------------------- sessionScope.jsp----------------------------
<%@ page language="java" import="java.util.*" pageEncoding="UTF-8"%>
<!-- 导入引用的 bean -->
<jsp:useBean id="circle2" class="com.eshore.pojo.Circle" scope="session"/>
<!DOCTYPE HTML PUBLIC "-//W3C//DTD HTML 4.01 Transitional//EN">
<html>
  <head>
    <title>测试 scope 为 page</title>
  </head>
  <body>
        sessionScope.jsp 页面信息：<br/>
         该 Bean 类对象为：<%=circle2.toString()%><br/>
         设置该 Bean 的半径为20：
        <%
           circle2.setRadius(20);
        %>
        <!-- 获取 Bean 的属性值 -->
       <br/>圆的半径为：<jsp:getProperty property="radius" name="circle2"/>
       <br/>圆的周长为:<jsp:getProperty property="circumference" name="circle2"/>
       <br/>圆的面积为：<jsp:getProperty property="circleArea" name="circle2"/>
       <!-- 跳转 sessionScope2.jsp 页面 -->
       <form action="sessionScope2.jsp" method="get">
            <input type="submit" value="跳转 sessionScope2.jsp 页面"/><br/>
       </form>
  </body>
</html>
```

pageScope2.jsp 中代码更改为如下：

```
----------------------- sessionScope2.jsp---------------------------
<%@ page language="java" import="java.util.*" pageEncoding="UTF-8"%>
<!-- 导入引用的 bean -->
<jsp:useBean id="circle2" class="com.eshore.pojo.Circle" scope="session"/>
<!DOCTYPE HTML PUBLIC "-//W3C//DTD HTML 4.01 Transitional//EN">
<html>
  <head>
    <title>测试 scope 为 page</title>
  </head>
  <body>
        sessionScope2.jsp 页面信息：<br/>
         该 Bean 类对象为：<%=circle2.toString()%><br/>
       <br/>圆的半径为：<jsp:getProperty property="radius" name="circle2"/>
       <br/>圆的周长为:<jsp:getProperty property="circumference" name="circle2"/>
```

```
        <br/>圆的面积为：<jsp:getProperty property="circleArea" name="circle2"/>
  </body>
</html>
```

上述两段代码中，主要更改的是 Bean 的范围以及跳转的页面。运行效果见图 8.48 所示。

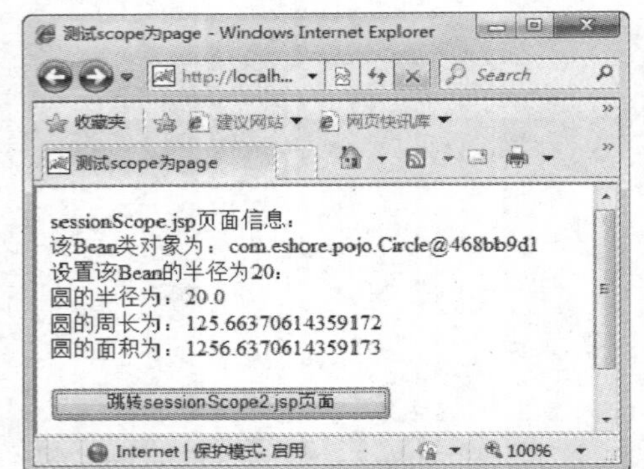

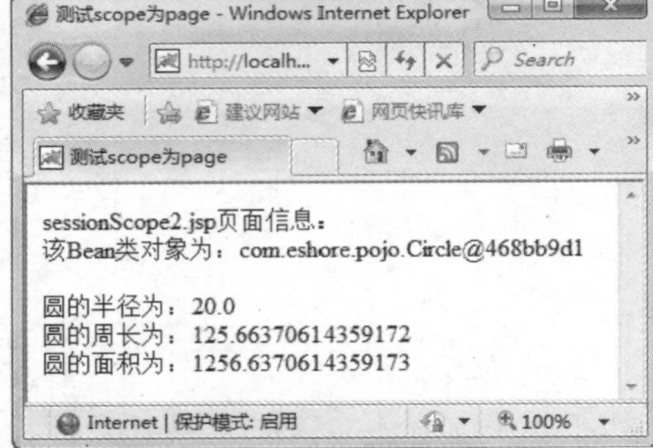

图 8.48　sessionScope.jsp 和 sessionScope2.jsp 运行效果图

从图中可以看出，引用的 Bean 是同一个对象，所以取得的属性值是一样的。

【例 8.16】演示 Bean 的 application 生命周期

applicationScope.jsp 页面设置 scope 的范围为 application，先获取系统默认的半径，然后更改半径值，页面再刷新一次，就能发现半径更改了。applicationScope.jsp 的源代码如下：

```
-----------------------applicationScope.jsp---------------------------
<%@ page language="java" import="java.util.*" pageEncoding="UTF-8"%>
<!-- 导入引用的 bean -->
<jsp:useBean id="circle2" class="com.eshore.pojo.Circle" scope="application"/>
<!DOCTYPE HTML PUBLIC "-//W3C//DTD HTML 4.01 Transitional//EN">
<html>
  <head>
    <title>测试 scope 为 page</title>
  </head>
  <body>
       applicationScope.jsp 页面信息：<br/>
        application 访问 Bean 类对象为：<%=circle2.toString()%><br/>
        获取 Bean 的半径：<jsp:getProperty property="radius" name="circle2"/><br/>
        设置该 Bean 的半径为30：
       <%
          circle2.setRadius(30);
       %>
       <!-- 获取 Bean 的属性值 -->
      <br/>圆的半径为：<jsp:getProperty property="radius" name="circle2"/>
      <br/>圆的周长为：<jsp:getProperty property="circumference" name="circle2"/>
      <br/>圆的面积为：<jsp:getProperty property="circleArea" name="circle2"/>
  </body>
```

```
</html>
```

上述代码的运行结果如图 8.49 所示。

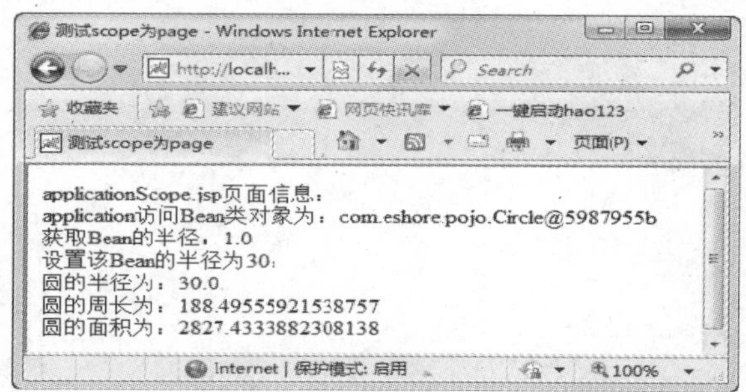

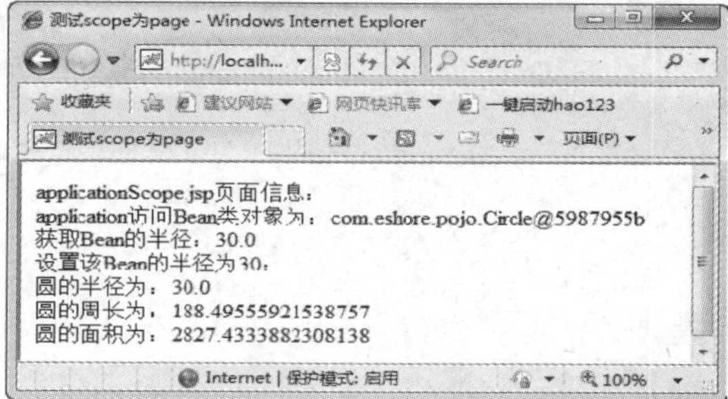

图 8.49　applicationScope.jsp 页面运行效果

从上述例子中可以发现，在同一范围内的 Bean 对象，其引用的是同一对象。

8.5 实例：利用 Java Bean 实现用户登录验证

前面章节曾经介绍过用户登录验证，本章介绍用 Java Bean 的方式进行登录验证。用户在表单中填入用户名和密码。如果存在用户则跳转到欢迎界面，否则显示错误信息。

在本程序中 JavaBean 是极其重要的一部分，不仅要完成数据的校验还要进行错误信息的显示。本例子为了简单，所有的错误信息存放在 Map 容器中。

【例 8.17】用 Java Bean 用户登录验证

类 User 包含基础的用户名、密码属性，并用 Map 来保存错误信息。其源代码如下：

```
-----------------------User.java---------------------------
package com.eshore.pojo;

import java.io.Serializable;
import java.util.HashMap;
import java.util.Map;

public class User implements Serializable{

    private String username="";           //用户名
    private String passwd="";                 //密码
    Map<String,String> userMap = null;      //存放用户
    Map<String,String> errorsMap = null;    //存放错误信息

    public User() {                       //无参的构造方法
        super();
```

```
        this.username = "";
        this.passwd="";
        userMap = new HashMap<String,String>();
        errorsMap = new HashMap<String,String>();
        //添加用户，模拟从数据库中查询出的数据库
        userMap.put("zhangsan", "123zs");
        userMap.put("lisi", "1234zs");
        userMap.put("wangwu", "1234ww");
        userMap.put("zhaoqi", "1234zq");
        userMap.put("zhengliu", "1234zl");
        // TODO Auto-generated constructor stub
    }
    //用户名和密码等数据验证
    public boolean isValidate(){
        boolean flag = true;
        //用户名验证
        if(!this.userMap.containsKey(this.username)){
            flag = false;
            errorsMap.put("username", "该用户不存在！");
            this.username = "";
        }
        //根据用户名进行密码验证
        String password = this.userMap.get(this.username);
        if(password==null||!password.equals(this.passwd)){
            flag = false;
            this.passwd = "";
            errorsMap.put("passwd", "密码错误，请输入正确密码！");
            this.username = "";
        }
        return flag;
    }
    //获取错误信息
    public String getErrors(String key){
        String errorV = this.errorsMap.get(key);
        return errorV==null?"":errorV;
    }
    //以下是属性的 get 和 set 方法，但必须是 public
    public String getUsername() {
        return username;
    }

    public void setUsername(String username) {
        this.username = username;
    }

```

```
61      public String getPasswd() {
62          return passwd;
63      }
64
65      public void setPasswd(String passwd) {
66          this.passwd = passwd;
67      }
68  }
```

上述代码中，第 21~25 行用 Map 保存用户数据。代码第 29~46 行验证页面输入的数据，如果验证失败，则会将错误信息保存在错误 Map 中。代码第 48~51 行根据错误 Key 获取错误信息。

登录页面 login.jsp，引用 User 类并用表单提交的方式设定 User 属性值。其源代码如下：

```
----------------------login.jsp--------------------------
<%@ page language="java" import="java.util.*" pageEncoding="UTF-8"%>
<jsp:useBean id="user" class="com.eshore.pojo.User" scope="session"/>
<!DOCTYPE HTML PUBLIC "-//W3C//DTD HTML 4.01 Transitional//EN">
<html>
  <head>
    <title>用户登录</title>
  </head>

  <body>
        <p>用户登录</p>
        <!-- 用 form 表单提交，用户名跟密码 -->
        <form action="check.jsp" method="post">
            <table border="1" width="250px;">
               <tr>
                   <td width="75px;">用户名：</td>
                   <td ><input name="username" value="<jsp:getProperty
                       name="user" property="username"/>"/>
                   <!-- 用户错误信息 -->
            <span style="color:red"> <%=user.getErrors("username") %></span><br/></td>
               </tr>
               <tr>
                   <td width="75px;">密  码：</td>
                   <td ><input type="password" name="passwd"
                       value="<jsp:getProperty name="user" property="passwd"/>"/>
                   <!-- 密码错误信息 -->
            <span style="color:red"> <%=user.getErrors("passwd") %></span><br/></td>
               </tr>
```

```
28                  <tr>
29                      <td colspan="2">
30                          <input type="submit" value="提交"/>  
31                          <input type="reset" value="重置"/>
32                      </td>
33                  </tr>
34              </table>
35          </form>
36      </body>
37  </html>
```

login.jsp 页面的主要功能是显示表单和错误信息，例如代码第 23 行和第 30 行所示。代码第 3 行定义 User 的有效范围为 session。

检验页面 check.jsp 页面，同样定义一个范围为 session 的 User，调用类的验证方法进行判断，其源代码如下：

```
-----------------------check.jsp--------------------------
01  <%@ page language="java" import="java.util.*" pageEncoding="UTF-8"%>
02  <jsp:useBean id="user" class="com.eshore.pojo.User" scope="session"/>
03  <!DOCTYPE HTML PUBLIC "-//W3C//DTD HTML 4.01 Transitional//EN">
04  <html>
05    <head>
06      <title>验证用户</title>
07    </head>
08    <body>
09      <!-- 设置 user 属性，判断是否合法
10           合法跳转成功，否则跳转到登录页面-->
11       <jsp:setProperty property="*" name="user"/>
12        <%
13            if(user.isValidate()){
14         %>
15         <jsp:forward page="success.jsp"/>
16         <%
17           }else{
18          %>
19          <jsp:forward page="login.jsp"/>
20          <%} %>
21    </body>
22  </html>
```

上述代码中，第 11 行设置 User 的属性值，第 12~20 行判断用户的合法性，如果验证通过则跳转到 success.jsp 页面，如果失败则跳转到 loging.jsp 页面提示错误信息。

success.jsp 页面显示欢迎界面，其源代码如下：

```
-----------------------success.jsp--------------------------
<%@ page language="java" import="java.util.*" pageEncoding="UTF-8"%>
```

```
02   <% request.setCharacterEncoding("UTF-8"); %>
03   <jsp:useBean id="user" class="com.eshore.pojo.User" scope="session"/>
04   <!DOCTYPE HTML PUBLIC "-//W3C//DTD HTML 4.01 Transitional//EN">
05   <html>
06     <head>
07       <title>登录成功</title>
08     </head>
09
10     <body>
11         <center>
12             <h4>欢迎您：
13                 <SPAN style="color: red">
14                 <jsp:getProperty property="username" name="user"/>
15                 </SPAN>用户！
16             </h4>
17         </center>
18     </body>
19   </html>
```

上述代码第 14 行显示登录成功的用户。本例子的运行结果如图 8.50 和图 8.51 所示。

图 8.50　登录界面和错误提示界面

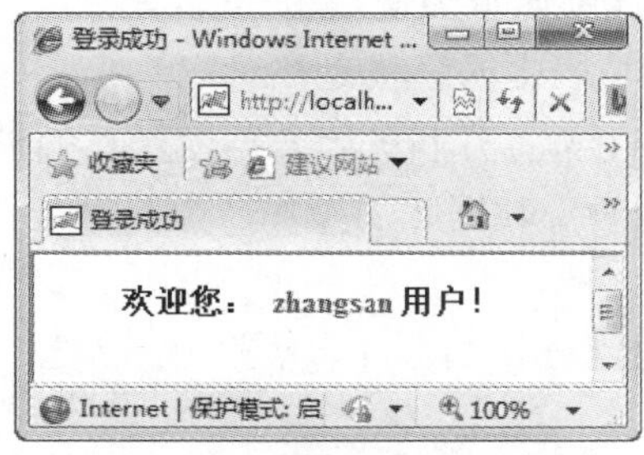

图 8.51　用户登录验证页面效果图

在实际开发中，应该灵活运用 Java Bean 开发。要根据具体的业务需要，尽量将代码解耦合。

8.6 DAO 设计模式

前面章节介绍了 Bean 属性的概念、属性以及如何使用 Java Bean。进而，聪明的大师们想提高开发效率，实现模块化开发。DAO 设计模式就这么应运而生，虽然现如今有很多成熟的框架，例如：Spring MVC 框架、Struts 框架等。但它仍然是一个值得大家学习的设计模式。本节将详细为大家介绍该设计模式，使得读者对 DAO 设计模式有个整体的框架概念，能编写出适合自己的简易框架。

8.6.1 DAO 设计模式简介

信息系统的开发架构如图 8.52 所示。

图 8.52 信息系统开发架构图

各层级的介绍分别如下：

- 客户层：实际上就是客户端浏览器；
- 显示层：用 JSP 和 Servlet 进行页面显示；
- 业务层：对数据层的原子性 DAO 操作进行整合；
- 数据层：对数据库进行原子操作，例如增加、删除、修改等；
- 数据库：顾名思义就是保存数据库的信息。

DAO 是（Data Access Object 的简称），主要是对数据的操作，对应上面的层级就是数据层。在数据操作过程中，主要是以面向接口编程为主。一般将 DAO 划分为以下几个部分。

（1）VO（Value Object）也称为 POJO：一个用于存放网页的一行数据即一条记录的类，比如网页要显示一条用户的信息，则这个类就是用户类。主要有属性、以及属性的 setter 和 getter 方法组成，VO 类中的成员变量与表中的字段是相对应的。

（2）DatabaseConnection：用于打开和关闭数据库操作的类。

（3）DAO 接口：用于声明对于数据库的操作，定义对数据库的原子性操作，例如增加、修改删除、修改等。

（4）DAOImpl：必须实现 DAO 接口，真实实现 DAO 接口的类，但是不负责数据库的打开和关闭。

（5）DAOProxy：也是实现 DAO 接口，主要完成数据库的打开和关闭。

（6）DAOFactory：工厂类，通过 getInstance()取得 DAO 的实例化对象。

8.6.2　DAO 命名规则

在开发中，命名规范非常重要。命名要具有可读性、可维护性。DAO 命名规则如下：

- DAO 命名为 XxxDao，有的开发人员喜欢在前加个 I 表示是接口类，例如 UserDao 或者 IUserDao。
- DAOImpl 命名为 XxxDaoImpl，表示是接口实现类，例如 UserDaoImpl。
- DAOProxy 命名为 XxxDaoProxy 或者 XxxService，例如 UserDaoProxy 或者 UserService。
- DAOFactory 命名为 XxxFactory，例如 UserDaoFactory。
- VO 的命名与表名一致，VO 中的属性与表字段一致。

8.6.3　DAO 开发

DAO 开发完全就是对数据的增、删、改、查的操作。在 MySQL 中新建一个产品表 product，建表脚本如下：

```
DROP TABLE IF EXISTS `product`;
CREATE TABLE `product` (
  `product_id` varchar(20) NOT NULL,
  `product_name` varchar(50) DEFAULT NULL,
  `price` decimal(6,2) DEFAULT NULL,
  `info` varchar(100) DEFAULT NULL,
  PRIMARY KEY (`product_id`)
) ENGINE=InnoDB DEFAULT CHARSET=utf8;
```

【例 8.18】VO 类使用例 8.10 中 Product 类，演示 DAO 开发

根据上述内容，要分别新建 ProductDao、ProductDaoImpl、ProductDaoProxy、DaoFactory、DBConnection。建完的目录结构如图 8.53 所示。

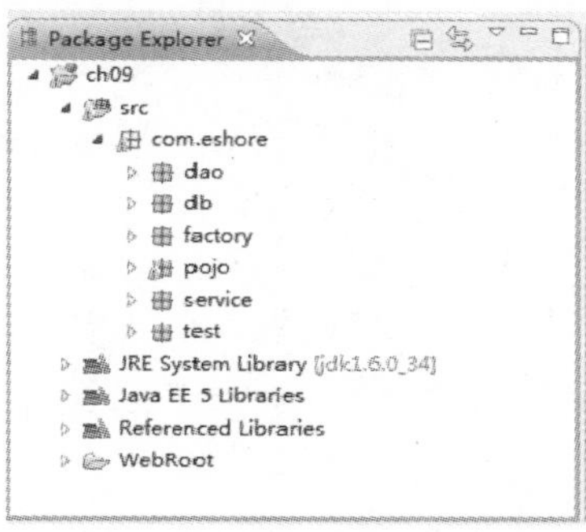

图 8.53　DAO 目录结构

首先，先新建 DBConnection 类。其源代码如下：

```
-----------------------DBConnection.java--------------------------
package com.eshore.db;
```

```
02
03  import java.sql.Connection;
04  import java.sql.DriverManager;
05
06  public class DBConnection {
07      private static final String Driver = "com.mysql.jdbc.Driver";
08      private static final String URL = "jdbc:mysql://localhost:3306/testweb";
09      private static final String USER = "root";
10      private static final String PASSWORD = "admin";
11      private Connection conn = null;
12
13      public DBConnection() throws Exception {    // 进行数据库连接
14          try {
15              Class.forName(Driver);           // 用反射加载数据库驱动
16              this.conn = DriverManager.getConnection(URL, USER, PASSWORD);
17          } catch (Exception e) {
18              throw e;                         // 抛出异常
19          }
20      }
21      public Connection getConnection() {
22          return this.conn;                    // 取得数据库的连接
23      }
24
25      public void close() throws Exception {      // 关闭数据库
26          if (this.conn != null) {
27              try {
28                  this.conn.close();
29              } catch (Exception e) {
30                  throw e;
31              }
32          }
33      }
34  }
```

上述代码中，主要是执行对数据库的连接配置。代码第 21 行获取当前的数据库连接。

接着，新建 DAO 接口类。接口类在 DAO 设计模式中，地位是极其重要的。在定义接口类之前，要分析业务的需求，要分析清楚系统需要哪些功能方法。在本例中只是完成新增、查询等简单功能。ProductDao 接口类的源码如下。

```
-----------------------ProductDao.java--------------------------
package com.eshore.dao;

import java.util.List;

```

```
import com.eshore.pojo.Product;

public interface ProductDao {
    /**
     * 数据库 新增数据
     *@param product 要增加的数据对象;
     *@return 是否增加成功的标记
     *@throws Exception 如果有异常交直接抛出
     */
   public boolean addProduct(Product product)throws Exception ;
    /**
     * 查询全部的 Product 数据
     *@param product_name 产品名称
     *@return 返回全部的查询结果，每一个 product 对象表示表的一行记录
     *@throws Exception 如果有异常交直接抛出
     */
   public List<Product> findAll(String product_name)throws Exception;
    /**
     * 根据产品编号查询产品
     *@param  product_id 产品编号
     *@return  产品的 vo 对象
     *@throws Exception 如果有异常交直接抛出
     */
   public Product findByProductId(String product_id)throws Exception;
}
```

上述 DAO 的接口类中，定义了 addProduct()、findAll()、findByProductId()等 3 个功能。addProduct()方法是执行数据库的插入操作；findAll()方法主要是完成数据的查询操作，用 List 来保存返回数据；findByProductId()方法根据产品编号返回单个 Product 对象。

DAO 接口定义完成后需要定义其实现类，为了降低耦合度，其实现类有两种：一种只是数据操作实现类，一种是业务操作实现类。

数据操作实现类主要是负责具体的数据库操作，其源代码如下：

```
----------------------ProductDaoImpl.java-------------------------
package com.eshore.dao;

import java.sql.Connection;
import java.sql.PreparedStatement;
import java.sql.ResultSet;
import java.util.ArrayList;
import java.util.List;

import com.eshore.pojo.Product;

public class ProductDaoImpl implements ProductDao {
```

```
    private Connection conn = null;              // 数据库连接对象
    private PreparedStatement pstmt = null;     // 数据库操作对象
    // 通过构造方法取得数据库连接
    public ProductDaoImpl(Connection conn) {
        this.conn = conn;
    }

    public boolean addProduct(Product product) throws Exception {
        boolean flag = false;                    // 定义标识
        String sql = "insert into product(product_id,product_name,
        price,info)
        values(?,?,?,?)";
        this.pstmt = this.conn.prepareStatement(sql);
         // 实例化 PrepareStatement 对象
        this.pstmt.setString(1,product.getProduct_id());   // 设置产品 id
        this.pstmt.setString(2,product.getProduct_name());// 设置产品名称
        this.pstmt.setDouble(3, product.getPrice());   // 设置产品价格
        this.pstmt.setString(4,product.getInfo());     // 设置产品信息

        if (this.pstmt.executeUpdate() > 0) {        // 更新记录的行数大于0
            flag = true;                             // 修改标识
        }
        this.pstmt.close();                    //关闭 PreparedStatement 操作
        return flag;
    }

    public List<Product> findAll(String product_name) throws Exception {
        List<Product> list = new ArrayList<Product>();
        // 定义集合，接受返回的数据
        String sql = "select product_id,product_name,price,info from product ";
        if(product_name!=null&&!"".equals(product_name)){
            sql = "select product_id,product_name,price,info
            from product where product_name like? ";
            this.pstmt.setString(1, "%" + product_name+"%");//设置查询产品名称
        }else{
            this.pstmt = this.conn.prepareStatement(sql);
            // 实例化 PreparedStatement
        }
        ResultSet rs = this.pstmt.executeQuery();  // 执行查询操作
        Product product = null;
        while (rs.next()) {
            product = new Product();              // 实例化新的 product 对象
            product.setProduct_id(rs.getString(1));
            product.setProduct_name(rs.getString(2));
```

```
53                product.setPrice(rs.getDouble(3));
54                product.setInfo(rs.getString(4));
55                list.add(product);                    // 向集合中增加 product 对象
56            }
57            this.pstmt.close();
58            return list;                              // 返回全部结果
59        }
60
61        public Product findByProductId(String product_id) throws Exception {
62            Product product = null;
63            String sql = "select product_id,product_name,price,info from
64            product where product_id=?";
65            this.pstmt = this.conn.prepareStatement(sql);
66            this.pstmt.setString(1, product_id);              // 设置产品编号
67            ResultSet rs = this.pstmt.executeQuery();
68            if (rs.next()) {
69                product = new Product();
70                product.setProduct_id(rs.getString(1));
71                product.setProduct_name(rs.getString(2));
72                product.setPrice(rs.getDouble(3));
73                product.setInfo(rs.getString(4));
74            }
75            this.pstmt.close();
76            return product; // 如果查询不到结果则返回 null,默认值为 null
77        }
78
79    }
```

上述代码中，代码第 13~14 行定义了数据库操作的接口对象 Connection 和 PreparedStatement。代码第 21 行定义一个成功标识，如果添加数据成功则返回 true，否则返回 false。代码第 23 行实例 PrepareStatement 对象，然后依次插入数据如代码第 24~27 行。代码第 29~31 行，判断是否插入数据成功，如果返回值大于 0 则表示插入成功否则失败。

代码第 35~53 行，是查询数据方法。首先定义 List 集合对象如代码第 36 行，在查询时定义产品名称为模糊查询条件如代码第 37~38 行。代码第 39~40 行执行数据库查询操作，并将查询出的结果进行循环遍历，保存在 List 集合中。

代码第 55~71 行根据产品 ID 值查询 Product 对象。第 57~61 行，设置查询条件语句并实例化 PrepareStatement 对象，executeQuery()执行查询，如果此产品编号存在，则实例化 Product 对象并设定属性值；若不存在此产品，返回 null。

上述代码中，在数据操作实现类中，没有针对数据库的打开和连接操作，只是由构造方法取得连接的数据库，而对数据库的打开和关闭由业务操作实现类完成。业务操作类 ProductService 源代码如下：

```
------------------------ProductService.java---------------------------
```

```java
package com.eshore.service;

import java.util.List;

import com.eshore.dao.ProductDao;
import com.eshore.dao.ProductDaoImpl;
import com.eshore.db.DBConnection;
import com.eshore.pojo.Product;

public class ProductService implements ProductDao{
    private DBConnection dbconn = null;                  // 定义数据库连接类
    private ProductDao dao = null;                   // 声明 DAO 对象
    // 在构造方法中实例化数据库连接，同时实例化 dao 对象
    public ProductService() throws Exception {
        this.dbc = new DBConnection();
        this.dao = new ProductDaoImpl(this. dbconn.getConnection());
        // 实例化 ProductDao 的实现类
    }

    public boolean addProduct(Product product) throws Exception {
        boolean flag = false;                  // 标识
        try {
            if(this.dao.findByProductId(product.getProduct_id()) == null) {
                // 如果要插入的产品编号不存在
                flag = this.dao.addProduct(product);// 新增一条产品信息
            }
        } catch (Exception e) {
            throw e;
        } finally {
            this. dbconn.close();
        }
        return flag;
    }

    public List<Product> findAll(String keyWord) throws Exception {
        List<Product> all = null;                   // 定义产品返回的集合
        try {
            all = this.dao.findAll(keyWord);          // 调用实现方法
        } catch (Exception e) {
            throw e;
        } finally {
            this. dbconn.close();
        }
        return all;
    }
```

```
46
47      public Product findByProductId(String product_id) throws Exception {
48          Product product = null;
49          try {
50              product = this.dao.findByProductId(product_id);
51          } catch (Exception e) {
52              throw e;
53          } finally {
54              this. dbconn.close();
55          }
56          return product;
57      }
58
59  }
```

在上述代码中，第 14~18 行实例化数据库连接类以及 ProductDao 的实现类，第 20~57 行也是分别实现 ProductDao 接口中的方法。在每个方法操作完成之后必须记得要关闭数据库。写完实现类，可以编写 DAO 工厂类用来获得业务操作类，在后续的客户端中就可以直接通过工厂类获得 DAO 接口的实例对象。DAOFactory 的源代码如下：

```
01  package com.eshore.factory;
02
03  import com.eshore.dao.ProductDao;
04  import com.eshore.service.ProductService;
05
06  public class DAOFactory {
07      public static ProductDao getIEmpDAOInstance()throws Exception {
08        return new ProductService();                    //取得业务操作类
09      }
10  }
```

上述代码中，第 7~8 行取得业务操作类。现在已经写好了所需的类，编写一个测试类测试方法是否可以用。测试添加产品类 TestInsertProduct.java，其原代码如下。

```
-----------------------TestInsertProduct.java--------------------------
package com.eshore.test;

import com.eshore.factory.DAOFactory;
import com.eshore.pojo.Product;

public class TestInsertProduct {
    public static void main(String[] args){
        Product product = null;
        try{
            for(int i=0;i<5;i++){
                product = new Product();
```

```
                    product.setProduct_id("350115001010"+i);
                    product.setProduct_name("水杯"+i);
                    product.setPrice(100+i);
                    product.setInfo("这是一个精美的杯子"+i);
                    DAOFactory.getIEmpDAOInstance().addProduct(product);
                }
            }catch(Exception e){
                e.printStackTrace();

            }
        }
    }
```

如上程序，循环添加 5 个产品信息，调用数据操作实例添加产品方法。运行后，在数据库表中添加了 5 条数据，如图 8.54 所示。从结果来看，程序是完全可以运行的。可以看出是用对象来操作数据库，而且插入的代码简单，易懂。

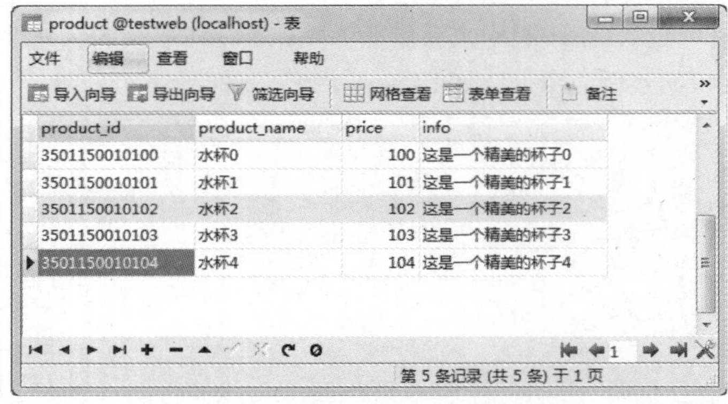

图 8.54　插入产品信息效果图

8.6.4　JSP 调用 DAO

当正确编写完一个 DAO 后，就可以结合 JSP 一起实现前台的显示。下面就演示下如何在 JSP 中使用 DAO。

【例 8.19】JSP 中增加产品信息

```
-----------------------product_add.jsp--------------------------
<%@ page language="java" import="java.util.*" pageEncoding="UTF-8"%>
<!DOCTYPE HTML PUBLIC "-//W3C//DTD HTML 4.01 Transitional//EN">
<html>
  <head>
    <title>添加产品信息</title>
  </head>
  <body>
     <form action="product_insert.jsp" method="post">
```

```
09                产品编号：<input name="product_id"/><br>
10                产品名称：<input name="product_name"/><br>
11                产品价格：<input name="price"/><br>
12                产品信息：<textarea rows="" cols="" name="info"></textarea><br/>
13                <input type="submit" value="添加">  
14                <input type="reset" value="重置">
15          </form>
16      </body>
17  </html>
```

上述代码中，第 8~15 行用表单 form 提供产品的提交信息，提交跳转到 product_insert.jsp 页面。product_insert.jsp 的原代码如下。

```
-----------------------product_ insert.jsp--------------------------
<%@ page  import="java.util.*,com.eshore.pojo.Product" pageEncoding=
"UTF-8"%>
<%@ page import="com.eshore.factory.DAOFactory" %>
<%
request.setCharacterEncoding("utf-8");                  //解决中文乱码
%>

<!DOCTYPE HTML PUBLIC "-//W3C//DTD HTML 4.01 Transitional//EN">
<html>
  <head>
    <title>执行添加产品</title>
  </head>
  <body>
      <%
        Product product = new Product();          //实例化 Product 对象
         product.setProduct_id(request.getParameter("product_id"));
         product.setProduct_name(request.getParameter("product_name"));
         product.setPrice(Double.parseDouble(request.getParameter("price")));
         product.setInfo(request.getParameter("info"));
         boolean flag = DAOFactory.getIEmpDAOInstance().
                        addProduct(product);    //执行添加操作
         if(flag){
      %>
         <h4>添加产品信息成功</h4>
        <%
        }else{
         %>
          <h4>添加产品信息失败</h4>
         <%} %>
  </body>
</html>
```

上述代码中，第 14 行定义了 Product 对象，然后将表单提交的参数依次设置到 Product 对象中，代码第 19 行调用 DAOFactory 的添加方法。代码第 21~28 行判断是否添加成功。其代码运行效果见图 8.55 所示。

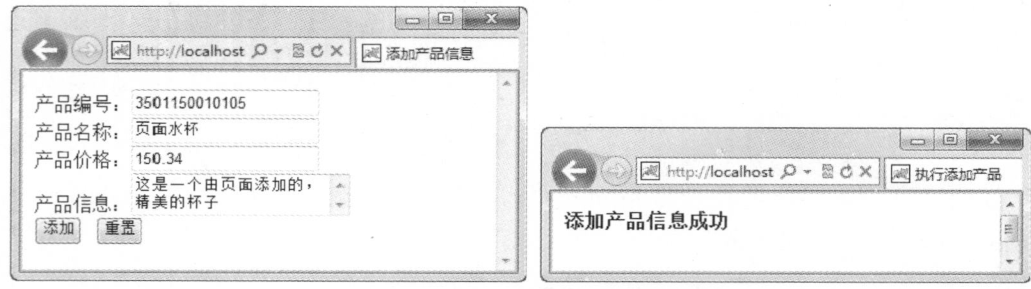

图 8.55 JSP 中增加产品信息效果图

【例 8.20】JSP 中查询产品信息

product_list.jsp 列出产品信息，其原代码如下。

```
-----------------------product_list.jsp--------------------------
<%@ page import="java.util.*,com.eshore.pojo.Product" pageEncoding="UTF-8"%>
<%@ page import="com.eshore.factory.DAOFactory" %>
<%
request.setCharacterEncoding("utf-8");//解决中文乱码
%>
<!DOCTYPE HTML PUBLIC "-//W3C//DTD HTML 4.01 Transitional//EN">
<html>
  <head>
    <title>查询产品列表</title>
  </head>

  <body>
      <%
        String product_name = request.getParameter("product_name");//
        List<Product> list = DAOFactory.getIEmpDAOInstance().findAll(product_name);
      %>
      <form action="product_list.jsp" method="post">
            请输入产品名称：<input name="product_name"/>
         <input type="submit" value="提交">
      </form>
      <table border="1">
         <tr>
            <td>产品编号</td>
            <td>产品名称</td>
            <td>产品价格</td>
            <td>产品信息</td>
```

```
27              </tr>
28              <%
29                  for(int i=0;i<list.size();i++){
30                          Product p = list.get(i);    //取出每一个产品
31               %>
32               <tr>
33                  <td><%=p.getProduct_id() %></td>
34                  <td><%=p.getProduct_name() %></td>
35                  <td><%=p.getPrice() %></td>
36                  <td><%=p.getInfo() %></td>
37              </tr>
38              <%} %>
39          </table>
40     </body>
41   </html>
```

在上述代码中，第 14~16 行取出产品列表，代码第 28~38 行循环遍历查询结果。如果输入产品名称可以实现模糊查询功能，如果不输入则输出所有的产品列表。程序的运行效果如图 8.56 所示。

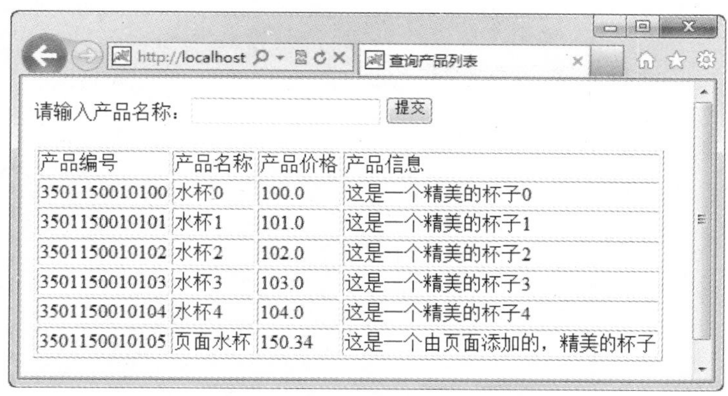

图 8.56　JSP 中查询产品信息效果图

8.7 上机实践

1. 创建、删除一个自定义的数据库，例如：Notice。

2. 创建数据库 School，在 School 中创建数据表 students，students 表结构如表 8.3 所示，按要求进行操作。

表 8.3　students表结构

字段名	数据类型	主键	外键	非空	唯一	自增
s_id	int(11)	是	否	是	是	是
s_name	varchar(20)	否	否	否	否	否
s_contact	varchar(20)	否	否	否	否	否
s_city	varchar(20)	否	否	否	否	否
s_birth	datatime	否	否	否	否	否

（1）创建数据库 School。

（2）创建数据包 students，在 s_id 字段上添加主键约束和自增约束。

（3）将 s_name 字段数据类型改为 vachar（30）。

（4）将 s_contact 字段名更改为 s_phone。

（5）将表名修改为 students_info。

（6）删除字段 s_city。

3. 用 Java Bean 技术实现网页的注册验证。

4. 应用 DAO 设计模式实现存储学生基本信息，并在页面中实现查询、新增、修改等操作。

第 9 章

JSTL标签库

标签的目的就是为 JSP 开发简化代码量，前面已经学过 EL 标签，本章将介绍另外一种标签：JSTL 标签。它不仅可以简化 JSP 代码量，而且使得 JSP 开发者维护工作更加轻松。读者要注意的就是，JSTL 标签经常与 EL 标签一起使用。本章将详细介绍 5 类标签库的使用：core 标签库、fmt 标签库、fn 标签库、XML 标签库和 SQL 标签库。

9.1 JSTL 标签概述

JSTL 标签的是一组与 HTML 标签相似，但又比 HTML 标签强大的功能标签，编程人员可通过它编写出动态的 JSP 页面。它包括 5 类标签库：core 标签库、fmt 标签库、fn 标签库、XML 标签库和 SQL 标签库。这 5 类标签库基本覆盖 Web 开发中所涉及到的技术展示。本节将通过一个实例介绍 JSTL 标签的基本用法，使读者对 JSTL 标签的使用有初步了解。

9.1.1 JSTL 的来历

JSTL（JSP Standard Tag Library）是一个开源的 JSP 标准标签库，是由 Apache 的 jakarta 小组开发并维护的，目前还在不断完善中。

使用 JSTL 标签具有以下优点：

- 接口统一，便于各服务器之间的移植；
- 简化了 Web 程序的开发。本需要大量的 Java 代码完成的功能，可以用少量的 JSTL 代码代替；
- JSTL 标签代码可读性强，易于理解；
- 用 JSTL 标签编写的 Web 程序维护相对简单。

JSTL 标签中的标签主要有以下 5 类：

- core 标签库（核心标签库），包括有通用标签（输出标签）、流控制标签和循环控制标签等；
- fmt 标签库，格式化、国际化标签库；
- fn 标签库，函数标签库；

- XML 标签库，关于 XML 操作的标签库；
- SQL 标签库，操作数据库的标签库。

9.1.2 一个标签实例带你入门

JSTL 最新版本的标准可以从下面的网址获得：

```
http://archive.apache.org/dist/jakarta/taglibs/standard/binaries/
```

从上面地址中选择 jakarta-taglibs-standard-1.1.2.zip 获得到 zip 文件。解压缩 zip 包，在 tld 目录下可以获得 c.tld、fmt..tld、fn.tld、sql.tld 和 x.tld，在 lib 目录下可以获得 jstl.jar 和 standard.jar 两个包，把这两个文件复制到 Web 应用的 lib 目录下，JSTL 就可以在当前的 Web 应用中使用；若想要 Tomcat 服务器中所有的 Web 应用都能使用 JSTL，就可以将这两个 jar 包复制到 Tomcat 安装目录的 lib 目录下。

下面通过一个简单例子，来初步了解 JSTL 如何应用。

【例 9.1】一个简单的 JSTL 示例，输出内容。

theFirstJSTL.jsp 是用户显示页面，源代码如下：

```
----------------- theFirstJSTL.jsp----------------
<%@ page language="java" import="java.util.*" pageEncoding="UTF-8"%>
<%@taglib prefix="c" uri="http://java.sun.com/jsp/jstl/core" %>
<!-- 引入标签 -->
<!DOCTYPE HTML PUBLIC "-//W3C//DTD HTML 4.01 Transitional//EN">
<html>
  <head>
    <title>一个简单的 JSTL 示例</title>
  </head>

  <body>
     <center>
          <c:out value="一个简单的 JSTL 示例"></c:out> <!-- 输出内容 -->
          <br/>
          <c:out value="《JSP 从零开始学》"></c:out>
     </center>
  </body>
</html>
```

如上程序中，页面用语句：<%@taglib prefix="c" uri="http://java.sun.com/jsp/jstl/core" %> 声明本页面要用到的标签库中的标签，prefix 是页面中标签的前缀，uri 是 JSTL 中 c.tld 文件声明的 uri 地址。这个 uri 地址可以修改。

程序中使用输出标签<c:out>，该标签的作用是将内容显示在页面。程序的运行结果如图 9.1 所示。

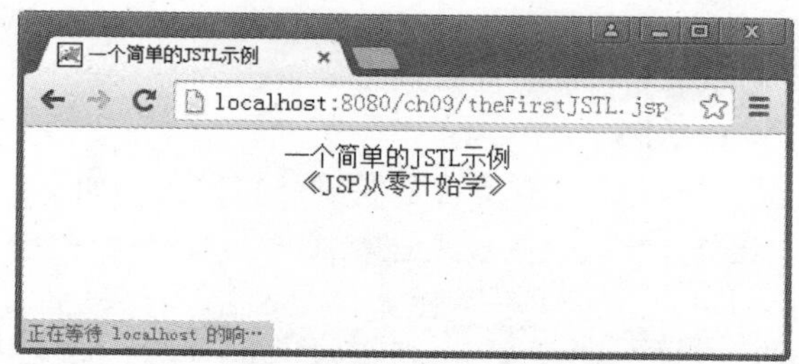

图 9.1　theFirstJSTL.jsp 页面的运行结果

9.2 JSTL 的 core 标签库

上一节介绍了 JSTL 的来源，以及简单的使用示例。本节开始介绍 JSTL 的核心标签库，核心标签库又可以分为表达式标签、流程控制标签和迭代标签。下面逐一进行介绍。

9.2.1　<c:set>标签、<c:out>标签

<c:set>标签用于在某个范围中设定某个值（可以是对象或者参数），这个范围可以是 request、page、session、application，其语法如下：

```
<c:set value="表达式"  var="varname"
[scope="request|page|session|application"]/>
```

在上述语法中，[]中的内容是可选项，“|”是或的意思。 scope 默认值是 page。

<c:set>标签语法也可以用<c:set var="varname">表达式</c:set>形式表示。

<c:out>标签用于把表达式中的结果输出到页面中，其语法如下：

```
<c:out value="表达式" [escapeXML="true|false"]/>
```

在上述语法中，[]中的内容是可选项，“|”是或的意思。escapeXML 默认值为 true，即将特殊字符转换，例如“<”转换为“<”、“>”转换为“>”等。

9.2.2　<c:if>标签

<c:if>标签用于条件判断，其语法如下：

```
<c:if  test="判断条件"  [var="varName"] [scope="{request|page|session|
application}"]>
条件为真时执行的语句
</c:if>
```

上述标签语法中，test 参数是<c:if>标签必须设置的，var 参数是条件的执行结果，scope

是 var 的有效范围。

【例 9.2】<c:if>、<c:out>示例

if.jsp 是<c:if>演示页面，验证输出数是否是偶数，其源代码如下：

```
----------------- if.jsp----------------
<%@ page language="java" import="java.util.*" pageEncoding="UTF-8"%>
<%@taglib prefix="c" uri="http://java.sun.com/jsp/jstl/core" %>
<!DOCTYPE HTML PUBLIC "-//W3C//DTD HTML 4.01 Transitional//EN">
<html>
  <head>
    <title>&lt;c:if&gt;标签使用例子</title>
  </head>

  <body>
      <center>
          <c:if test="${2%2==0}" var="num">
              输出数为：偶数！
          </c:if>
          <br/>该数为偶数的检查结果为：
          <c:out value="${num}"/>
          <br/>
          <c:if test="${3%2==0}" var="num">
              输出数为：奇数！
          </c:if>
          <br/>该数为偶数的检查结果为：
          <c:out value="${num}"/>
      </center>
  </body>
</html>
```

程序的运行结果如图 9.2 所示。

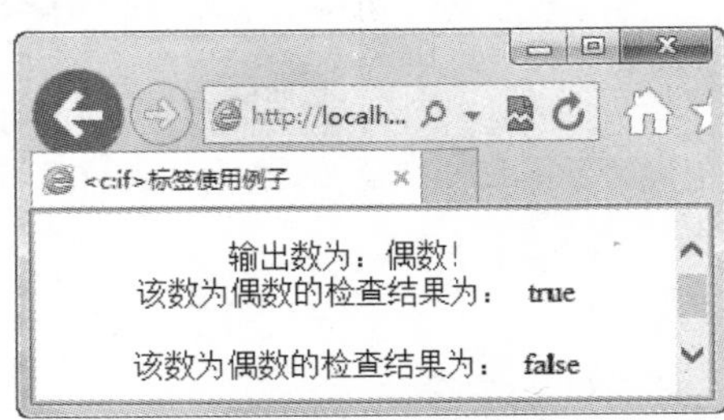

图 9.2　if.jsp 页面运行结果

程序中可以看出用<c:if>标签可以非常简单的进行判断，但要注意别名的使用，别名相当于 Java 中的局部变量。别名的默认范围是 page。

9.2.3　<c:choose>、<c:when>、<c:otherwise>标签

<c:choose>、<c:when>、<c:otherwise>标签是另外一组 JSTL 的流程控制标签。其语法形式如下：

```
<c:choose>
    <c:when  test="表达式">
         表达式为真时执行的语句
    </c:when>
    [<c:otherwise>
     表达式为假时执行的语句
]
</c:choose>
```

在上述的语法中，<c:choose>是父标签，<c:when>和<c:otherwise>是子标签，<c:when>标签可以有 0 个或者多个，同样<c:otherwise>标签也可以用 0 个或者多个，但是<c:when>标签必须在<c:otherwise>标签之前。当<c:when>标签中的判断语句为假时，才会执行<c:otherwise>中的内容。

若<c:when>标签中有多少条件都是真时，只会执行条件最先为真中的<c:when>中的内容。

【例 9.3】<c:choose>、<c:when>、<c:otherwise>示例

choose.jsp 是<c:choose>演示页面，验证输出数是否是偶数，其源代码如下：

```
----------------- choose.jsp----------------
<%@ page language="java" import="java.util.*" pageEncoding="UTF-8"%>
<%@taglib prefix="c" uri="http://java.sun.com/jsp/jstl/core" %>
<!DOCTYPE HTML PUBLIC "-//W3C//DTD HTML 4.01 Transitional//EN">
<html>
  <head>
    <title>&lt;c:choose&gt;标签使用例子</title>
  </head>

  <body>
      <center>
                   输出数检查结果：
          <c:set value="2" var="num"/><!-- set 标签 设定输入值为2-->
          <c:choose>
              <c:when test="${num%2==0}">
                    偶数
              </c:when>
              <c:otherwise test="${num%2!=0}">
                    奇数
              </c:otherwise >
          </c:choose>
```

```
21          </center>
22      </body>
23  </html>
```

代码第 12 行是<c:set>标签，为页面设定值。

程序的运行结果如图 9.3 所示。

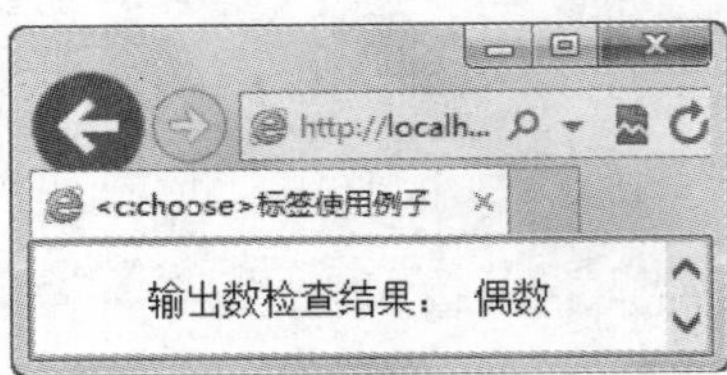

图 9.3　choose.jsp 页面运行结果

程序中第一个<c:when>条件为真，所有输出第一标签内容。

<c:otherwise>标签可以是 0 个或者是多个，只有当<c:when>标签的条件判断都为假时才执行<c:otherwise>中的内容。

9.2.4　<c:set>标签

<c:set>标签用于在某个范围中设定某个值（可以是对象或者参数），这个范围可以是 request、page、session、application，其语法如下：

```
<c:set value="表达式" var="varname" [scope="request|page|session|application"]/>
```

在上述语法中，[]中的内容是可选项，“|”是或的意思。 scope 默认值是 page。

<c:set>标签语法也可以用<c:set var="varname">表达式</c:set>形式表示。例如上述代码第 14 行所示。

9.2.5　<c:forEach>标签

<c:forEach>标签是核心标签中的迭代标签，它的功能类似于 Java 中的 for 循环语句。其语法如下：

```
    <c:forEach [var="varname"]  [varStatus="varstatusName"] [begin="开始"] [end="
结束"] [step="step"]]
      Java 程序或者 HTML 代码
    </c:forEach>
```

或者

```
    <c:forEach item="collection" [var="varname" [varStatus="varstatusName"]
[begin="开始"] [end="结束"] [step="step"]]
```

```
    Java 程序或者 HTML 代码
</c:forEach>
```

在上述语法中，varname 用来存放当前迭代到的成员值，collection 用来迭代集合，varstatusName 中存放当前迭代成员的状态信息；begin 是迭代开始，end 是迭代结束，step 是迭代的步长。collection 集合可以是数组、Java 集合（List 容器和 Map 容器）。varstatusName 中存放的信息有：index（当前迭代的索引号）、count（当前迭代的次数）、first（是否第一次迭代）、last。

【例 9.4】<c:forEach>示例，遍历输出 List 集合

secondForEach.jsp 是测试页面，输出 List 中集合对象，其源代码如下：

```
------------------secondForEach.jsp----------------
<%@ page language="java" import="java.util.*" pageEncoding="UTF-8"%>
<%@taglib prefix="c" uri="http://java.sun.com/jsp/jstl/core" %>
<!DOCTYPE HTML PUBLIC "-//W3C//DTD HTML 4.01 Transitional//EN">
<html>
  <head>
    <title>&lt;c:forEach&gt;标签使用例子</title>
  </head>

  <body>
      <%
            List<String> nameLists = new ArrayList<String>();
            nameLists.add("Toms");
            nameLists.add("Smith");
            nameLists.add("John");
            nameLists.add("Anna");
            nameLists.add("James");
            nameLists.add("Roses");
            nameLists.add("Bruce");
            request.setAttribute("nameLists",nameLists);
       %>
      <center>
                 输出集合中的内容：<hr/>
            <c:forEach items="${nameLists}" var="name" varStatus=
"currentStatus">
                当前元素为：<c:out value="${name}"/>  
                当前元素索引号为：<c:out value="${currentStatus.index}"/>
  
                当前迭代数为：<c:out value="${currentStatus.count}"/>
  
               <c:if test="${currentStatus.first}">第一次循环操作</c:if>
               <c:if test="${currentStatus.last}">最后一次循环操作</c:if>
                  <hr/>
```

```
30              </c:forEach>
31          </center>
32      </body>
33  </html>
```

代码 10~20 行在 JSP 页面中组合 List 集合，第 19 行代码向页面传递参数 nameLists 集合，第 23 行代码用 EL 标签获得参数 nameLists 值进行遍历。其运行结果如图 9.4 所示。

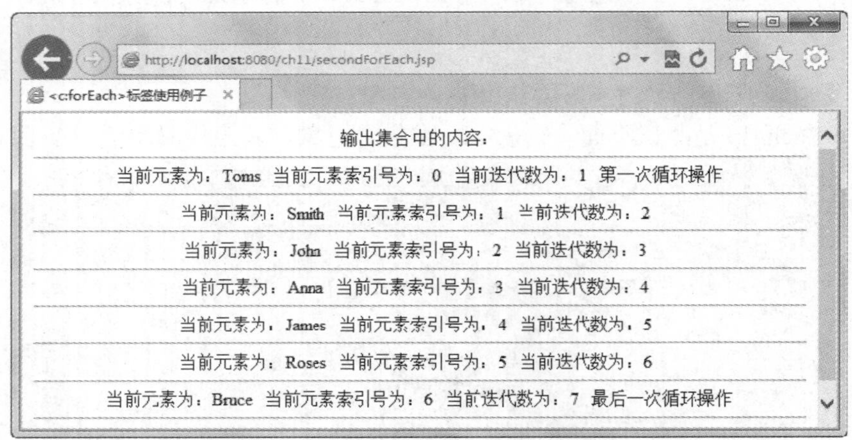

图 9.4　secondForEach 页面运行结果

9.2.6　<c:forTokens>标签

<c:forTokens>标签用于对字符串进行分隔，类似于 Java 中的 split 方法。语法形式如下：

```
<c:forTokens items="字符串" delims="分隔符" [var="别名"] [varStatus="
varstatusName"] [begin="开始"] [end="结束"] [step="步长"]>
Java 代码，HTML 代码等
</c:forTokens>
```

在上述的语法中，items 中放的是进行处理的字符串，delims 是进行拆分处理的分隔符，var 用来指明迭代值的别名，varStatus 是指当前迭代的状态信息，begin 是迭代开始，是一个可选项，end 是迭代结束值，setp 是迭代的步长。

9.2.7　<c:remove>标签

<c:remove>标签，从标签命名中可以看出，该标签主要功能是删除作用，它可以删除某个范围中设定的值。其中范围可以是 page、request、session、application；值可以是某对象或者系统中某参数。其语法形式如下：

```
<c:remove var="varname" [scope="page|request|sessioin|applicatioin"]/>
```

在上述语法中，varname 是指要删除的元素名，scope 是指删除的范围，默认为 page。

9.2.8 <c:catch>标签

<c:catch>标签用于捕获嵌套在<c:catch>标签中的代码抛出的异常，并作相应的处理，类似于 Java 中的 try...catch 方法。其语法形式如下：

```
<c:catch [var="varname"]>
    需要捕获异常的程序代码
</c:catch>
```

上述语法中，参数 varname 用于标识捕获异常信息。

9.2.9 <c:import>标签与<c:param>标签

<c:import>标签的作用是把当前 JSP 页面之外的静态或者动态文件导入进来，甚至可以是其他网站的文件，这功能类似于<jsp:include>JSP 动作指令。但是它们有所不同，<jsp:include>只能导入当前 JSP 页面同一 Web 应用下的文件而<c:import>标签不是。其语法形式有两种，分别如下：

```
<c:import url="url" [context="context"] [var="varname"]
  [scope=page|request|session|application] [charEncoding="charencoding"]
    [<c:param/>标签语句]
</c:param>
```

或

```
<c:import url="url" [context="context"] [varReader="readerName"]
      [charEncoding="coding"]>
    [<c:param/>标签语句]
</c:import>
```

上述语法中，第一种导入文件时，把文件的内容以字符串形式存入 varname 中，第二种导入文件时，把文件的内容以 Reader 的方式向外提供读取。

url 是输入的地址，它可以是网址、FTP 服务的文件地址，也可以是 Web 应用中的文件。

<c:param>标签用于向导入的页面中传入参数，其语法形式如下：

```
<c:param name="paramName" value="paramValue"/>
```

paramName 是要传入的参数名称，paramValue 为传入的参数值。

9.2.10 <c:redirect>标签

<c:redirect>标签用于把 client 端发来的请求重定向到另一个页面。其语法如下：

```
<c:redirect url="url" [context="context"]>
  [<c:param/>标签语句]
</ c:redirect>
```

在上述语法中，url 是重定向的目标网页，<c:param>标签是负责带入参数，是可选项，参数 context 的作用是当要重定向目标网址为其他 Web 应用的网页时指出其应用名。例如，当前 Web 应用为 testJstl，要重定向到 Web 应用 targetJstl 中的 taget.jsp 文件时，语句如下：

```
<c:redirect url="/target.jsp" context=" targetJstl"/>
```

如果在传递传递参数时页面显示中文的为乱码，可以按照第三章中乱码的方式解决，也可以修改 Tomcat 服务器中的 server.xml 配置。

在 Tomcat 的安装目录 conf 子目录中找到 server.xml 文件并打开，找到如下内容：

```
<Connector port="8080" protocol="HTTP/1.1"
           connectionTimeout="20000"
           redirectPort="8443" />
```

在上述内容中添加 URIEncoding="utf8"，即修改为：

```
<Connector port="8080" protocol="HTTP/1.1"
           connectionTimeout="20000"
           redirectPort="8443" URIEncoding="utf8" />
```

重新启动 Tomcat，就可以发现中文显示正常。

9.2.11 <c:url>标签

<c:url>标签用于生成一个 URL。其语法形式如下：

```
<c:url  value="url" [context="context"] var="varname"
[scope="page|request|session|application"]>
  [<c:param/>标签语句]
</ c:url>
```

上述语法中，url 是要生成的 URL，参数 context 是用于 URL 为其他应用中的应用名字跟<c:import>标签中的 context 属性一样是可选项，参数 varname 是生成 URL 字符串的变量名称，scope 是其作用范围，默认是 page。

如果生成的 URL 中中文是乱码，可以对其转换字符集。

9.3 JSTL 的 fmt 标签库

上一节介绍 JSTL 的核心标签库，让读者初步了解其标签库中各标签的用法。本节介绍 JSTL 的格式化标签库。格式化标签库在 JSP 开发中也是经常使用到的，它可以将日期，数字用很简单的方式进行转换。 fmt 标签大致有以下几种：

- 国际化标签：<fmt:requestEncoding>和<fmt:setLocale>。

- 消息标签：<fmt:bundle>、<fmt:message>、<fmt:setBundle>和<fmt:param>。
- 数字和日期格式化标签：<fmt:formatNumber>、<fmt:formatDate>、<fmt:parseDate>、<fmt:parseNumber>、<fmt:setTimeZone>和<fmt:timeZone>。

9.3.1　<fmt:requestEncoding>设置编码

<fmt:requestEncoding>标签用于设置请求中数据的字符集，它的作用跟 JSP 中设定字符集语句：request.setCharacterEncoding("charsetName")是一样的。其语法形式如下：

```
<fmt:requestEncoding value="charsetName"/>
```

上述语句中，参数 charsetName 是要设置的字符集名称。其语句如下：

```
<fmt:requestEncoding value="utf-8"/>
```

设置完字符集，就不必再用 request 为每个参数作字符集编码转换。

9.3.2　<fmt:setLocale>显示所有地区的数据格式

<fmt:setLocale>标签用于设置用户的语言国家或地区。其语法如下：

```
<fmt:setLocale value="localcode" [scope=page|request|session|application]
[variant="variant"] />
```

上述语法中，参数 localcode 代表语言代码，例如：zh、en，也可以在后面加上国家或者地区的两位数的代码符号，中间用"_"连接，例如：zh_TW（中国台湾地区）、zh_HK（中国香港）。参数 variant 是设置浏览器类型的。例如：win 代表 Windows。Scope 是设置其有效范围，默认是 page。

9.3.3　<fmt:bundle>、<fmt:message>、<fmt:param>资源国际化

<fmt:bundle>、<fmt:message>、<fmt:setBundle>、<fmt:param>这 4 个标签 fmt 标签库中的消息标签，其数据来源都是.properties 文件。

（1）<fmt:bundle>标签的作用是绑定数据源.properties 文件。其调用方式如下：

```
<fmt:bundle basename="resourceName" prefix="pre">
  代码块
</fmt:bundle>
```

上述方式中，参数 basename 是要绑定的数据源.properties 文件的文件名，参数 prefix 是要获取.properties 文件的前缀。若设置 prefix 属性，则嵌套的<fmt:message>标签中 key 属性就可以省略 prefix 属性设置的前缀部分，这功能主要是针对有相同前缀的多个关键字情况。

【例 9.5】<fmt:bundle>示例

bundle.jsp 页面是测试绑定源文件页面，其作用是读取源文件信息，源代码如下：

```
------------------ bundle.jsp-----------------
<%@ page language="java" import="java.util.*" pageEncoding="UTF-8"%>
<%@taglib prefix="c" uri="http://java.sun.com/jsp/jstl/core" %>
<%@taglib prefix="fmt" uri="http://java.sun.com/jsp/jstl/fmt" %>
<fmt:requestEncoding value="utf-8"/>
<!DOCTYPE HTML PUBLIC "-//W3C//DTD HTML 4.01 Transitional//EN">
<html>
  <head>
    <title>&lt;fmt:bundle&gt;标签使用例子</title>
  </head>

  <body>
        <c:out value="读取资源文件(myresource.properties)" />
        <br>
        <fmt:bundle basename="myresource"  prefix="my.">
            <fmt:message key="author" var="author" />
            <fmt:message key="teacher" var="teacher" />
        </fmt:bundle>
        作者:<c:out value="${author}" />
        老师:<c:out value="${ teacher }" />
        <br>
   </body>
</html>
```

参数 basename 中的文件名不能带扩展名。

代码中，用标签<fmt:bundle/>绑定数据源 myresource.properties 文件，设定前缀是“my.”，用标签<fmt:meaage>取出值，其运行结果如图 9.5 所示。

图 9.5　bundle.jsp 页面运行结果

如果.properties 文件中有中文的，那么在读取的时候会显示乱码，必须对.properties 文件进行编码转换。可以用 JDK 中提供的转换工具进行转换，转换命令如下：

```
native2ascii -encoding utf8 源.properties 文件名  目标.properties 文件名
```

执行这个命令时，要确定是否设置了 JDK 环境变量。

myresource.properties 文件的源文件信息如下：

```
my.author=linl
my.teacher=\u8C2D\u5DE5
```

（2）<fmt:message>标签的作用是从指定的资源文件。其调用方式如下：

```
<fmt:message  key="messageName"  [var="varname"]  [bundle="resourceName"]
[scope="page|request|session|applicatioin"]>
   [<fmt:param>标签]
</fmt:bundle>
```

上述语法中，参数 key 是要从.properties 文件中取出的键名称；参数 bundle 是要绑定的数据源.properties 文件的文件名，参数 varname 用来保存取出的键值，scope 是设置标签的有效范围。

（3）<fmt:param>标签需要与<fmt:message>表联合使用，其用于设定<fmt:message>标签指定键的动态值。其调用方式如下：

```
<fmt:param value="keyvalue"/>
```

若<fmt:message>标签对应的键有多个参数，可以用多个<fmt:param>标签来设置其动态值。若.properties 文件中有如下的键值对：

```
messageT=Welcome  {0},today is {1,date}
```

{0}表示第 1 个参数，{1,date}表示第 2 个参数，参数的格式为日期类型。

参数是从 0 开始编号的。

若数据已经绑定，则设定的标签语句为：

```
<fmt:message key="messageT">
   <fmt:param>linl</fmt:param>
   <fmt:param value="${dateT}"/>
</fmt:message>
```

上述语句中，第一个<fmt:param>对应.properties 中的{0}，第二个<fmt:param>对应.properties 中的{1,date}，变量 dateT 是设定好的日期变量。

9.3.4 <fmt:setBundle>标签

<fmt:setBundle>标签与<fmt:bundle>标签类似都是用于设置默认的数据源。其调用方法如下：

```
<fmt:setBundle basename="resourceName" [var="绑定数据源别名"]
[scope="page|session|request|application"] >
   代码块
</fmt:setBundle>
```

上述方式中，参数 basename 是要绑定的数据源.properties 文件的文件名，参数 var 是代表了绑定的数据源，scope 设置 var 参数的有效范围，默认是 page。

<fmt:setBundle>示例可以参考例 9.5 所示，只需将标签<fmt:bundle>更改成<fmt:setBundle>即可，其运行结果是一样的。

9.3.5　<fmt:formatNumber>显示不同地区的各种数据格式

<fmt:formatNumber>标签用于显示不同地区的各种数据格式。其调用方法如下：

```
<fmt: formatNumber  value="numberValue" [type="number|percent|currency"]
      [pattern="pattern"] [currencyCode="currenCyCode"]
[currencySymbol="currencySymbol']
      [groupingUsed="true|false"] [maxIntegerDigits="maxIntegerDigits"]
       [minIntegerDigits="minIntegerDigits"] [maxFractionDigits="
maxFractionDigits"]
      [minFractionDigits=" minFractionDigits"] [var="varName"]
   [scope="page|session|request|application"] >
   </fmt: formatNumber >
```

参数说明见表 9.1 所示。

表 9.1　formatNumber 参数说明

参数	说明
参数 value	要格式化的数字
参数 type	设定数字的单位，有 3 种：number、currency、percent
参数 pattern	设定显示的模式
参数 currencyCode	设置 ISO-4217 编码
参数 currencySymbol	设置货币符号
参数 groupingUsed	设置是否在显示数字时，隔开显示
参数 maxIntegerDigits	设置最多的整数位，若设定的数值少于数字的实际位数时，数字的左边位数会被截去响应的位数，例如，123456，maxIntegerDigits 设定为 4，则结果为 3456
参数 minIntegerDigits	设置最少的整数位，若设定的数值多于数字的实际位数时，会在数字的左边补 0，例如，123，minIntegerDigits 设定为 4，则结果为 0123
参数 maxFractionDigits	设置最多小数位数，若设定的数值小于数字的实际小数位数，则会从右边戒掉多于位数，例如，123.56，maxFractionDigits 设定为 1，则结果为 123.5
参数 minFractionDigits	设置最少小数位数，若设定的数值大于数字的实际小数位数，则会从右边补 0，例如，123.56，maxFractionDigits 设定为 4，则结果为 123.5600
参数 var	代表格式化后的数字，若设定了该参数，需用<c:out>标签输出
参数 scope	设定参数 var 的有效范围，默认为 page

9.3.6　<fmt:parseNumber>解析数字

<fmt:parseNumber>标签用于把字符串中表示的数字、货币、百分比转换成数字数据类型。其调用方法如下：

```
    <fmt:parseNumber value="number" [ integerOnly="true|false"]
[parseLocale="parseLocal"]
             pattern="customPattern"
scope="page|request|session|application"
    type="number|currency|percent" var="varname"/>
```

参数说明见表 9.2 所示。

表 9.2　parseNumber参数说明

参数	说明
参数 value	要解析的数字
参数 type	设定数字的单位，有 3 种：number、currency、percent　.
参数 pattern	设定显示的模式
参数 integerOnly	设置是否只输出整数部分
参数 parseLocale	解析数字时所用的区域
参数 var	代表格式化后的数字，若设定了该参数，需用<c:out>标签输出
参数 scope	设定参数 var 的有效范围，默认为 page

9.3.7　<fmt:formatDate>格式化日期

<fmt:formatDate>标签用于格式化日期，使得可以用不同方式输出日期和时间。其调用方法如下：

```
    <fmt:formatDate value="datevalue"
[dateStyle="default|short|medium|long|full"]
    [pattern="customPattern"]  [scope="page|request|session|application"]
     [timeStyle=" default|short|medium|long|full "]  [timeZone="timeZone"]
   [type="time|date|both" ] [var="varname"]/>
```

参数说明见表 9.3 所示。

表 9.3　formateDate参数说明

参数	说明
参数 value	要显示的日期
参数 type	设置输出的类别，time、date 和 both
参数 dateStyle	设定日期输出的格式
参数 pattern	自定义格式模式
参数 timeStyle	设定日期的输出风格
参数 timeZone	设定日期的时区
参数 var	代表格式化后的数字，若设定了该参数，需用<c:out>标签输出
参数 scope	设定参数 var 的有效范围，默认为 page

formateDate 输出模式说明见表 9.4 所示。

表 9.4　formateDate常用的日期和时间输出模式

模式	示例
yyyyMMdd	20140120
HH:mm	11:50
HH:mm:ss	11:50:20
yyyy.mm.dd G 'at' HH:mm:ss z	2014.01.20 公元 at 11:30:10 CST
yyMMddHHmmssZ	140120112030+0800

9.3.8　<fmt:parseDate>解析日期

<fmt:parseDate />标签用于把字符串类型的日期转换成日期数据类型，其调用方法如下：

```
<fmt:parseDate  [dateStyle="default|short|medium|long|full"]
[parseLocale="parseLocale" ]  [pattern="customPattern"]
[scope="page|request|session|application"]
[timeStyle=" default|short|medium|long|full "]
[timeZone="timeZone"]  [type="time|date|both" ]
        value="parseDateValue"  [var="varname"]/>
```

参数说明见表 9.5 所示。

表 9.5　parseDate参数说明

参数	说明
参数 value	要显示的日期
参数 type	设置输出的类别，time、date 和 both
参数 dateStyle	设定日期输出的格式
参数 pattern	自定义格式模式
参数 timeStyle	设定日期的输出风格
参数 timeZone	设定日期的时区
参数 var	代表格式化后的数字，若设定了该参数，需用<c:out>标签输出
参数 scope	设定参数 var 的有效范围，默认为 page

9.3.9　<fmt:setTimeZone>标签和<fmt:timeZone>标签

（1）<fmt:setTimeZone>标签用于设定默认的时区，其调用方法如下：

```
<fmt:setTimeZone value="timeValue"
 [scope="page|request|session|application"]   var="varname"/>
```

在上述语法中，参数 value 为要设置的时区；var 为存储新时区的变量名；scope 为变量 var 的有效范围。

（2）<fmt:timeZone>标签用来指定时区，供其他标签使用，其调用方法如下：

```
<fmt:timeZone value="timeValue" >
代码块
```

```
</fmt:timeZone>
```

9.4 JSTL 的 fn 方法库

上一节介绍 JSTL 的 fmt 标签库，让读者初步了解 JSTL 中格式化标签库的使用方法。本节介绍 JSTL 的函数标签库。函数标签库在 JSP 开发中经常使用到的，它可以将字符串处理的变的简单，就像在页面中 Java 方法处理的一样。

要使用 fn 方法库，需在 JSP 页面的头部加入下面语句：

```
<%@taglib prefix="fn" uri="http://java.sun.com/jsp/jstl/functions" %>
```

9.4.1 fn:contains()函数与 fn: containsIgnoreCase()函数

（1）fn:contains()函数用于判断一个字符串是否包含指定的字符串。其语法如下：

```
boolean contains(sourceStr, testStr)
```

在上述语法中，sourceStr 是源字符串，testStr 是指定的字符串。

（2）fn: containsIgnoreCase ()函数用于判断一个字符串是否包含指定的字符串，忽略大小写敏感。其语法如下：

```
boolean containsIgnoreCase (sourceStr, testStr)
```

在上述语法中，sourceStr 是源字符串，testStr 是指定的字符串。

9.4.2 fn:startsWith()函数与 fn:endsWith()函数

（1）fn:startsWith()函数用于判断一个字符串是否以指定的前缀开始。其语法如下：

```
boolean startsWith (sourceStr, startPrefix)
```

在上述语法中，sourceStr 是源字符串，startPrefix 是指定的开始前缀。跟 Java 中 startsWith 方法类似。

（2）fn:endsWith()函数用于判断一个字符串是否以指定的后缀结尾。其语法如下：

```
boolean endsWith (sourceStr, endPrefix)
```

在上述语法中，sourceStr 是源字符串，endPrefix 是指定的结束后缀。跟 Java 中 endsWith 方法类似。

9.4.3 fn:escapeXml()实现 HTML 编码

fn: escapeXml()函数用于忽略 XML 标记的字符，其语法如下：

```
java.lang.String escapeXml(spcialString)
```

在上述语法中，spcialString 指定字符串。

9.4.4 fn:indexOf()函数与 fn:length()函数

（1）fn:indexOf()函数用于返回指定子字符串在此字符串中第一次出现处的索引，其用法跟 Java 中 indexOf()方法是一样的，语法如下：

```
int indexOf(sourceStr, specialStr)
```

在上述语法中，sourceStr 是源字符串，specialStr 是指定的字符串。

（2）fn:length()函数用于返回指定字符串的长度，其用法和 Java 中 length()方法是一样的，语法如下：

```
int length(sourceStr)
```

在上述语法中，sourceStr 是要测试的字符串。

【例 9.6】fn: indexOf ()函数和 fn:length()函数示例

indexOf.jsp 页面是测试页面，其功能是输出 indexOf 函数中指定的字符串索引和字符串长度，其源代码如下：

```
----------------- indexOf.jsp-----------------
<%@ page language="java" import="java.util.*" pageEncoding="UTF-8"%>
<%@taglib prefix="c" uri="http://java.sun.com/jsp/jstl/core" %>
<%@taglib prefix="fn" uri="http://java.sun.com/jsp/jstl/functions" %>
<!DOCTYPE HTML PUBLIC "-//W3C//DTD HTML 4.01 Transitional//EN">
<html>
  <head>
    <title>fn:indexOf()函数和 fn:length()函数使用例子</title>
  </head>
  <body>
    <h3>fn:indexOf()函数和 fn:length()函数使用例子</h3>
    <c:set var="string1" value="This is the First my test String."/>
     <c:set var="string2" value="This <abc>is the Second my test String.</abc>"/>

     <p><h4>使用 fn:indexOf()函数:</h4></p>
     <p>字符串1 : ${fn:indexOf(string1,"Fir")}</p>
     <p>字符串2 : ${fn:indexOf(string2,"my")}</p>

     <p><h4>使用 fn:length()函数:</h4></p>
     <p>字符串1的长度 : ${fn:length(string1)}</p>
  </body>
</html>
```

上述代码中第 14~15 行是输出指定字符串的索引，代码第 18 行是输出字符串 1 的长度，其运行结果如图 9.6 所示。

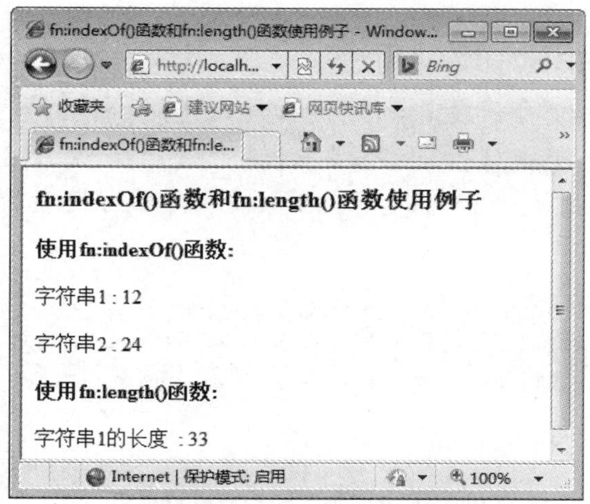

图 9.6　例 9.6 运行结果

9.4.5　fn:split()函数与 fn:join()函数

（1）fn:split()函数用于将字符串用指定的分隔符分隔为一个子串数组。其用法和 Java 中 split()方法是一样的，语法如下：

```
java.lang.String[] split(sourceStr, regex)
```

在上述语法中，sourceStr 是源字符串，regex 是指定的分隔符。

（2）fn:join()函数用于将一数组中的所有元素用指定的分隔符来连接成一字符串，其语法如下：

```
String join (array[], regex)
```

在上述语法中，array 是源数组，regex 是指定的分隔符。

9.5　JSTL 的 SQL 标签库

上一节介绍 JSTL 的部分 fn 方法库，让读者初步了解 JSTL 中方法库的使用方法。本节介绍 JSTL 的 SQL 标签库即数据库标签库。有了 SQL 标签，操作数据库就很方便了，虽然在大型的网站中不建议使用这种标签，但是在小型网站中经常会用到。

要使用 SQL 标签库，需在 JSP 页面的头部加入下面语句：

```
<%@taglib prefix="fn" uri="http://java.sun.com/jsp/jstl/sql" %>
```

SQL 标签库主要有<sql:setDateSource>、<sql:query>、<sql:update>、<sql:dateParam>、<sql:param>等标签，下面会逐一介绍。

9.5.1 <sql:setDateSource>标签

<sql:setDateSource>标签用于设定操作的数据源，其调用方法如下：

```
   <sql:setDataSource dataSource="dataSource"
[scope="page|session|request|application"]   [var="varname"]/>
```

或者

```
   <sql:setDataSource driver="drivername" [password="password"]  url="jdbcURL"
user="username"  var="varname"
   [scope="page|session|request|application"] />
```

参数说明见表 9.6 所示。

表 9.6 setDateSource 参数说明

参数	说明
参数 driver	注册的 JDBC 驱动
参数 url	数据库连接的 JDBC URL
参数 user	连接数据库时使用的用户名
参数 password	连接数据库时使用的密码
参数 dataSource	已经存在的数据源
参数 var	代表数据源的变量
参数 scope	设定参数 var 的有效范围，默认为 page

设置<sql:setDateSource>标签的参数并不复杂，若要连接到本地 MySQL 数据库，用标签连接语句如下：

```
   <sql:setDateSource
       var="test" driver="com.mysql.jdbc.Driver"
url="jdbc:mysql://localhost:3306/test"
       user="lin" password="123456" />
```

上述语句中，lin 是数据库用户名，password 是用户密码，driver 是数据库驱动

9.5.2 <sql:query>标签

<sql:query>标签用于查询数据库中的数据，其调用方法如下：

```
<sql:query var="varname"  [dataSource="dataSource"]  [maxRows="maxRows" ]
       [scope="page|session|request|application"]  sql="sqlQuery"
       [startRow="startRow"]/>
```

参数说明见表 9.7 所示。

表 9.7　query参数说明

参数	说明
参数 maxRows	设置最多可存放的记录条数
参数 sql	查询的 SQL 语句
参数 startRow	设置结果集从查询结果的第几条记录开始
参数 dataSource	连接的数据源
参数 var	代表 SQL 查询的结果
参数 scope	设定参数 var 的有效范围，默认为 page

查询出的结果是存放在 var 变量中，要输出结果就要通过其属性来输出。属性说明见表 9.8 所示。

表 9.8　var属性说明

参数	说明
参数 rows	以字段名称读取 query 结果记录集
参数 rowsByIndex	以数字作为索引的查询结果
参数 columnNames	设置结果集从查询结果的第几条记录开始
参数 rowCount	结果集中记录的条数
参数 limitedByMaxRows	查询结果记录数是否因为 maxRows 而受到限制，若超过 maxRows，返回 true，否则为 false

9.5.3　<sql:update>标签

<sql:update>标签用于更新数据库中的数据，其调用方法如下：

```
<sql:update var="varname"  [dataSource="dataSource"]
        [scope="page|session|request|application"]  sql="sqlUpdate"/>
```

参数说明见表 9.9 所示。

表 9.9　<sql:update>标签参数说明

参数	说明
参数 sql	更新的 SQL 语句，可以是 insert、update、delete 语句
参数 dataSource	连接的数据源
参数 var	用来存储所影响行数的变量
参数 scope	设定参数 var 的有效范围，默认为 page

9.5.4　<sql:dateParam>标签与<sql:param>标签

（1）<sql:dateParam>标签与<sql:query>标签、<sql:update>标签结合使用，用来提供日期和时间的动态值。其调用方法如下：

```
<sql:dateParam value="value" type="DATE|time|timestamp" />
```

参数说明见表 9.10 所示。

表 9.10　dataParam 参数说明

参数	说明
参数 value	代表要设置的动态参数值
参数 type	设置日期数据种类，有 date、time、timestamp 3 种类型

（2）<sql:param>标签用来提供设置 SQL 语句中的动态值。其调用方法如下：

```
<sql:param value="value" />
```

【例 9.7】sql 标签综合示例

sqlZonghe.jsp 是 sql 标签综合示例，其源代码如下：

```
------------------sqlZonghe.jsp-----------------
<%@ page import="java.io.*,java.util.*,java.sql.*" pageEncoding="UTF-8"%>
<%@ page import="javax.servlet.http.*,javax.servlet.*" %>
<%@ page import="java.util.Date,java.text.*" %>
<%@ taglib uri="http://java.sun.com/jsp/jstl/core" prefix="c"%>
<%@ taglib uri="http://java.sun.com/jsp/jstl/sql" prefix="sql"%>
<%@taglib prefix="fmt" uri="http://java.sun.com/jsp/jstl/fmt"%>
<html>
<head>
   <title>sql 标签综合示例</title>
</head>
<body>
<!-- 设置数据源-->
<sql:setDataSource var="test" driver="com.mysql.jdbc.Driver"
     url="jdbc:mysql://localhost:3306/TEST"
     user="root"  password="root"/>
<!-- 将用户的年龄增加2岁 -->
<sql:update dataSource="${test}"  var="updatecount">
   update users set user_age=user_age+?
   <c:set value="2" var="count"/>
   <sql:param value="${count}"/>
</sql:update>
<!-- 给 id 为1的用户设置日期 -->
<%
Date nowdate = new Date();
int userId = 1;
%>
<sql:update dataSource="${test}"  var="updatecount2">
   UPDATE users SET createtime = ? WHERE Id = ?
   <sql:dateParam value="${nowdate}" type="timestamp" />
   <sql:param value="${userId}" />
</sql:update>
```

```
31  <!-- 查询数据-->
32  <sql:query dataSource="${test}" var="result"
33    sql="SELECT * from users;" >
34  </sql:query>
35  <!--显示数据  -->
36  <table border="1" width="100%">
37  <tr>
38    <td colspan="6" align="center">
39       共查询${result.rowCount}条用户记录
40    </td>
41  </tr>
42  <tr>
43    <th>用户 ID</th>
44    <th>用户姓名</th>
45    <th>用户性别</th>
46    <th>用户年龄</th>
47    <th>联系电话</th>
48    <th>出身地</th>
49    <th>创建日期</th>
50  </tr>
51  <c:forEach var="user" items="${result.rows}">
52  <tr>
53    <td><c:out value="${user.id}"/></td>
54    <td><c:out value="${user.user_name}"/></td>
55    <td><c:out value="${user.user_sex}"/></td>
56    <td><c:out value="${user.user_age}"/></td>
57    <td><c:out value="${user.user_phone}"/></td>
58    <td><c:out value="${user.user_address}"/></td>
59    <td><fmt:formatDate type="time" value="${user.createtime}"
var="formatUsertime"/>
60       <c:out value="${formatUsertime}"></c:out></td>
61  </tr>
62  </c:forEach>
63  </table>
64
65  </body>
66  </html>
```

上述程序第 15~20 行给每个用户的年龄都增加 2 岁，第 26~30 行给指定用户设置创建日期，程序运行结果如图 9.7 所示。

图 9.7　sqlZonghe.jsp 页面运行结果

9.5.5　<sql:transaction>标签事务管理

<sql:transaction>标签是事务标签，它用来将<sql:query>标签、<sql:update>标签封装在单一事务中，确保事务的一致性。<sql:transaction>标签调用方法如下：

```
<sql:transaction dataSource="" isolation="">
  <sql:query>标签语句或者<sql:update>标签语句
 </sql:transaction>
```

参数说明见表 9.11 所示。

表 9.11　transaction 参数说明

参数	说明
参数 dataSource	代表要设置好的数据源
参数 isolation	设置事务的隔离级别，有 4 个取值：READ_COMMITTED，READ_UNCOMMITTED，REPEATABLE_READ 或 SERIALIZABLE

9.6 JSTL 的 XML 标签库

上一节介绍 JSTL 的部分 SQL 标签库，让读者了解 JSTL 中 SQL 标签库的使用方法，可以让读者了解如何在 JSP 中访问数据库。本节介绍 JSTL 的 XML 标签库，有了 XML 标签，可以轻松处理 XML 文件。 XML 标签库与核心标签库功能很相似，读者使用时可以参考核心标签库。

要使用 XML 标签库，需在 JSP 页面的头部加入下面语句：

```
<%@taglib prefix="x" uri="http://java.sun.com/jsp/jstl/xml" %>
```

XML 标签库主要有<x:out>、<x:parse>、<x:set>、<x:choose>、<x:when>、<x:othersise>、

<x:forEach>、<x:if>等标签，下面会逐一介绍。

9.6.1　<x:parse>获取新浪 RSS 新闻

<x:parse>标签用来解析 XML 文件，其调用方法如下：

```
<x:parse {doc="XMLDocument"| xml="XMLDocument"} [filter="filter" ]
     [systemId="systemId"] {var="varname"
[scope="request|page|session|application"]
     |varDom="" [scopeDom="request|page|session|application"] }  />
```

参数说明见表 9.12 所示。

表 9.12　<x:parse>参数说明

参数	说明
参数 var	代表已解析 XML 数据的变量
参数 varDom	代表已解析 XML 数据的变量
参数 xml	需要解析的 XML 文档的文本内容
参数 doc	需要解析的 XML 文档的文本内容
参数 filter	文档过滤器
参数 systemId	XML 文档的 URI
参数 scope	参数 var 的作用范围
参数 scopeDom	参数 varDom 的作用范围

<x:parse>标签常与<c:import>标签结合使用，解析完的节点需要用<x:out>标签输出。

【例 9.8】<x:parse>标签获取新浪 RSS 新闻

xparse.jsp 页面是获取新浪 RSS 新闻的测试页面，输出第一个 outline 中的xmlUrl 属性值，其源代码如下：

```
-----------------sqlZonghe.jsp----------------
<%@ page import="java.io.*,java.util.*" pageEncoding="UTF-8"%>
<%@ taglib uri="http://java.sun.com/jsp/jstl/core" prefix="c"%>
<%@taglib prefix="x" uri="http://java.sun.com/jsp/jstl/xml"%>
<html>
<head>
   <title>&lt;x:parse&gt;标签示例</title>
</head>
<body>
    <h3>新浪 RSS 节点信息:</h3>
    <!-- 导入新浪 RSS 的 xml 信息 -->
     <c:import var="xinlangInfo"
url="http://rss.sina.com.cn/sina_all_opml.xml "/>
    <x:parse xml="${xinlangInfo}" var="output"/>
    <b>第一个 outline 的 xmlUrl 属性值</b>:
```

```
14        <x:out select="$output/opml/body/outline/outline[1]/@xmlUrl" />
15    </body>
16    </html>
```

上述代码第 11 行是导入新浪 RSS 的 xml 信息，第 12 行解析 xml 文件，第 14 行输出指定的内容，其运行结果如图 9.8 所示。

图 9.8　xparse.jsp 页面的运行结果

9.6.2　<x:out>输出指定元素

<x:out>标签用于输出 XML 文件中指定的内容，其调用方法如下：

```
<x:out select="expression"  {escapeXml="true|false"}/>
```

上述语句中，select 属性是 XPath 表达式，escapseXml 代表是否忽略 XML 特殊字符，默认值是 true，即转换特殊字符为实体代码。<x:out>标签的使用参见例 9.33 所示。

9.6.3　<x:forEach>遍历新浪 RSS 新闻

<x:forEach>标签的功能跟<c:forEach>相似，但<x:forEach>是针对 XML 文件内容的。<x:forEach>标签调用方法如下：

```
<x:forEach  select="expression" [var="varname"] [varStatus="varstatusName"]
[begin="开始"] [end="结束"] [step="step"]]
    Java 程序或者 HTML 代码
</x:forEach>
```

其参数的说明可以参见<c:forEach>标签。

9.6.4　<x:if>标签

<x:if>标签与<c:if>标签相似，但<x:if>标签是判断一个 XPath 表达式的值为真或假。<x:if>调用方法如下：

```
<x:if  select="expression"  [var="varName"]
[scope="{request|page|session|application}"]>
条件为真时执行的语句
```

```
</x:if>
```

其中，参数 select 是要判断的 XPath 表达式，参数 var 代表条件结果的变量，参数 scope 是 var 的有效范围，默认值是 page。

9.6.5 <x:choose>、<x:when>、<x:otherwise>标签

<x:choose>、<x:when>、<x:otherwise>标签与<c:choose>、<c:when>、<c:otherwise>标签相似，但前 3 者标签是判断的是 XPath 表达式。调用方法如下：

```
<x:choose>
    <x:when  select="expression">
          表达式为真时执行的语句
    </x:when>
    [<x:otherwise>
     表达式为假时执行的语句
]
</x:choose>
```

其中，参数 select 是要判断的 XPath 表达式。

9.6.6 <x:set>标签

<x:set>标签用于把 XML 文件中 XPath 表达式的值设置一个变量，调用方法如下：

```
  <c:set select="expression"  var="varname"
[scope="request|page|session|application"]/>
```

在上述语句中，select 是 XPath 表达式，参数 var 代表 XPath 表达式值的变量，scope 是参数 var 的有效范围即作用域。

用<x:set>标签设定值之后，用<c:out>标签输出。

9.6.7 <x:transform>转化 XML 为 HTML

<x:transform>标签可以将 XML 文档转化为 HTML 格式，其调用方法如下：

```
  <x:transform {doc="XMLDocument" docSystemId="docSystemId" | xml="XMLDocument"
xmlSystemId="xmlSystemId"}
  [result="result"] [scope="request|page|session|application"] [var="varname"]
xslt="xslt" [xsltSystemId=""]/>
```

参数说明见表 9.13 所示。

表 9.13　<x:transform>参数说明

参数	说明
参数 xml	需要转化的 XML 文档
参数 doc	需要转化的 XML 文档
参数 docSystemId	XML 文档的 URI
参数 xmlSystemId	XML 文档的 URI
参数 xslt	XSLT 样式表
参数 xsltSystemId	源 XSLT 文档的 URI
参数 result	转换结果的对象
参数 var	代表被转换的 XML 文档的变量
参数 scope	参数 var 的作用范围

xsl 是可扩展样式表语言（EXtensible Stylesheet Language），它包括 3 个部分：XSLT（一种用于转换 XML 文档的语言）、XPath、XSL-FO（一种用于格式化 XML 文档的语言）。

9.7 上机实践

1. 现有一报文信息如下：

```
username=Smith|password=aaaaa|age=30|number=15920578XXX
```

编写一个 JSP 程序，并应用 JSTL 相关标签对上述报文进行解析并显示解析结果。

2. 使用数据库标签对数据库表进行增、删、改、查操作。

3. 编写一个 XML 文件，应用 XML 标签对其解析遍历。

第 10 章

实现自定义标签

在 JSP 页面中，最为理想的代码结构是页面中不含有 Java 代码，只含有 HTML 代码和部分标签代码，Java 代码只存在于业务逻辑处理的后台中。在上一章节中介绍了 JSTL 标签使得 JSP 中的 Java 代码得到简化，页面逻辑更加清晰，本章介绍 JSP 的自定义标签，通过本章的介绍学习，可以做到 JSP 页面有标签组成，不留下 Java 代码。

在实际的应用中，自定义标签应用最多的是自定义分页标签，读者可以根据分页原理自定义出一份分页标签，这样在实现应用时可以减少大量的开发工作。如果读者需要进行更为高端的开发（比如开发自己的框架或者中间件等）时，那么自定义标签就显得尤为重要了。

10.1 编写自定义标签

所谓自定义标签就是有开发者自己定义的标签，该标签具有某些特殊功能。在 JSP 页面中使用自定义标签，可以实现在页面中无任何 Java 代码但却能实现业务的作用，它可以使 JSP 页面成为一个完完全全有各种标签组成的 HTML 文件。

本节开始介绍如何自定义标签，以便读者对自定义标签有个感性的认识。

10.1.1 版权标签

版权标签就是在 JSP 页面中显示出版权信息。新建一个新的自定义标签步骤如下：

（1）编写自定义标签实现类

CopyRightTag 类实现了生成版权信息并输出的功能：

```
-----------------------CopyRightTag.java---------------------------
public class CopyRightTag extends TagSupport{

    @Override
    public int doStartTag(){
     String copyRight = "清华版权所有  &copy2015";
     try {
            pageContext.getOut().print(copyRight);
        } catch (IOException e) {
```

```
                e.printStackTrace();
            }
        return EVAL_PAGE;
        }

    public int doEndTag(){

        return EVAL_PAGE;
        }
    }
```

CopyRightTag 类若不是在编译器中编写，则需要进行编译，然后将编译好的 class 文件放在项目的 WEB-INF\classes 下面。

（2）配置标签

自定义标签实现类编译完成后，还需要配置标签，这跟配置 JSTL 是一致的，只不过自定义标签需要自己编写 tld 文件。自定义标签的 tld 文件需要保存在项目的 WEB-INF\目录下，为了管理方便可以在目录下新建文件夹 tlds 文件夹。

新建 copyright.tld 文件如下：

```
-----------------------copyright. tld--------------------------
<?xml version="1.0" encoding="UTF-8" ?>
<!DOCTYPE taglib
  PUBLIC "-//Sun Microsystems, Inc.//DTD JSP Tag Library 1.2//EN"
  "http://java.sun.com/dtd/web-jsptaglibrary_1_2.dtd">
<taglib>
 <tlib-version>1.0</tlib-version>
 <jsp-version>1.2</jsp-version>
 <short-name>lms</short-name>
    <uri>/copyr-tags</uri>

 <tag>
        <!-- 自定义标签的的名称,在页面中通过它来使用标签 -->
     <name>copyright</name>
     <!-- 自定义标签的实现类路径 -->
     <tag-class>com.eshore.CopyRightTag</tag-class>
     <!-- 正文内容类型  没有正文内容用 empty 表示 -->
     <body-content>empty</body-content>
     <!-- 自定义标签的功能描述 -->
     <description>自定义版权</description>
 </tag>
</taglib>
```

上述文件中有且只有一对 taglib 标签，而 taglib 标签下可以有多个 tag 标签，每个 tag 标签代表一个自定义标签含义。

上述自定义标签的含义是：定义一个 copyrght 的自定义标签，其对应的实现类是 com.eshore.CopyRightTag，该标签不需要标签内容，并且没有指定属性。

标签文件的后缀为.tld，新建完成后要保存在 WEB-INF 文件夹下。

（3）使用自定义标签

最后一步就是如何使用标签，实际上使用自定义标签跟使用 JSTL 标签是一样的。

以下就是使用上述 copyright 标签的示例。

【例 10.1】一个简单的 copyright 标签示例，输出版本内容。

copyright.jsp 是引用标签示例，源代码如下：

```
-----------------------copyright.jsp--------------------------
01  <%@ page language="java" import="java.util.*" pageEncoding="UTF-8"%>
02  <%@ taglib prefix="linl" uri="/copyr-tags" %>
03  <!DOCTYPE HTML PUBLIC "-//W3C//DTD HTML 4.01 Transitional//EN">
04  <html>
05    <head>
06      <title>自定义版本标签示例</title>
07    </head>
08
09    <body>
10            <p>这里是正文的内容</p>
11            <linl:copyright/>
12    </body>
13  </html>
```

上述代码第 02 行代码就是引用自定义标签 copyright，引用 copyright.tld 标签库时通过 prefix 属性指定了标签前缀 linl，那么在页面中使用自定义标签时就可以直接使用标签<linl:copyright>来引用标签库 copyright.tld 中的 copyright 标签。页面运行结果如图 10.1 所示。

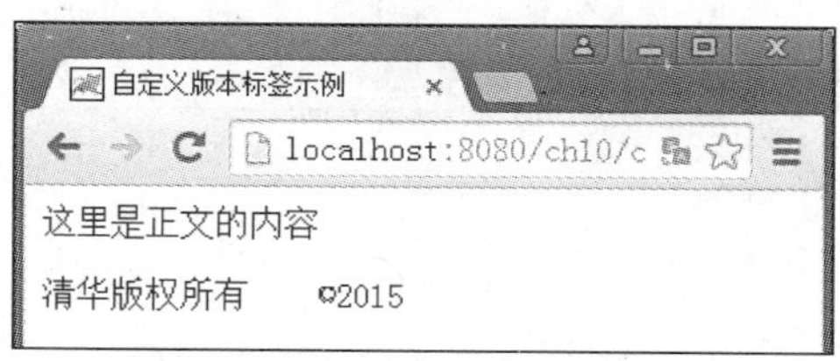

图 10.1　copyright.jsp 页面运行结果

经过以上 3 个步骤，一个完整的自定义标签就设置完成并成功应用于页面中。在后续的例子中也是按照这 3 个步骤进行的，但是不会进行详细的描述标签的属性等内容。

10.1.2　tld 标签库描述文件

tld 标签库描述文件实质是采用 XML 文件格式进行描述的，tld 文件中常用的元素有 taglib、

tag、attribute 和 variable。以下以 copyright.tld 为例逐个描述其含义。

（1） 标签库元素<taglib>

<taglib>元素是用来设置整个标签库信息的，其属性说明见表 10.1 所示。

表 10.1 标签库元素中属性说明

属性	说明
tlib-version	标签库版本号
jsp-version	JSP 版本号
short-name	当前标签库的前缀
uri	页面引用的自定义标签 uri 地址
name	自定义标签名称
tag-class	自定义标签实现类路径
body-content	自定义标签正文内容（也称标签体）类型，若无则写 empty
description	自定义标签的功能描述
attribute	自定义标签功能指定属性，可以有多个

（2）标签元素<tag>

<tag>元素用来定义标签具体的内容，其说明见表 10.2 所示。

表 10.2 tag 元素属性说明

属性	说明
name	自定义标签名称
tag-class	自定义标签实现类
body-content	自定义标签正文内容（也称标签体）类型，有 3 个值：empty（表示无标签体）、JSP（表示标签体可以加入 JSP 程序代码）、tagdependent（表示标签体中的内容由标签自己处理）
description	自定义标签的功能描述
attribute	自定义标签功能指定属性，可以有多个
variable	自定义标签的变量属性

（3）标签属性元素<attribute>

<attribute>元素用来定义标签<tag>中的属性，其说明见表 10.3 所示。

表 10.3 attribute 元素属性说明

属性	说明
name	属性名称
description	属性描述
required	属性是否是必须的，默认为 false
rtexprvalue	属性值是否支持 JSP 表达式
type	定义该属性的 Java 类型。默认为 String
fragment	如果声明了该属性,属性值将被视为一个 JspFragment

在编写 attribute 属性时还要注意元素顺序

（4） 标签变量元素<variable>

<variable >元素用来定义标签<tag>中的变量属性，其说明见表 10.4 所示。

表 10.4　variable 元素属性说明

属性	说明
declare	变量声明
description	变量描述
name-from-attribute	指定的属性名称，其值为变量，在调用 JSP 页面可以使用的名字
name-given	变量名（标签使用时的变量名）
scope	变量的作用范围，NESTED（开始和结束标签之间）、AT_BEGIN（从开始标签到页面结束）、AT_END（在结束标签之后到页面结束）
variable-class	变量的 Java 类型，默认 String

10.1.3　TagSupport 类简介

在 JSP 1.0 中，自定义标签库的实现类大多继承 TagSupport 类来实现自身的方法，其实现了 Tag 接口，有 4 个重要的方法见表 10.5 所示。

表 10.5　TagSupport 类的 4 个重要方法

方法	说明
int doStartTag()	遇到自定义标签开始时调用该方法， 有 2 个可选值：SKIP_BODY(表示不用处理标签体，直接调用 doEndTag()方法)、EVAL_BODY_INCLUDE(正常执行标签体，但不对标签体做任何处理)
int doAfterBody()	重复执行标签体内容的方法， 有 2 个可选值：SKIP_BODY（表示不用处理标签体，直接调用 doEndTag()方法）、EVAL_BODY_AGAIN（重复执行标签体内容）
int doEndTag()	遇到自定义标签结束时调用该方法， 有 2 个可选值：SKIP_PAGE（忽略标签后面的 JSP 内容，中止 JSP 页面执行）、EVAL_PAGE（处理标签后，继续处理 JSP 后面的内容）
void release()	释放获得的资源

TagSupport 类中各方法执行过程如下：

- 当页面中遇到自定义标签的开始标记，先建立一个标签处理对象，JSP 容器回调 setPageContext()方法，然后初始化自定义标签的属性值；
- JSP 容器运行 doStartTag()方法，如果该方法返回 SKIP_BODY，则表示 JSP 忽略此标签主体的内容；如果返回 EVAL_BODY_INCLUDE，则表示 JSP 容器会执行标签主体的内容，接着运行 doAfterBody()方法；
- 运行 doAfterBody()方法若返回 EVAL_BODY_AGAIN，则表示 JSP 容器再次执行标

签主体的内容；返回 SKIP_BODY，JSP 容器将会运行 doEndTag()方法；

- 运行 doEndTag()方法，返回 SKIP_PAGE，则表示 JSP 容器会忽略自定义标签之后的 JSP 内容；返回 EVAL_PAGE，则运行自定义标签以后的 JSP 内容。

TagSupport 类的生命周期见图 10.2 所示。

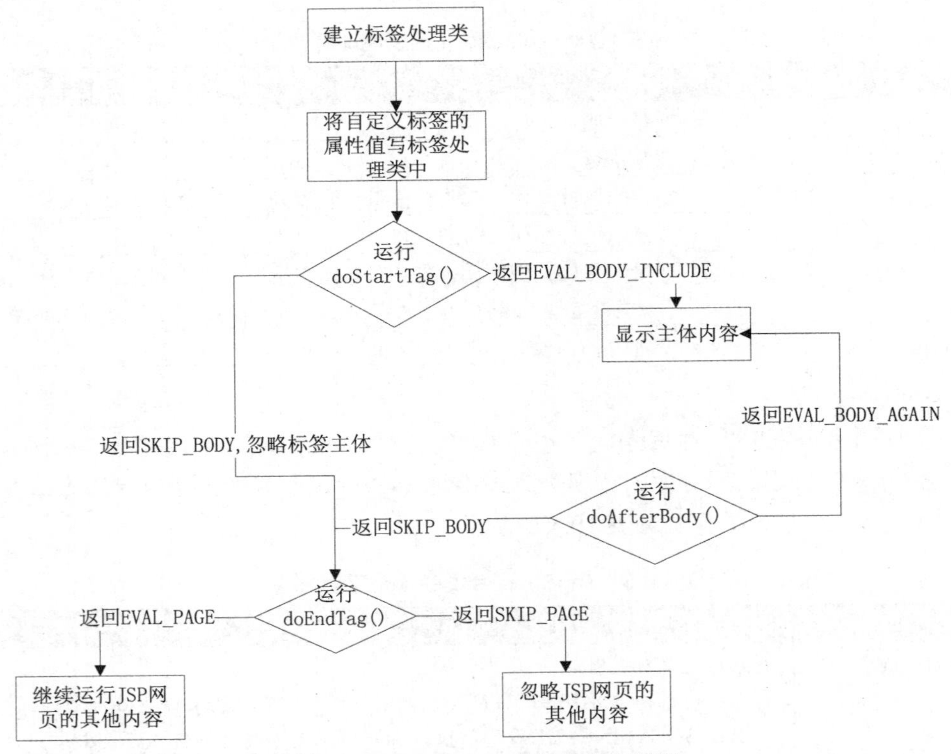

图 10.2　TagSupport 类的生命周期图

10.1.4　带参数的自定义标签

在自定义标签中，如果可以提供参数支持，那么自定义标签的应用范围就会大幅度提高。例如，对于上面显示版权标签的示例，如果能通过参数配置来指定版权所有人的名字，那么这个标签库就可以被任意使用，它将具有通用性和可配置性。

类似的，如果版权信息中的年份也可以进行指定，那么这个版权标签就不需要每年进行更新。假设满足上述需求的自定义标签的名称为 copyrightV，其属性 user 用来指定版权所有人，startY 用来指定版权开始年份，那么在 JSP 页面中使用标签 copyrightV 的代码如下：

```
<%@ page language="java" import="java.util.*" pageEncoding="UTF-8"%>
<%@ taglib prefix="linl" uri="/copyr-tags" %>
<!DOCTYPE HTML PUBLIC "-//W3C//DTD HTML 4.01 Transitional//EN">
<html>
  <head>
```

```
        <title>自定义版本标签示例</title>
    </head>

    <body>
            <p>这里是正文的内容</p>
            <linl:copyrghtV user="清华" startY="2014"/>
    </body>
</html>
```

1. 定义自定义标签的参数

显然，自定义标签如果需要支持参数，那么必须在标签定义时添加参数。并在 TLD 文件中添加参数属性。

当前要实现的 copyrightV 标签要有 user 和 startY 两个属性，因此在 tag 标签中必须添加两个 attribute 属性标签，而每个 attribute 属性标签则通过 name 标签指定属性的名字。

copyrightV.tld 文件的完整定义如下：

```
----------------------- copyrightV. tld---------------------------
<?xml version="1.0" encoding="UTF-8" ?>
<!DOCTYPE taglib
  PUBLIC "-//Sun Microsystems, Inc.//DTD JSP Tag Library 1.2//EN"
  "http://java.sun.com/dtd/web-jsptaglibrary_1_2.dtd">
<taglib>
    <tlib-version>1.0</tlib-version>
    <jsp-version>1.2</jsp-version>
    <short-name>linl</short-name>
   <uri>/copyright-tags</uri>

    <tag>
        <name>copyright</name>
        <!-- 自定义标签的实现类路径 -->
        <tag-class>com.eshore.CopyRightTag</tag-class>
        <!-- 正文内容类型  没有正文内容用 empty 表示 -->
        <body-content>empty</body-content>
        <!-- 自定义标签的功能描述 -->
        <description>自定义版权</description>
        <!-- 添加 attribute 属性 -->
        <attribute>
           <name>user</name>
           <required>true</required>
           <rtexprvalue>true</rtexprvalue>
           <type>java.lang.String</type>
           <description>版权拥有者</description>
        </attribute>
         <attribute>
```

```
28              <name>startY</name>
29              <required>true</required>
30              <rtexprvalue>true</rtexprvalue>
31              <type>java.lang.String</type>
32              <description>开始时间</description>
33          </attribute>
34      </tag>
35  </taglib>
```

上述代码中第 20~32 行是添加属性的代码，并配置其属性的各个元素，它们都是必须的元素，type 设置成 String 类型。至此，带参数的自定义标签的 TLD 文件配置完成。

2. 定义带参数的自定义标签实现类

要实现带参数的自定义标签，还要在实现类中添加相应的属性代码。在处理类代码中，通过增加属性的 set 方法，系统就可以自动将标签中的属性值传递给标签类的实例。CopyRightTag2 类相对于 CopyRightTag 类增加了两个属性，因此需要增加两个私有成员变量，并分别给它们实现 set 方法。处理类 CopyRightTag2 的完整代码如下：

```
----------------------CopyRightTag2.java---------------------------
package com.eshore;

import java.io.IOException;
import java.text.SimpleDateFormat;
import java.util.Date;
import javax.servlet.jsp.tagext.TagSupport;

public class CopyRightTag2 extends TagSupport{

     private String user;          //用户名
     private String startY;        //开始月份
    @Override
    public int doStartTag(){      //开始标签
     SimpleDateFormat sdf = new SimpleDateFormat("yyyy");
     String endY = sdf.format(new Date());
     String copyRight = user+" 版权所有  "+startY+"-"+endY;
     try {
            pageContext.getOut().print(copyRight);
        } catch (IOException e) {
            e.printStackTrace();
        }
     return EVAL_PAGE;
    }
    public int doEndTag(){  //结束标签

```

```
        return EVAL_PAGE;
      }
      public void setUser(String user) {
          this.user = user;
      }
      public void setStartY(String startY) {
          this.startY = startY;
      }

  }
```

完成标签处理类的编写后，经过编译，项目部署后，程序的运行结果如图 10.3 所示，页面调用代码如下。

```
<linl:simpleTagcopyright startY="2014" user="清华"/>
```

处理类未实现变量的 get 方法，是因为自定义标签在实际项目中多是将标签的属性值传递给标签实现类处理，get 方法很少被调用，所以一般不实现 get 方法。

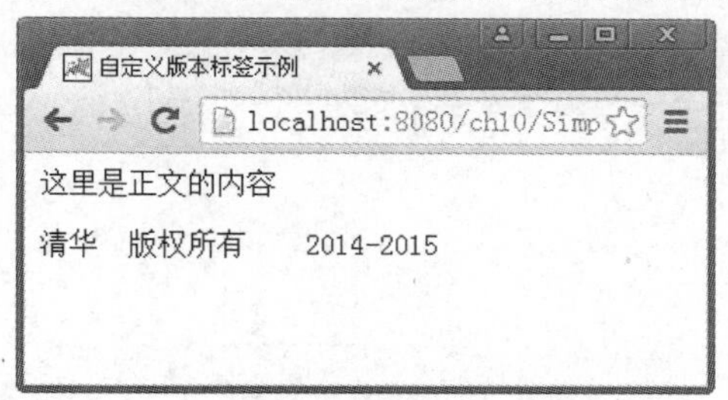

图 10.3　copyrightV.jsp 页面运行结果

10.1.5　带标签体的自定义标签

从 tld 标签库描述文件中，可以看出标签除了名称和属性外，还可以有标签，因为有了标签体，所以自定义标签的灵活性更高了。

标签体的含义是指标签起始标记和标签结束标记之间的内容。若无标签体内容，是将起始标记和结束标记合二为一。

对于带标签体的自定义标签，它的定义和应用方法分步骤说明如下。

（1）定义包含标签体的 tld 文件

要实现自定义标签可以包含标签体，则必须修改标签库中的 tld 定义。如 tld 描述文件中所述，标签定义中的 bodycontent 属性用来说明当前自定义标签的标签体情况，在版权标签中它的值为 empty，它表明版权标签无标签体。如果希望标签体中可以包含页面代码，则可以将其值设置为 JSP。

在标签库中添加版权标签 copyrightBodycontent，将 bodycontent 的值设定为 JSP，其源代码如下：

```
-----------------------copyrightBodycontent. tld----------------------------
<?xml version="1.0" encoding="UTF-8" ?>
<!DOCTYPE taglib
  PUBLIC "-//Sun Microsystems, Inc.//DTD JSP Tag Library 1.2//EN"
  "http://java.sun.com/dtd/web-jsptaglibrary_1_2.dtd">
<taglib>
    <tlib-version>1.0</tlib-version>
    <jsp-version>1.2</jsp-version>
    <short-name>linl</short-name>
   <uri>/copyrightBodycontent-tags</uri>
    <tag>
        <name>copyright</name>
        <!-- 自定义标签的实现类路径 -->
        <tag-class>com.eshore.CopyRightTag3</tag-class>
        <!-- 正文内容类型  允许有 JSP 代码-->
        <body-content>JSP</body-content>
        <!-- 自定义标签的功能描述 -->
        <description>自定义版权</description>
        <!-- 添加 attribute 属性 -->
        <attribute>
          <name>user</name>
          <required>true</required>
          <rtexprvalue>true</rtexprvalue>
          <type>java.lang.String</type>
          <description>版权拥有者</description>
        </attribute>
         <attribute>
          <name>startY</name>
          <required>true</required>
          <rtexprvalue>true</rtexprvalue>
          <type>java.lang.String</type>
          <description>开始时间</description>
        </attribute>
    </tag>
</taglib>
```

将 tld 配置文件放在 WEB-INF/tlds 目录下，标签体的版权标签就定义完成。

（2）定义自定义标签处理类处理标签体

要使得自定义标签能够处理标签体，还需要修改标签处理类。从 TagSupport 类的生命周期可以看出，如果要处理标签体，则需要重写 doAfterBody 方法并将类继承 BodyTagSupport 类。

添加了 doAfterBody 方法的版权标签处理类源代码如下：

```
----------------------CopyRightTag3.java--------------------------
package com.eshore;

import java.io.IOException;
import java.text.SimpleDateFormat;
import java.util.Date;

import javax.servlet.jsp.JspWriter;
import javax.servlet.jsp.tagext.BodyContent;
import javax.servlet.jsp.tagext.BodyTagSupport;

public class CopyRightTag3 extends BodyTagSupport{

    private String user;        //用户名
    private String startY;      //开始月份
   @Override
   public int doStartTag(){     //开始标签
    SimpleDateFormat sdf = new SimpleDateFormat("yyyy");
    String endY = sdf.format(new Date());
    String copyRight = user+" 版权所有  "+startY+"-"+endY;
    try {
            pageContext.getOut().print(copyRight);
        } catch (IOException e) {
            e.printStackTrace();
        }
    return EVAL_PAGE;
   }
   public int doAfterBody(){              //取得标签体

    BodyContent bc = getBodyContent();
    JspWriter out = getPreviousOut();
    try{
        //将标签体中的内容写入到 JSP 页面中
        out.write(bc.getString());
    }catch(IOException e){
        e.printStackTrace();
    }
    return SKIP_BODY;
   }
   public int doEndTag(){           //结束标签
    return EVAL_PAGE;
   }
    public void setUser(String user) {
        this.user = user;
```

```
44      }
45      public void setStartY(String startY) {
46          this.startY = startY;
47      }
48
49  }
```

上述代码第 27~38 行，添加 doAfterBody()方法，该方法获取标签体内容，将标签体内容写到 JSP 页面中。至此，处理类的程序逻辑编写完成。

（3）使用带标签体的自定义标签

【例 10.2】使用带标签体的自定义标签

copyrightBodycontent.jsp 是使用带标签体的版本标签，源代码如下：

```
----------------------- copyrightBodycontent.jsp--------------------------
01  <%@ page language="java" import="java.util.*" pageEncoding="UTF-8"%>
02  <%@ taglib prefix="linl" uri="/copyrightBodycontent-tags" %>
03  <!DOCTYPE HTML PUBLIC "-//W3C//DTD HTML 4.01 Transitional//EN">
04  <html>
05    <head>
06      <title>自定义版本标签示例</title>
07    </head>
08
09    <body>
10            <p>这里是正文的内容</p>
11        < linl:copyright startY="2014" user="清华">
12            <a href="http://www.sina.com">新浪网</a>
13          </linl:copyright>
14    </body>
15  </html>
```

上述代码第 12 行加入一个超链接，还可以加入其他 JSP 代码，它们都能被浏览器解析，并在页面上很好的展示出来。其运行结果如图 10.4 所示。

图 10.4 copyrightBodycontent.jsp 页面运行结果

10.1.6　多次执行的循环标签

自定义标签中的循环标签是指当标签执行 doAfterBody()方法的时候，其返回值是 EVAL_BODY_AGAIN（重复执行标签体内容）。

【例 10.3】自定义标签的循环标签

循环标签的实现类源代码如下：

```
---------------------- CopyRightTag3.java---------------------------
package com.eshore;
import java.io.IOException;
import java.text.SimpleDateFormat;
import java.util.Date;
import javax.servlet.jsp.JspWriter;
import javax.servlet.jsp.tagext.BodyContent;
import javax.servlet.jsp.tagext.BodyTagSupport;

public class CopyRightTag3 extends BodyTagSupport{
     private static final long serialVersionUID = 1L;
     private int time;
    @Override
    public int doStartTag(){          //开始标签
     return EVAL_BODY_INCLUDE;
    }
    public int doAfterBody(){         //标签体内容
     if(time>1){
         time--;
         return EVAL_BODY_AGAIN;
     }else{
         return SKIP_BODY;
     }
    }
    public int doEndTag(){       //结束标签
     JspWriter out = pageContext.getOut();
     try {
             out.print("");
         } catch (IOException e) {
             e.printStackTrace();
         }
     return EVAL_PAGE;
    }
```

```
37
38      public void setTime(int time) {
39          this.time = time;
40      }
41  }
```

上述代码中第 18~26 行是当次数大于 5 时，就循环执行页面的内容，否则忽略标签体的内容。

将下面代码加入到 copyrightBodycontent.tld 标签文件中。

```
01  <tag>
02          <name>loop</name>
03          <!-- 自定义标签的实现类路径 -->
04          <tag-class>com.eshore.CopyRightTag3</tag-class>
05          <!-- 正文内容类型  允许有 JSP 代码-->
06          <body-content>JSP</body-content>
07          <!-- 自定义标签的功能描述 -->
08          <description>自定义版权</description>
09          <!-- 添加 attribute 属性 -->
10          <attribute>
11             <name>time</name>
12             <required>true</required>
13             <rtexprvalue>true</rtexprvalue>
14             <type>java.lang.Integer</type>
15             <description>循环次数</description>
16          </attribute>
17      </tag>
```

上述代码中第 10~16 行是增加 time 属性值参数，因为实现类中的 time 是 int 类型，所以 type 用 Integer 类。

copyright3.jsp 页面是输出循环标签页面，源代码如下：

```
-----------------------copyright3.jsp--------------------------
<%@ page language="java" import="java.util.*" pageEncoding="UTF-8"%>
<%@ taglib prefix="linl" uri="/copyrightBodycontent-tags" %>
<!DOCTYPE HTML PUBLIC "-//W3C//DTD HTML 4.01 Transitional//EN">
<html>
  <head>
    <title>自定义版本标签示例</title>
  </head>

  <body>
        <p>这里是正文的内容</p>
        <linl:loop time="5" >
            <a href="http://www.sina.com">新浪网</a><br/>
        </linl:loop>
```

```
    </body>
  </html>
```

上述代码中表示循环 5 次输出标签体的内容，其运行结果如图 10.5 所示。

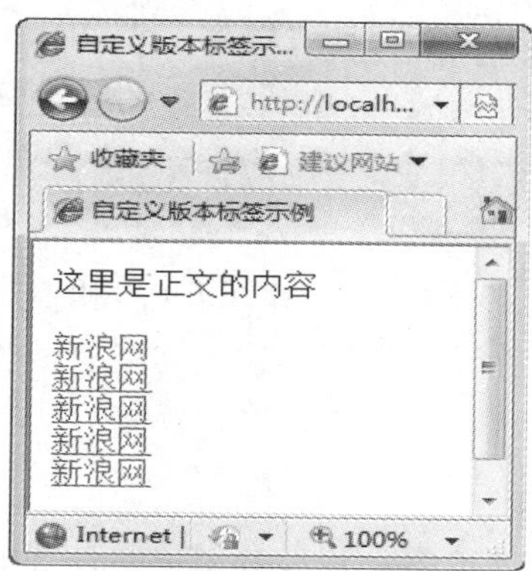

图 10.5　copyright.jsp 页面运行结果

10.1.7　带动态属性的自定义标签

前面都是在标签中直接输入属性值，如果这个属性值可以动态的输入，这样更符合实际的开发需求。以例 10.3 为例在页面 copyright.jsp 中将标签<linl:loop>中的 time 值变成动态的输入，源代码如下：

```
01  <%@ page language="java" import="java.util.*" pageEncoding="UTF-8"%>
02  <%@ taglib prefix="linl" uri="/copyrightBodycontent-tags" %>
03  <!DOCTYPE HTML PUBLIC "-//W3C//DTD HTML 4.01 Transitional//EN">
04  <html>
05    <head>
06      <title>自定义版本标签示例</title>
07    </head>
08    <%
09
10        int num = (int)(Math.random()*6)+1;
11     %>
12    <body>
13           <p>这里是正文的内容</p>
14           <linl:loop time="<%=num %>" >
15                <a href="http://www.sina.com">新浪网</a><br/>
16           </linl:loop>
17    </body>
18  </html>
```

上述代码第 8~11 行是产生 1~6 的随机数，那么循环的次数就根据随机数而定。

10.2 嵌套的自定义标签

嵌套的自定义标签是指自定义的标签相互嵌套，例如以下形式：

```
<linl:table var="item" items="${users}">
 <linl:showUserInfo user="${item}" />
</linl:table>
```

从上述的形式可以看出有个迭代标签<linl:table>，和输出标签<linl:showUserInfo>，那么要分别建立这两个标签，下面来叙述下如何创建上述嵌套标签。

10.2.1 实例：表格标签

先建立实体类和对应的标签处理类，假设有用户实体类如下：

```
public class UserInfo {
 private String userName;                                    //用户名
 private int age;                                            //年龄
 private String email;                                       //用户邮箱

 public UserInfo(String userName, int age, String email) {  //构造函数
     super();
     this.userName = userName;
     this.age = age;
     this.email = email;
 }
 public UserInfo() {
     super();

 }
 //省去属性的 get 和 set 方法
}
```

上述代码中省略了属性的 get 和 set 方法，为了篇幅大小原因在不是必要的情况下本文都省略 get 和 set 方法，请读者在编辑的代码是自行加上。

表格标签的处理类 UserInfoTag.java 如下：

```
-----------------------UserInfoTag.java--------------------------
import javax.servlet.jsp.JspException;
import javax.servlet.jsp.JspWriter;
import javax.servlet.jsp.tagext.TagSupport;

public class UserInfoTag extends TagSupport{
```

```
06
07      private static final long serialVersionUID = 1L;
08      private UserInfo user;
09
10     @Override
11     public int doStartTag() throws JspException {        //开始标签
12
13         try {
14             if(user == null) {
15                 out.println("No UserInfo Found...");
16                 return SKIP_BODY;
17             }
18             String content = "<td>"+user.getUserName()+"</td>";
19             content+="<td>"+user.getAge()+"</td>";
20             content+="<td>"+user.getEmail()+"</td>";
21             this.pageContext.getOut().println("<tr>"+content+"</tr>");
22
23         } catch(Exception e) {
24             throw new JspException(e.getMessage());
25         }
26         return SKIP_BODY;
27     }
28
29     @Override
30     public int doEndTag() throws JspException {      //结束标签
31         return EVAL_PAGE;
32     }
33
34     @Override
35     public void release() {                          //释放资源
36
37         super.release();
38         this.user = null;
39
40     }
41     //省去 user 属性的 get 和 set 方法
42 }
```

上述代码中第 14~21 行获得用户对象的值，并将值输出到页面中。

循环迭代标签 TableTag.java 的处理类如下：

```
----------------------TableTag.java--------------------------
public class TableTag extends TagSupport {

    private static final long serialVersionUID = 1L;
    private Collection items;
```

```
05      private Iterator it;
06      private String var;
07
08      @Override
09      public int doStartTag() throws JspException {        //开始标签
10
11          if(items == null || items.size() == 0) return SKIP_BODY;
12          it = items.iterator();
13          if(it.hasNext()) {
14              pageContext.setAttribute(var, it.next());
15          }
16          return EVAL_BODY_INCLUDE;
17      }
18      @Override
19      public int doAfterBody() throws JspException {       //标签体内容
20          if(it.hasNext()) {
21              pageContext.setAttribute(var, it.next());
22              return EVAL_BODY_AGAIN;
23          }
24          return SKIP_BODY;
25      }
26      @Override
27      public int doEndTag() throws JspException {      //结束标签
28          return EVAL_PAGE;
29      }
30      //省去属性的 get 和 set 方法
31  }
```

上述代码中第 9~16 行标签开始是，将集合容器中的内容输出到页面中，第 20~22 行代码中，如果集合还有数据则继续遍历标签体的内容否则返回页面。items 属性值是遍历的集合对象，var 是存放对象的别名。

10.2.2　嵌套标签的配置

嵌套标签的配置与一般的自定义标签配置基本一致，依次配置标签的名称、实现类、标签体、实现值等配置。

表格标签的配置如下：

```
<?xml version="1.0" encoding="UTF-8" ?>
<!DOCTYPE taglib  PUBLIC "-//Sun Microsystems, Inc.//DTD JSP Tag Library 1.2//EN"
  "http://java.sun.com/dtd/web-jsptaglibrary_1_2.dtd">
<taglib>
    <tlib-version>1.0</tlib-version>
    <jsp-version>1.2</jsp-version>
```

```
07      <short-name>linl</short-name>
08      <uri>/table-tags</uri>
09      <!-- 显示表格标签内容 -->
10      <tag>
11          <name>showUserInfo</name>
12          <tag-class>com.eshore.UserInfoTag</tag-class>
13          <body-content>empty</body-content>
14          <attribute>
15              <name>user</name>
16              <required>false</required>
17              <rtexprvalue>true</rtexprvalue>
18          </attribute>
19      </tag>
20      <!-- 表格标签遍历 -->
21      <tag>
22          <name>table</name>
23          <tag-class>com.eshore.TableTag</tag-class>
24          <body-content>JSP</body-content>
25          <!-- 属性 -->
26          <attribute>
27              <name>items</name>
28              <required>false</required>
29              <rtexprvalue>true</rtexprvalue>
30          </attribute>
31          <attribute>
32              <name>var</name>
33              <required>true</required>
34              <rtexprvalue>true</rtexprvalue>
35          </attribute>
36      </tag>
37  </taglib>
```

上述代码中第 26~35 行配置表格标签的 items 和 var 属性，分别设置它们的属性值。

10.2.3　嵌套标签的运行效果

经过上面两小节的介绍，嵌套标签的主体步骤已经完成，剩下如何使用它。实际上，使用嵌套标签也是十分简单的，跟 JSTL 标签是一样的。

【例 10.4】用表格的方式输出用户信息

页面 selfTableTag.jsp 用表格输出用户，其源代码如下：

```
-----------------------selfTableTag.jsp--------------------------
<%@ page language="java" import="java.util.*,com.eshore.*" pageEncoding="UTF-8"%>
<%@ taglib prefix="linl" uri="/table-tags" %>
```

```
03  <!DOCTYPE HTML PUBLIC "-//W3C//DTD HTML 4.01 Transitional//EN">
04  <html>
05    <head>
06      <title>自定义表格标签示例</title>
07    </head>
08    <%
09      //模拟从数据库中取出数据
10      List<UserInfo> users = new ArrayList<UserInfo>();
11      users.add(new UserInfo("张三", 20, "Zhangsan@163.com"));
12      users.add(new UserInfo("李四", 26, "Lisi@sina.com"));
13      users.add(new UserInfo("王五", 33, "Wangwu@qq.com"));
14      pageContext.setAttribute("users", users);
15     %>
16    <body>
17     <center>
18        用户信息<br/>
19     </center>
20      <table width='400px' border='1' align='center'>
21          <tr>
22              <td width='20%'>用户名</td>
23              <td width='20%'>年龄</td>
24              <td>邮箱</td>
25          </tr>
26          <!-- 使用标签输出用户信息 -->
27          <linl:table var="item" items="${users}">
28             <linl:showUserInfo user="${item}" />
29          </linl:table>
30      </table>
31
32    </body>
33  </html>
```

上述代码中第 2 行引入自定义标签，第 9~15 行是模拟从数据库中取得用户信息数据，第 26~29 行使用表格标签。其页面的运行结果如图 10.6 所示。

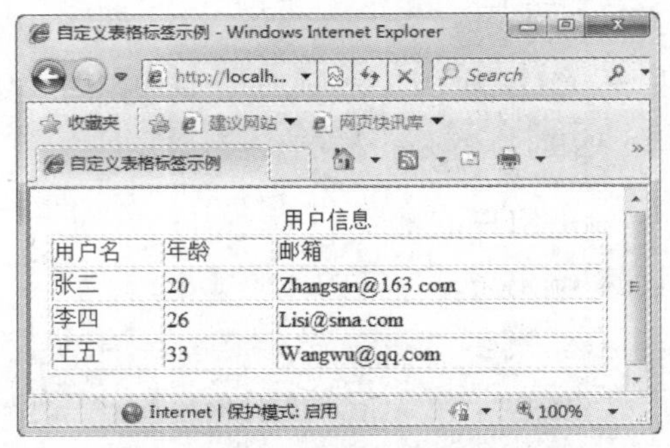

图 10.6　selfTableTag.jsp 页面运行结果

本示例的缺点是表格的列数输出是固定的，实际上应该做成类似 HTML 标记中的 table 标签中的嵌套，有待改善。

10.3 JSP 2.x 标签

JSP 2.X 标签库新增了一新的自定义标签接口类：SimpleTag 接口。该接口极其简单的提供 doTag()方法去处理自定义标签中的逻辑过程、循环体以及标签体的过程。而不像在 JSP1.x 中那么复杂，需要重写 doStartTag()、doAfterBody ()、doEndTag()方法，逻辑处理也简单，实现的接口也相对较少。

在 SimpleTag 接口中还提供 setJspBody()和 getJspBody()方法用于设置 JSP 的相关内容。JSP 容器会依据 setJspBody()方法产生一个 JspFragment 对象，它的基本特点是可以使处理 JSP 的容器推迟评估 JSP 标记属性，一般 JSP 容器是先设定 JSP 标记的属性，然后在处理 JSP 标签时使用这些属性，而 JspFragment 提供了动态属性。这些属性在 JSP 处理标记体时是可以被改变的。JSP 将这样的属性定义为 javax.servlet.jsp.tagext.JspFragment 类型。当 JSP 标记设置成这种形式时，这种标记属性的处理方法类似于处理标记体。

【例 10.5】用 SimpleTag 接口改写版本标签

首先，编写自定义标签处理类 SimpleTagCopyRight，其源代码如下：

```
------------------------SimpleTagCopyRight.java---------------------------
package com.eshore;

import java.io.IOException;
import java.text.SimpleDateFormat;
import java.util.Date;

import javax.servlet.jsp.JspException;
import javax.servlet.jsp.tagext.SimpleTagSupport;

public class SimpleTagCopyRight extends SimpleTagSupport{

    private String user;                                    //用户名
    private String startY;                                  //开始月份
    @Override
   public void doTag() throws JspException, IOException { //开始标签
        SimpleDateFormat sdf = new SimpleDateFormat("yyyy");
        String endY = sdf.format(new Date());
        String copyRight = user+" 版权所有  "+startY+
        "-"+endY;
```

```
        getJspContext().getOut().write(copyRight);
    }
    public void setUser(String user) {
        this.user = user;
    }
    public void setStartY(String startY) {
        this.startY = startY;
    }
}
```

编写 SimpleTag 的自定义标签需继承 SimpleTagSupport 类。

上述代码中，自定义标签只需重写一个 doTag()方法，其余和继承 TagSupport 类相同。

其次，编写 TLD 文件，编写的方法跟前面的例子一样，下面给出其标签文件内容

```
<tag>
    <name>simpleTagcopyright</name>
    <!-- 自定义标签的实现类路径 -->
    <tag-class>com.eshore.SimpleTagCopyRight</tag-class>
    <!-- 正文内容类型  没有正文内容用 empty 表示 -->
    <body-content>empty</body-content>
    <!-- 自定义标签的功能描述 -->
    <description>SimpleTag 版权</description>
    <!-- 添加 attribute 属性 -->
    <attribute>
      <name>user</name>
      <required>true</required>
      <rtexprvalue>true</rtexprvalue>
      <type>java.lang.String</type>
      <description>版权拥有者</description>
    </attribute>
     <attribute>
      <name>startY</name>
      <required>true</required>
      <rtexprvalue>true</rtexprvalue>
      <type>java.lang.String</type>
      <description>版权拥有者</description>
    </attribute>
</tag>
```

将上述内容添加到相应的标签文件中。

SimpleTagcopyright.jsp 页面的源代码如下：

```
<%@ page language="java" import="java.util.*" pageEncoding="UTF-8"%>
<%@ taglib prefix="linl" uri="/copyright-tags" %>
```

```
02  <!DOCTYPE HTML PUBLIC "-//W3C//DTD HTML 4.01 Transitional//EN">
03  <html>
04    <head>
05      <title>自定义版本标签示例</title>
06    </head>
07    <body>
08        <p>这里是正文的内容</p>
09        <linl:simpleTagcopyright startY="2014" user="清华"/>
10    </body>
11  </html>
```

上述代码第 9 行使用版本标签，其实跟原来的方法一样，其运行效果跟图 10.3 一样。综上所述，JSP 2.X 中使用标签的方法更为简单，只需实现 doTag()方法即可，逻辑处理也更为简单。

10.4 上机实践

1. 编写一个自定标签显示自己的签名。
2. 编写处理带标签体的自定义标签。
3. 编写自定义方法。

第 11 章

使用JDBC连接数据库

本章介绍 Web 应用程序如何与数据库通信，包括对数据的 CRUD 的操作。目前，主流的数据库都支持 JDBC，使用 JDBC 连接某个数据库时，必须找到对应数据库的 JDBC 驱动包，这样就能连接到数据库。本文主要是讨论连接 MySQL 数据库，读者可以去 MySQL 官网下载 JDBC 驱动包。

本章将通过示例讲解具体的操作过程，包括对数据库的查询、数据录入、修改、删除等操作，还会介绍对结果集的操作，包括查询多个结果集、可以滚动的结果集、带条件的查询以及元数据的显示等。这些方面基本覆盖 Web 开发中的基本问题和对数据的处理流程。

11.1 JDBC 简介

Java 中主要是使用 JDBC 来访问数据库的，JDBC（Java Database Connectivity)api 是 Java 语言访问数据库的一种规范，是 Java 数据库编程接口，是一组标准的 Java 接口和类，使用这些接口和类，可以访问各种不同的数据库。

JDBC 的接口类位于 java.sql 包中，其常用的类有 java.sql.Connection、java.sql.Statement、java.sql.ResultSet 等。一般的 JDBC 建立分为 4 个步骤：

- 建立数据库的一个连接;
- 执行 SQL 语句;
- 处理数据库返回结果;
- 关闭数据库的连接。

下面以一个查询人员列表实例说明操作 JDBC 过程。

11.1.1 查询实例：列出人员信息

（1）首先建立一张人员信息表，并插入数据。打开 MySQL 控制台或者 MySQL 工具，并执行如下语句：

```
drop database if exists testweb;
create database testweb character set utf8;
use testweb;
drop table if exists person;
```

```
05  create table person(
06     id integer auto_increment comment 'id',
07     name varchar(20) comment '姓名',
08     age integer comment '年龄',
09     sex varchar(10) comment '性别',
10     birthday date comment '出生日期',
11     description text comment '备注',
12     create_time timestamp default current_timestamp(),
13     primary key(id)
14  );
15  insert into person(name,age,sex,birthday,description)
16    values('公孙胜','30','男','1984-02-18','绰号入云龙');
17  insert into person(name,age,sex,birthday,description)
18    values('李逵','26','男','1988-03-18','绰号铁牛');
19  insert into person(name,age,sex,birthday,description)
20    values('柴进','30','男','1984-01-18','绰号小旋风');
21  insert into person(name,age,sex,birthday,description)
22    values('秦明','24','男','1990-06-18','绰号霹雳火');
23  insert into person(name,age,sex,birthday,description)
24    values('林冲','30','男','1984-010-18','绰号豹子头');
```

上述代码中第 1 行建立以 UTF8 为字符集的 testweb 数据库，第 5~14 行建 person 表，第 15~24 行向表中插入数据。

（2）其次，建立 Web 项目，编写相应的 Servlet 类，例如编写 PersonServlet.java 类，源代码如下：

```
public void doPost(HttpServletRequest request, HttpServletResponse
response)
            throws ServletException, IOException {
        Connection con = null;
        Statement st = null;
        ResultSet rs = null;
        try {
            Class.forName("com.mysql.jdbc.Driver");          //注册数据库

         } catch (ClassNotFoundException e) {
            e.printStackTrace();
            System.out.println("驱动程序加载错误");
         }
         try {
            con = DriverManager.
getConnection("jdbc:mysql://localhost:3306/testWeb","root","admin");
                                                         //获取数据库连接
            st = con.createStatement();                     //获取 Statement
```

```
            rs = st.executeQuery("select * from person");
           //执行查询，返回结果集

            response.setContentType("text/html;charset=utf-8");
            PrintWriter out = response.getWriter();
            out .println("<!DOCTYPE HTML PUBLIC \"-//W3C//DTD HTML
                  4.01 Transitional//EN\">");
            out.println("<HTML>");
            out.println("  <HEAD><TITLE>列出人员信息表</TITLE></HEAD>");
            out.println("  <BODY>");
            out.println("<center><h4>人员信息列表</h4>");
            out.println("  <table border=\"1\" width=\"100%\"
                    cellpadding=\"2\" cellspacing=\"1\">");
            out.println("<tr>");
            out.println("<td>选择</td>");
            out.println("<td>姓名</td>");
            out.println("<td>年龄</td>");
            out.println("<td>性别</td>");
            out.println("<td>生日</td>");
            out.println("<td>备注</td>");
            out.println("</tr>");
            while(rs.next()){
                      //遍历结果集 ResultSet
                int id = rs.getInt("id");                          //获取 ID
                String name  = rs.getString("name");              //获取姓名
                int age = rs.getInt("age");                        //获取年龄
                String sex  = rs.getString("sex");                 //获取性别
                Date birthday = rs.getDate("birthday");    //获取生日
                String description = rs.getString("description");
                //获取备注
                out.println("<tr>");
                out.println("<td><input type=\"checkbox\"
                        name=\"id\" value=\""+id+"\"></td>");
                out.println("<td >"+name+"</td>");
                out.println("<td >"+age+"</td>");
                out.println("<td >"+sex+"</td>");
                out.println("<td >"+birthday+"</td>");
                out.println("<td >"+description+"</td>");
               out.println("</tr>");
            }
            out.println("</table></center>");
            out.println("  </BODY>");
            out.println("</HTML>");
            out.flush();
```

```
61                out.close();
62            } catch (SQLException e) {
63
64                e.printStackTrace();
65            }finally{
66                try { //记住关闭连接
67                    rs.close();
68                    st.close();
69                    con.close();
70                } catch (SQLException e) {
71
72                    e.printStackTrace();
73                }
74            }
75        }
```

上述代码中第 7 行注册数据库连接，说明是 MySQL 数据库类型，第 14 行获得 MySQL 驱动，第 19 行获得数据集，第 39~56 行遍历数据结果集，第 66~72 行是关闭数据库连接，数据库连接一定要记得关闭，否则它会一直占着连接池。其运行结果如图 11.1 所示。

上述源代码中，只给出 doPost 方法，其余的 doInit()、doDestory()方法可以为默认的方法。

选择	姓名	年龄	性别	生日	备注
	公孙胜	30	男	1984-02-18	绰号入云龙
	李逵	26	男	1988-03-18	绰号铁牛
	柴进	30	男	1984-01-18	绰号小旋风
	秦明	24	男	1990-06-18	绰号霹雳火
	林冲	30	男	1984-10-18	绰号豹子头

图 11.1　人员信息表效果图

11.1.2　各种数据库的连接

不同数据库的类型虽然不同，但是使用 JDBC 连接的步骤是一样的，只是在获取驱动的 URL 上有所不同。主流数据库的连接见表 11.1 所示。

表 11.1 主流数据库的连接

数据库类型	反射注册	连接 url	备注
MySQL	com.mysql.jdbc.Driver	jdbc:mysql://IP:3306/database	IP 为连接的 IP 地址，默认端口是 3306, database 是连接的数据库
Oracle	oracle.jdbc.driver.OracleDriver	jdbc:oracle:thin:@IP:1521:database	IP 为连接的 IP 地址，默认端口是 1521, database 是连接的数据库
DB2	com.ibm.db2.jdbc.app.DB2Driver	jdbc:db2://IP:6789/database	IP 为连接的 IP 地址，默认端口是 6789, database 是连接的数据库
PostgreSQL	org.postgresql.Driver	jdbc:postgresql://IP:5432/database	IP 为连接的 IP 地址，默认端口是 5432, database 是连接的数据库
Sysbase	com.sybase.jdbc.SybDriver	jdbc:sybase:Tds:IP:5000/ database	IP 为连接的 IP 地址，默认端口是 5000, database 是连接的数据库
SQLServer 2005	com.microsoft.sqlserver.jdbc.SQLServerDriver	jdbc:sqlserver://IP:1433;databaseName=database	IP 为连接的 IP 地址，默认端口是 1433, database 是连接的数据库

11.2 MySQL 的乱码解决

在 MySQL 控制台，显示数据和插入数据有时会出现乱码，出现乱码一般情况都是字符集设定的有问题。通过设定数据库的字符集可以解决这个问题，下面介绍 3 种修改乱码的方法。

11.2.1 MySQL 的乱码解决

MySQL 从 MySQL4 开始支持 UTF-8 等几十种编码方式，其安装的时候默认是 latin1 编码方式，它不支持中文，因此需要修改编码方式。

UTF-8 字符集能够编码目前世界上所有的语言，一般的网站也是采用 UTF-8 编码方式，它的缺点就是比较占空间，解析也比较复杂。但是为了能统一编码格式，都采用 UTF-8 进行编码。

支持中文的编码方式还有 GBK、GB2312、GB18030，GB2312 是专门针对中文的编码方式，但是它支持的中文字符有限；GBK 比 GB2312 支持更多的中文和中文字符；GB18030 不仅可以支持中文字符还可以支持少数民族语言。

11.2.2 从控制台修改编码

从控制台修改编码，可以修改数据库编码、数据表编码、表字段编码。

（1）修改数据库编码，输入如下命令：

```
ALTER DATABASE testweb CHARACTER SET utf8;
```

运行效果如图 11.2 所示。

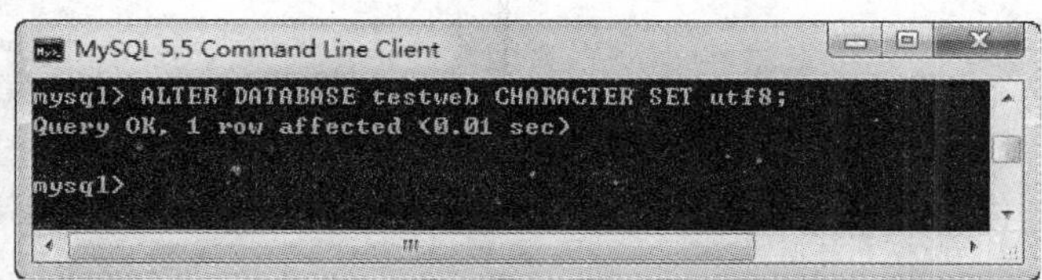

图 11.2 修改数据库编码效果图

（2）修改数据表编码，输入如下命令：

```
ALTER TABLE  person  CHARACTER SET utf8;
```

运行效果如图 11.3 所示。

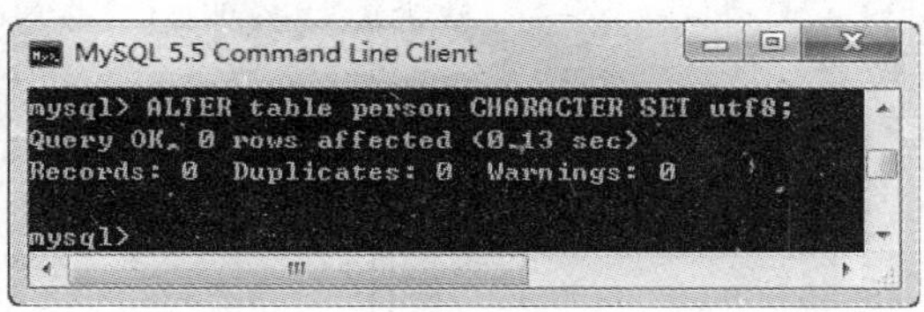

图 11.3 修改数据表编码效果图

（3）修改数据表字段编码，输入如下命令：

```
ALTER TABLE person CHANGE name name VARCHAR(100) CHARACTER SET utf8
```

运行效果如图 11.4 所示。

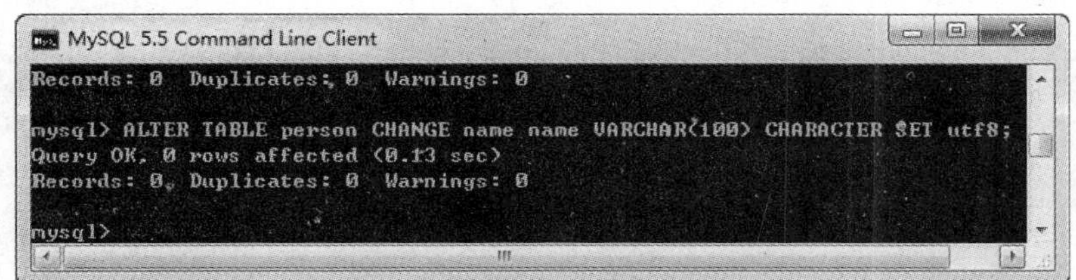

图 11.4 修改数据表字段编码效果图

修改完成可以查看数据库字符集是否修改成功，表字段的字符集是否修改成功。

查看系统字符集命令如下：

```
show variables like 'char%';
```

查看数据库字符集命令如下：

```
show  create database databasename;
```

查看数据表字符集命令如下：

```
show  full columns from tablename;
```

例如：show full columns from person，其效果如图 11.5 所示。

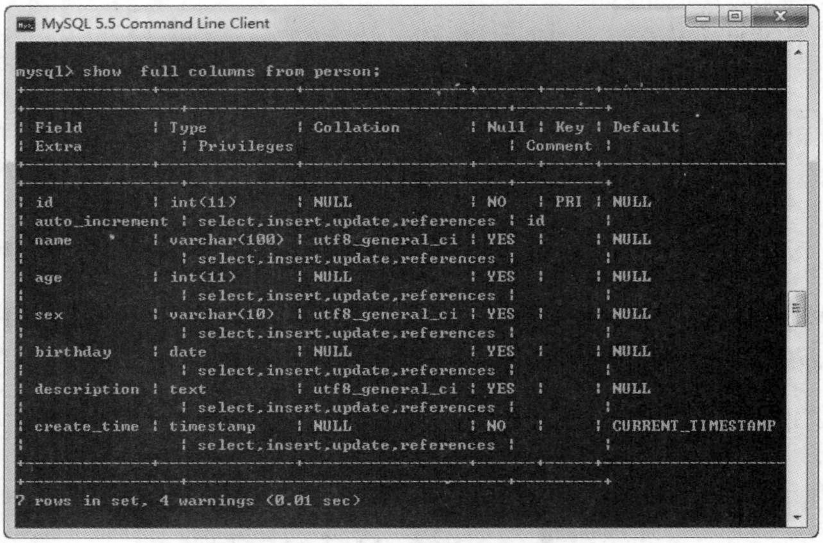

```
MySQL 5.5 Command Line Client

mysql> show  full columns from person;
| Field       | Type         | Collation       | Null | Key | Default
| Extra          | Privileges                      | Comment |
| id          | int(11)      | NULL            | NO   | PRI | NULL
| auto_increment | select,insert,update,references | id      |
| name        | varchar(100) | utf8_general_ci | YES  |     | NULL
|                | select,insert,update,references |         |
| age         | int(11)      | NULL            | YES  |     | NULL
|                | select,insert,update,references |         |
| sex         | varchar(10)  | utf8_general_ci | YES  |     | NULL
|                | select,insert,update,references |         |
| birthday    | date         | NULL            | YES  |     | NULL
|                | select,insert,update,references |         |
| description | text         | utf8_general_ci | YES  |     | NULL
|                | select,insert,update,references |         |
| create_time | timestamp    | NULL            | NO   |     | CURRENT_TIMESTAMP
|                | select,insert,update,references |         |
7 rows in set, 4 warnings (0.01 sec)
```

图 11.5　查看数据表字符集效果图

11.2.3　从配置文件修改编码

上一小节介绍在控制台中修改编码，那么也可以在配置文件中进行永久性的修改数据库编码。用 UE 工具或者记事本打开 MySQL 安装目录下的 my.ini 文件，找到如下语句：

```
default-character-set=latin1
```

和

```
character-set-server= latin1
```

将编码方式 latin1 都更改为 utf8，然后重启 MySQL 服务。

11.2.4　利用图形界面工具修改

利用 MySQL 自带的图形界面工具修改。用图形界面可以修改数据库各种参数。在图形界面中右击表名，在弹出的对话框中选择 Alter Table 属性，在 Collation 中选择字符集，如图 11.6 所示。

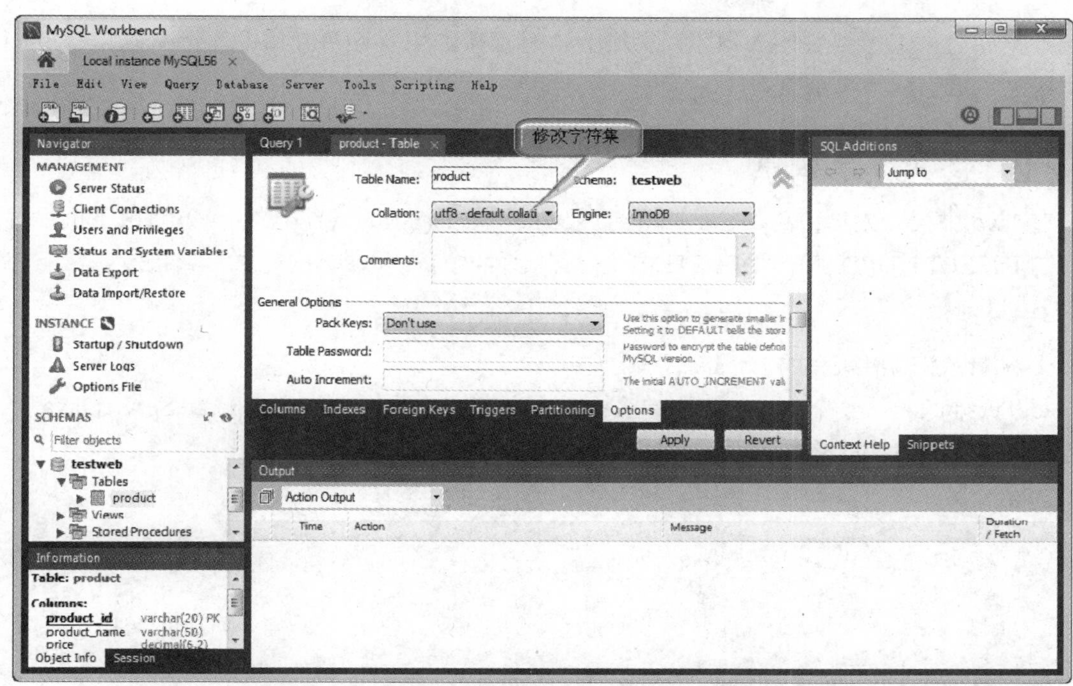

图 11.6　利用图形界面工具修改字符集效果图

11.2.5　URL 中指定编码方式

上面介绍了修改数据库的编码方式，在用 JDBC 连接时，也可以指定编码方式，方法是在连接 URL 后面添加 unicode=true&characterEncoding=UTF-8 就可以了，例如：

```
Jdbc:mysql://localhost:3306/database?unicode=true&characterEncoding=UTF-8
```

一般情况下经过上述步骤是不会有乱码出现了，如果在程序显示中还有乱码，就检查在请求中是否进行了编码转换。如果是在控制台中还出现乱码或者插入不了中文数据，建议重装数据库，重装数据库时一定要删除干净 MySQL 文件夹。

11.3　JDBC 基本操作：CRUD

数据库的操作常被称为 CRUD，其主要是指对数据库进行创建（Create）、查询（Read）、更新（Update）、删除（Delete）等操作。本节将主要介绍利用 JDBC 进行数据库的 CRUD 的基本操作。

11.3.1　查询数据库

在第 11.1.1 中介绍了一个查询人员信息列表的例子，它说明了查询数据库的一般步骤。在查询数据库时，记得要关闭数据库，否则会占有连接，引发连接异常。示例代码请看 11.1.1 节中所述，这里就不再重复叙述。

11.3.2 插入人员信息

在 Java 中插入人员信息，主要是用到 executeUpdate(String sql)方法，该方法用于执行对数据库的 INSERT、UPDATE、DELETE 操作，它返回执行语句影响的行数，如果没有影响的行数则返回 0。

【例 11.1】输出所有的 Locale 代码

addPerson.jsp 是一个人员信息表单页面，在 action 中页面跳转到 OperateServlet.java 类中执行插入数据的后台操作，成功后返回列表信息。addPerson.jsp 页面如下：

```
-----------------------addPerson.jsp---------------------------
<%@ page language="java" import="java.util.*" pageEncoding="UTF-8"%>
<!DOCTYPE HTML PUBLIC "-//W3C//DTD HTML 4.01 Transitional//EN">
<html>
  <head>
    <title>增加人员列表</title>
    <!-- 调用日期控件的js -->
    <script language="javascript"
        type="text/javascript"
     src="${pageContext.request.contextPath}/My97DatePicker/WdatePicker.js"></script>
    <script type="text/javascript"
     src="${pageContext.request.contextPath}/js/jquery-1.8.1.js"></script>
  </head>
  <body>
      <form action="${pageContext.request.contextPath}/servlet/OperateServlet"
         method="post">
           <table>
                 <tr>
                     <td>姓名</td>
                     <td><input name="name"/></td>
                 </tr>
                 <tr>
                     <td>性别</td>
                     <td><select name="sex">
                             <option value="男">男</option>
                             <option value="女">女</option>
                       </select></td>
                 </tr>
                 <tr>
                     <td>年龄</td>
```

```
30                      <td><input name="age"/></td>
31                  </tr>
32                  <tr>
33                      <td>生日</td>
34                      <!-- 调用日期控件 -->
35                      <td><input id="d11" name="datePicker"
36                         type="text" onClick="WdatePicker()"/></td>
37                  </tr>
38                  <tr>
39                      <td>描述</td>
40                      <td><textarea name="description"></textarea></td>
41                  </tr>
42                  <tr>
43                      <td colspan="2">
44                         <input type="submit" value="提交"/>
45                      </td>
46                  </tr>
47              </table>
48          </form>
49      </body>
50  </html>
```

上述代码中，就是叙写基本的表单页面，第 14 行填写跳转的页面或者 servlet，在第 35 行中填写日期控件。

OperateServlet.java 类的 doPost 方法如下：

```
-----------------------OperateServlet.java---------------------------
public void doPost(HttpServletRequest request, HttpServletResponse response)
            throws ServletException, IOException {
        request.setCharacterEncoding("UTF-8");
        String name = request.getParameter("name");
        String age = request.getParameter("age");
        String sex = request.getParameter("sex");
        String birthday = request.getParameter("birthday");
        String description = request.getParameter("description");
        String sql = "insert into person(name,age,sex,birthday,description) "+

         "values('"+name+"','"+age+"','"+sex+"','"+birthday+"','
         "+description+"')";
        Connection con = null;
        Statement st = null;
        int  result = 0;
        try {
            Class.forName("com.mysql.jdbc.Driver");          //注册数据库
            con = DriverManager.
```

```
18                  getConnection("jdbc:mysql://localhost:3306/testWeb","root",
                    "admin");
19                                                          //获取数据库连接
20                  st = con.createStatement();             //获取 Statement
21                  result = st.executeUpdate(sql);         //执行查询，返回结果集
22                  response.setContentType("text/html;charset=utf-8");
23                  PrintWriter out = response.getWriter();
24                  out .println("<!DOCTYPE HTML PUBLIC \"-//W3C//DTD HTML
25                  4.01 Transitional//EN\">");
26                  out.println("<HTML>");
27                  out.println("  <HEAD><TITLE>列出人员信息表</TITLE></HEAD>");
28                  out.println("  <BODY>");
29                  out.println("");
30                  out.println("<a href=\""+request.getContextPath()+
                    "/listPerson.jsp\">
31                   返回人员列表</a>");
32                  out.println("  </BODY>");
33                  out.println("</HTML>");
34                  out.flush();
35                  out.close();
36              } catch (SQLException e) {
37
38                  e.printStackTrace();
39              }catch (ClassNotFoundException e) {
40
41                  e.printStackTrace();
42              }finally{
43                  try {                                               //记住关闭连接
44                      st.close();
45                      con.close();
46                  } catch (SQLException e) {
47
48                      e.printStackTrace();
49                  }
50              }
51      }
```

上述代码中，跟 PersonServlet.java 类差别不大，主要是修改了第 21 行，执行插入方法 executeUpdate()方法。

查询列表的 listPerson.jsp 页面代码如下：

```
-----------------------listPerson.jsp--------------------------
<%@ page language="java" contentType="text/html; charset=UTF-8"%>
<%@ page import="java.sql.*" %>
<!DOCTYPE HTML PUBLIC "-//W3C//DTD HTML 4.01 Transitional//EN">
<HTML>
```

```
	<HEAD>
		<TITLE>人员信息列表</TITLE>
		<script type="text/javascript"
			src="${pageContext.request.contextPath}/js/jquery-
			1.8.1.js"></script>
	</HEAD>
	<BODY>
		<center>
			<h4>人员信息列表</h4>
		</center>
		<%
			Connection con = null;
			Statement st = null;
			ResultSet rs = null;
			try {
				Class.forName("com.mysql.jdbc.Driver");     //注册数据库
				con = DriverManager.
				getConnection("jdbc:mysql://localhost:3306/testWeb",
				"root","admin");
				                                            //获取数据库连接
				st = con.createStatement();                 //获取 Statement
				rs = st.executeQuery("select * from person");
				//执行查询，返回结果集
		%>
		<a href="addPerson.jsp">新增人员信息</a>
		<br/>
		<br/>
		<table border="1" width="100%" cellpadding="2" cellspacing="1">
			<tr>
				<td>选择</td>
				<td>姓名</td>
				<td>年龄</td>
				<td>性别</td>
				<td>生日</td>
				<td>备注</td>
				<td>操作</td>
			</tr>
			<%
				while(rs.next()){
				                                            //遍历结果集 ResultSet
				int id = rs.getInt("id");                   //获取 ID
				String name  = rs.getString("name");        //获取姓名
				int age = rs.getInt("age");                 //获取年龄
				String sex  = rs.getString("sex");          //获取性别
				Date birthday = rs.getDate("birthday");     //获取生日
```

```
                    String description = rs.getString("description");
                    //获取备注
                    out.println("<tr>");
                    out.println("<td><input type=\"checkbox\"
                    name=\"checkPerson\"
                        value=\""+id+"\"></td>");
                        out.println("<td >"+name+"</td>");
                        out.println("<td >"+age+"</td>");
                        out.println("<td >"+sex+"</td>");
                        out.println("<td >"+birthday+"</td>");
                        out.println("<td >"+description+"</td>");
                        out.println("<td><a href=\"modify.jsp?id="+id+"\">修改
                        </a>  "+
                        "<a href=\"delete.jsp?id="+id+"\"
                        onclick=\"return confirm('确定删除该记录？')\">
                        删除</a></td>");
                       out.println("</tr>");
                    }
                %>
        </table>
        <table>
                <tr>
                   <td>
                       全选<input  type="checkbox" onclick="selectPerson(this);"/>

                   </td>
                </tr>
        </table>
        <%
                } catch (SQLException e) {

                    e.printStackTrace();
                }finally{
                    try {//记住关闭连接
                        rs.close();
                        st.close();
                        con.close();
                    } catch (SQLException e) {

                        e.printStackTrace();
                    }
                }
         %>
    </BODY>
    <script type="text/javascript">
```

```
89          function selectPerson(obj){
90              $('input[name="checkPerson"]').attr("checked",obj.checked);
91          }
92      </script>
93  </HTML>
```

在上述代码中，实际上是将 PersonServlet.java 改写成 JSP 页面，并新增加操作列分别有修改、删除操作，代码第 89~91 行 js 中应用 jQuqrry 技术对数据的全选进行选择，在页面中输入访问地址 http://localhost:8080/ch13/addPerson.jsp，并填写人员基本信息，然后提交，执行成功返回 OperateServlet 输出的页面，点击返回人员信息列表超链接，即可显示是否插入数据成功。页面效果如图 11.7 所示。

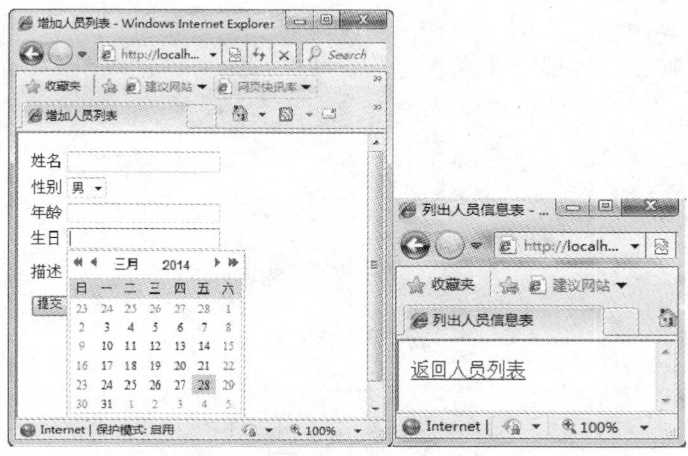

图 11.7 表单页面与 servlet 效果图

11.3.3 注册数据库驱动

注册数据库驱动有两种方式，分别如下：

（1）直接调用 DriverManager 注册：

```
DriverManager.registerDriver(new com.mysql.jdbc.Driver());
```

（2）用 Java 反射机制注册：

```
Class.forName("com.mysql.jdbc.Driver");
```

这两种方法注册数据库驱动都是可行的，在目前的开发过程中运用第 2 种方法的比较普遍。

11.3.4 获取自动插入的 ID

在上述例子中，建数据库表的主键是采用自动增长的数值类型，它由数据库自己去计算与维护，开发者无需关心，但是在对数据进行修改、删除操作时，需要用到主键 ID，那么要获取这些 ID 值，方便对数据的操作。

Statement 提供了 getGeneratedKeys()方法来获取数据库主键的方法，遍历其结果集就可以获得自动插入的 ID。

11.3.5 删除人员信息

删除人员与插入人员信息的 Statement 执行方法是一样，都是用 executeUpdate(sql)方法，只是 SQL 语句不一样，删除的语句大致为 delete from tableName where id=1，其中 where 条件可以自定义，如果不指定条件，则是全表删除。

【例 11.2】删除指定的人员信息

在查询人员信息的 JSP 页面中，有个删除操作，单击删除操作连接，单击“确定”按钮，程序根据用户删除的是哪一条或者哪些条数据，执行删除数据操作。其运行效果如图 11.8 所示。

图 11.8 删除操作选择效果图

delete.jsp 的页面源代码如下

```
-----------------------delete.jsp---------------------------
<%@ page language="java" contentType="text/html; charset=UTF-8"%>
<%@ page import="java.sql.*" %>
<!DOCTYPE HTML PUBLIC "-//W3C//DTD HTML 4.01 Transitional//EN">
<HTML>
    <HEAD>
        <TITLE>删除人员信息</TITLE>
    </HEAD>
    <BODY>
        <%
            Connection con = null;
            Statement st = null;
            try {
             Class.forName("com.mysql.jdbc.Driver");     //注册数据库
```

```
                    con = DriverManager.getConnection(      //获取数据库连接
                    "jdbc:mysql://localhost:3306/testWeb","root","admin");
                    st = con.createStatement();             //获取 Statement
                    request.setCharacterEncoding("UTF-8");
                String id = request.getParameter("id");   //获取页面参数 id
                String sql = "delete from Person where id='"+id+"'";
                //删除指定人员
                 int  result = st.executeUpdate(sql);
              } catch (SQLException e) {

                  e.printStackTrace();
              }finally{
                                                        //记住关闭连接
                  try {
                      st.close();
                      con.close();
                  } catch (SQLException e) {

                      e.printStackTrace();
                  }
              }
          %>
          <a href="listPerson.jsp">返回人员信息列表</a>
          <br/>
      </BODY>
  </HTML>
```

从上述代码中，可以发现删除代码比插入语句简单，只有 SQL 语句发生了变化，其余代码雷同，在页面中执行删除操作，然后返回人员信息列表会发现系统少了一条记录，当然也可以删除多条记录，那么可以编写删除多条记录的方法。

11.3.6　修改人员信息

修改操作是 CRUD 操作中最复杂的操作，一个完整的修改过程需要多个子过程来协助完成，首先它需要先从数据库中查询出数据呈现给用户进行查看修改信息，然后用户提交保存到数据库中。

在本例中，修改人员信息列表的流程如图 11.9 所示。首先，单击页面 listPerson.jsp 的修改操作，提交到 modify.jsp 页面从数据库中查询出数据显示到 update.jsp 上，执行保存方法。

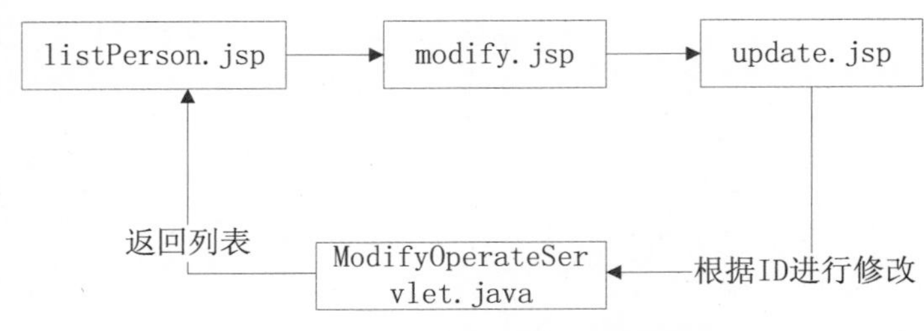

图 11.9　修改人员信息流程图

【例 11.3】修改指定的人员信息

listPerson.jsp 页面是显示人员信息列表，其源代码前面例子已经给出，这里就不再叙述。modify.jsp 是选择某人员，并从查询数据库有无该人员的页面，其源代码如下：

```
------------------------ modify.jsp---------------------------
<%@ page language="java" contentType="text/html; charset=UTF-8"%>
<%@ page import="java.sql.*" %>
<!DOCTYPE HTML PUBLIC "-//W3C//DTD HTML 4.01 Transitional//EN">
<HTML>
    <HEAD>
        <TITLE>修改人员信息</TITLE>
    </HEAD>
    <BODY>
        <%
            Connection con = null;
            Statement st = null;
            ResultSet rs = null;
            try {
                Class.forName("com.mysql.jdbc.Driver");     //注册数据库
                con = DriverManager.getConnection(         //获取数据库连接
                "jdbc:mysql://localhost:3306/testWeb","root","admin");
                st = con.createStatement();          //获取 Statement
                request.setCharacterEncoding("UTF-8");
                                                     //获得修改人员的主键 ID
                String id = request.getParameter("id");
                //查询该人员的 SQL 语句
                String sql = "select *  from Person where id='"+id+"'";
                rs = st.executeQuery(sql);
                if(rs.next()){
                    //往页面传递人员信息参数
                    request.setAttribute("id",rs.getInt("id"));
                    request.setAttribute("name",rs.getString("name"));
                    request.setAttribute("sex",rs.getString("sex"));
                    request.setAttribute("age",rs.getInt("age"));
                    request.setAttribute("birthday",rs.getString
                    ("birthday"));
                    request.setAttribute("description",rs.getString
```

```
                        ("description"));
32                      RequestDispatcher rd = request.getRequestDispatcher
                          ("update.jsp");
33                      rd.forward(request,response);
34                  }else{
35                   out.println("没有找到 id 为"+id+"的人员记录");
36                  }
37              } catch (SQLException e) {
38
39                   e.printStackTrace();
40              }finally{
41                   try {//记住关闭数据库连接
42                       rs.close();
43                        st.close();
44                        con.close();
45                   } catch (SQLException e) {
46
47                        e.printStackTrace();
48                   }
49              }
50          %>
51          <br/>
52      </BODY>
53  </HTML>
```

上述代码中第 14~17 行跟前面例子是一样的，都是连接数据库，第 22 行是组合 SQL 语句，第 23 行执行查询结果语句，如果数据库存在该数据，则往修改页面传递人员信息，如果不存在打印出该人员不存在信息。同样要记得关闭数据库连接。

update.jsp 修改页面和 addPerson.jsp 页面差不多，主要的区别在与修改页面需要将传递的参数显示在页面中，以便客户修改，其源代码如下：

```
----------------------- update.jsp--------------------------
<%@ page language="java" import="java.util.*" pageEncoding="UTF-8"%>
<!DOCTYPE HTML PUBLIC "-//W3C//DTD HTML 4.01 Transitional//EN">
<html>
  <head>
    <title>修改人员</title>
    <!-- 调用日期控件的 js -->
   <script language="javascript" type="text/javascript"
     src="${pageContext.request.contextPath}/My97DatePicker/WdatePicker.
     js"></script>
  </head>
  <%
     //获取传递的参数信息
     Integer id = (Integer)request.getAttribute("id");
```

```
        String name = (String)request.getAttribute("name");
        Integer age = (Integer)request.getAttribute("age");
        String sex = (String)request.getAttribute("sex");
        String birthday = (String)request.getAttribute("birthday");
        String description = (String)request.getAttribute("description");
    %>
  <body>
     <form action="${pageContext.request.contextPath}/servlet
      /ModifyServlet?
         id=<%=id%>" method="post">
         <table>
             <tr>
                 <td>姓名</td>
                 <td><input name="name" value="<%=name %>" />
                 </td>
             </tr>
             <tr>
                 <td>性别</td>
                 <td><select name="sex">
                         <%
                          if("男".equals(sex)){
                        %>
                        <option value="男" selected>男</option>
                        <%
                          }else{
                         %>
                         <option value="男" >男</option>
                         <%
                          }
                          %>
                        <%
                          if("女".equals(sex)){
                        %>
                        <option value="女" selected>女</option>
                        <%
                          }else{
                         %>
                         <option value="女" >女</option>
                         <%
                          }
                          %>
                   </select></td>
             </tr>
             <tr>
                 <td>年龄</td>
```

```
57                      <td><input name="age" value="<%=age%>"/></td>
58                  </tr>
59                  <tr>
60                      <td>生日</td>
61                      <!-- 调用日期控件 -->
62                      <td><input id="d11" name="birthday" type="text"
63                      onClick="WdatePicker()" value="<%=birthday %>"/></td>
64                  </tr>
65                  <tr>
66                      <td>描述</td>
67                      <td><textarea name="description" ><%=description
                        %></textarea>
68                     </td>
69                  </tr>
70                  <tr>
71                      <td colspan="2">
72                         <input  type="submit" value="提交"/>
73                      </td>
74                  </tr>
75              </table>
76          </form>
77    </body>
78  </html>
```

上述代码中，第 10~18 行代码嵌入 Java 代码获取人员参数信息，第 22~69 行在 HTML 中嵌入 JSP 表达式，将人员信息赋值给 HTML 中相应的值，最后用 submit 提交 form 到指定的 ModifyServlet 中。

ModifyServlet.java 是对提交的人员信息更新到数据库操作的 servlet，其 post 方法源代码如下：

```
----------------------- ModifyServlet.java--------------------------
public void doPost(HttpServletRequest request, HttpServletResponse response)
            throws ServletException, IOException {
        request.setCharacterEncoding("UTF-8");
        String id = request.getParameter("id");
        String name = request.getParameter("name");
        String age = request.getParameter("age");
        String sex = request.getParameter("sex");
        String birthday = request.getParameter("birthday");
        String description = request.getParameter("description");
        String sql = "update Person
        set name='"+name+"',age='"+age+"',sex='"+sex+"',birthday='"
        +birthday+"',description='"+description+"' where id='"+id+"'";
        Connection con = null;
        Statement st = null;
```

```
15          int  result = 0;
16          try {
18              Class.forName("com.mysql.jdbc.Driver");          //注册数据库
19              //DriverManager.registerDriver(new com.mysql.jdbc.
                Driver());
20              con =DriverManager.getConnection(           //获取数据库连接
21              "jdbc:mysql://localhost:3306/testWeb","root","admin");
22                                                   //获取 Statement
23                  st = con.createStatement();
24                  System.out.println(sql);
25                  result = st.executeUpdate(sql);     //执行查询，返回结果集
26                  response.setContentType("text/html;charset=utf-8");
27                  PrintWriter out = response.getWriter();
28                  out .println("<!DOCTYPE HTML PUBLIC \"-//W3C//DTD HTML
29                  4.01 Transitional//EN\">");
30                  out.println("<HTML>");
31                  out.println("  <HEAD><TITLE>列出人员信息表</TITLE></HEAD>");
32                  out.println("  <BODY>");
33                  out.println("");
34                  out.println("<a href=\""+request.getContextPath()+
                    "/listPerson.jsp\">
35                 返回人员列表</a>");
36                  out.println("  </BODY>");
37                  out.println("</HTML>");
38                  out.flush();
39                  out.close();
40            } catch (SQLException e) {
41
42                  e.printStackTrace();
43            }catch (ClassNotFoundException e) {
44
45                  e.printStackTrace();
46            }finally{
47                  //记住关闭连接
48                  try {
49                      st.close();
50                      con.close();
51                  } catch (SQLException e) {
52
53                      e.printStackTrace();
54                  }
55            }
56        }
```

上述代码中第4~9行是获取页面提交的参数值，第10~12行组合更新的SQL语句，第18~21

行注册数据库连接，第 29~41 行输出页面信息，第 49~52 行关闭数据库连接。执行成功后返回列表信息查看是否更新成功，update.jsp 的运行效果如图 11.10 所示。

图 11.10　update.jsp 更新页面效果

提交完，返回列表页面发现该人员已经修改成功。

修改操作是 CRUD 操作中最复杂的，因此需要注意流程的顺序以及信息的传递。

11.3.7　使用 PreparedStatement

上述代码中使用 Statement 来处理数据库的操作，还可以用 Statement 子类 PreparedStatement 来处理数据的 CRUD。

PreparedStatement 优点在于允许使用不完整的 SQL 语句，可以用"？"来代替数据列值，在执行前再将值设置进去。因为 PreparedStatement 使用"？"来代替 SQL 中所有可变的部分，剩下了不变的内容，所以 JDBC 将 PreparedStatement 的 SQL 中不变的内容进行预先编译，这样下次执行相同的 SQL 语句时效率就会提高。PreparedStatement 与 Statement 的主要区别如表 11.2 所示。

表 11.2　PreparedStatement与Statement的主要区别

区别点	PreparedStatement	Statement
预编译	可以	不可以
大文本数据	支持	不支持
二进制数据	支持	不支持
执行效率	相对较高	相对较低

【例 11.4】使用 PreparedStatement 修改指定的人员信息

针对例 11.3，将 ModifyServlet.java 修改为 PreparedStatement 实现的程序如下：

```
----------------------ModifyServlet.java--------------------------
public void doPost(HttpServletRequest request, HttpServletResponse
```

```
response)
            throws ServletException, IOException {
        request.setCharacterEncoding("UTF-8");
        String id = request.getParameter("id");
        String name = request.getParameter("name");
        String age = request.getParameter("age");
        String sex = request.getParameter("sex");
        String birthday = request.getParameter("birthday");
        SimpleDateFormat sdf = new SimpleDateFormat("yyyy-MM-dd");
        String description = request.getParameter("description");
        String sql = "update Person set name=?,age=?,sex=?,birthday=?,
                     description=? where id=?";
        Connection con = null;
        PreparedStatement prest = null;
        int  result = 0;
        try {
            Class.forName("com.mysql.jdbc.Driver");       //注册数据库
            con = DriverManager.getConnection(           //获取数据库连接
                "jdbc:mysql://localhost:3306/testWeb","root","admin");
            prest = con.prepareStatement(sql);
            //获取 PreparedStatement,并且预编译 SQL 语句
            prest.setString(1, name);
            prest.setInt(2, Integer.parseInt(age));      //设定第2个参数
            prest.setString(3, sex);                     //设定第3个参数
            Date date = new Date(sdf.parse(birthday).getTime());
            prest.setDate(4, date);                      //设定第4个参数
            prest.setString(5, description);             //设定第5个参数
            prest.setInt(6, Integer.parseInt(id));       //设定第6个参数
            result = prest.executeUpdate(sql);       //执行查询，返回结果集
            response.setContentType("text/html;charset=utf-8");
            PrintWriter out = response.getWriter();
            out .println("<!DOCTYPE HTML PUBLIC \"-//W3C//DTD HTML
            4.01 Transitional//EN\">");
            out.println("<HTML>");
            out.println("  <HEAD><TITLE>列出人员信息表</TITLE></HEAD>");
            out.println("  <BODY>");
            out.println("");
            out.println("<a href=\""+request.getContextPath()+
            "/listPerson.jsp\">
           返回人员列表</a>");
            out.println("  </BODY>");
            out.println("</HTML>");
            out.flush();
            out.close();
        } catch (SQLException e) {
```

```
45
46                  e.printStackTrace();
47              }catch (ClassNotFoundException e) {
48
49                  e.printStackTrace();
50              }catch (ParseException e) {
51
52                  e.printStackTrace();
53              }finally{
54                  try {//记住关闭连接
55                      prest.close();
56                      con.close();
57                  } catch (SQLException e) {
58
59                      e.printStackTrace();
60                  }
61              }
62      }
```

从上面代码，可以看出主体的代码是没有大的变化，代码第 20 行产生 PreparedStatement 对象，第 22~28 行开始为 SQL 语句中的参数赋值，执行效果跟 Statement 是一样的。

11.3.8　Statement 与 PreparedStatement 批处理 SQL

所谓的批处理 SQL，就是批量执行 SQL 语句，PreparedStatement 与 Statement 都可以进行批处理 SQL，且都是通过 addBatch()方法添加 SQL 语句，通过 executeBatch()方法批量执行 SQL 语句，并返回 int 类型数组，表示 SQL 语句执行情况。

能够进行批量的 SQL 语句，必须是 INSERT、UPDATE、DELETE 这样的语句，因为它们返回的是 int 类型。

下面演示 PreparedStatement 批处理 SQL 的例子。

【例 11.5】使用 PreparedStatement 批量插入人员信息

```
String sql = "insert into Person(name,age,sex,birthday,description)
values(?,?,?,?,?)";
        Connection con = null;
        PreparedStatement prest = null;
    try {
            Class.forName("com.mysql.jdbc.Driver");     //注册数据库
            con = DriverManager.getConnection(          //获取数据库连接
"jdbc:mysql://localhost:3306/testWeb","root","admin");
            prest = con.prepareStatement(sql);
```

```
09            //获取 PreparedStatement,并且预编译 SQL 语句
10              for(int i=0;i<=10;i++){
11                  prest.setString(1, "李四"+i);            //设定第1个参数
12                  prest.setInt(2, 30);                     //设定第2个参数
13                  prest.setString(3, "男");                //设定第3个参数
14                  prest.setDate(4,                         //设定第4个参数
15                          new Date(System.currentTimeMillis()));
16                  prest.setString(5, "PreparedStatement 批量插入");
                    //设定第5个参数
17                  prest.addBatch();                        //添加 SQL 语句
18              }
19              int[] result = prest.executeBatch();  /执行批量插入，返回结果集
20          } catch (SQLException e) {
21
22              e.printStackTrace();
23          }catch (ClassNotFoundException e) {
24
25              e.printStackTrace();
26          }finally{
27              //记住关闭连接
28              try {
29                  prest.close();
30                  con.close();
31              } catch (SQLException e) {
32
33                  e.printStackTrace();
34              }
35          }
```

从上述代码第 11~19 可以看出，跟执行单条 SQL 语句相比，在执行多条时，只要将 SQL 语句添加进 addBatch 方法中，然后再执行 executeBatch 方法，批处理 SQL 就完成。如果有异常错误，则进行事务回滚，执行都不成功。

11.4 处理结果集

上一节中，详细介绍了 JDBC 的 CRUD 基本操作，使得读者能清楚地了解 Java 操作数据的基本流程。在查询结果数据中，数据都是返回在 ResultSet 结果集中，遍历结果集就能得到数据。本节介绍结果集的处理的相关内容。

11.4.1 查询多个结果集

查询多个结果集是指查询多张表，返回同一个 ResultSet 对象，在执行第一次查询时，

Statement 会返回一个结果集对象，第二次查询时，它会返回一个全新的结果集对象。中间不要关闭 Statement。因为它会自动关闭上一次查询的结果集，下次返回全新表的数据。例如：

```
ResultSet rs = stmt.executeQuery("select * from 表一");
//遍历 ResultSet 对象
while(rs.next()){
}
rs = stmt.executeQuery("select * from 表二");
//遍历 ResultSet 对象
while(rs.next()){
}
```

11.4.2　可以滚动的结果集

在创建结果集对象时，可以设定可以滚动的结果集，形式如下：

```
conn.createStatement(ResultSet.TYPE_SCROLL_INSENSITIVE,
ResultSet.CONCUR_UPDATABLE);
```

ResultSet.TYPE_SCROLL_INSENSITIVE 指定结果集是可以滚动的，ResultSet.CONCUR_UPDATABLE 指定结果集是可以直接修改的。

11.4.3　带条件的查询

一般的 Web 网站中，带条件的查询是非常普遍的，因此从实用方面考虑，复杂的查询条件是网站所必备的，它的实现原理是将页面中的条件转化为 SQL 语句中的查询条件。

【例 11.6】带条件的查询

```
-----------------------searchPerson.jsp--------------------------
<%@ page language="java" contentType="text/html; charset=UTF-8"%>
<%@ page import="java.sql.*" %>
<!DOCTYPE HTML PUBLIC "-//W3C//DTD HTML 4.01 Transitional//EN">
<HTML>
    <HEAD>
        <TITLE>查询人员信息列表</TITLE>
        <script type="text/javascript"
 src="${pageContext.request.contextPath}/js/jquery-1.8.1.js"></script>
        <!-- 调用日期控件的 js -->
<script language="javascript" type="text/javascript"
src="${pageContext.request.contextPath}/My97DatePicker/WdatePicker
.js"></script>
    </HEAD>
    <BODY>
        <user>
```

```
            <name>张三</name>
            <english-name>zhangsan</english-name>
            <age>20</age>
            <sex>男</sex>
            <address>广东省广州市</address>
            <description>他是一个工程师</description>
        </user>
        <user>
            <property name="name" value="张三"/>
            <property name="english-name" value="zhangsan"/>
            <property name="age" value="20"/>
            <property name="sex" value="男"/>
            <property name="address" value="广东省广州市"/>
            <property name="description" value="他是一个工程师"/>
        </user>
        <center>
            <h4>人员信息列表</h4>
        </center>
        <%
            //获取页面查询条件
            request.setCharacterEncoding("UTF-8");
            String name = request.getParameter("name");
            String sex = request.getParameter("sex");
            String age = request.getParameter("age");
            String description = request.getParameter("description");
            String startTime = request.getParameter("startTime");
            String endTime = request.getParameter("endTime");
            Connection con = null;
           PreparedStatement st = null;
           ResultSet rs = null;
           //组合 SQL 的 where 条件
           String sql = "select * from person where 1=1 ";
           if(name!=null&&!"".equals(name)){
            sql+="and name like '%"+name+"%'";
           }
           if(sex!=null&&!"".equals(sex)){
            sql+="and sex ='"+sex+"'";
           }
           if(age!=null&&!"".equals(age)){
            sql+="and age ='%"+age+"'";
           }
           if(description!=null&&!"".equals(description)){
            sql+="and description like '%"+description+"%'";
```

```
            }
            if(startTime!=null&&!"".equals(startTime)){
             sql+="and birthday >= '"+startTime+"'";
            }
            if(endTime!=null&&!"".equals(endTime)){
             sql+="and birthday <= '%"+endTime+"'";
            }
            try {
                Class.forName("com.mysql.jdbc.Driver");     //注册数据库
                con = DriverManager.getConnection(        //获取数据库连接
                "jdbc:mysql://localhost:3306/testWeb","root","admin");
                st = con.prepareStatement(sql);     //获取 Statement
                rs = st.executeQuery(sql);              //执行查询，返回结果集
        %>
        <form action="searchPerson.jsp" method="post">
            <table>
                <tr>
                    <td>姓名：</td>
                    <td><input name="name"/></td>
                    <td>性别：</td>
                    <td><select name="sex" style="width:100">
                            <option value="">无限制</option>
                            <option value="男">男</option>
                            <option value="女">女</option>
                      </select></td>
                </tr>
                <tr>
                    <td>年龄：</td>
                    <td><input name="age"/></td>
                    <td>备注：</td>
                    <td><input name="description"/></td>
                </tr>
                <td colspan="4">出生日期：
                <label> 从：</label><input class="Wdate"
name="startTime" id="startBeginTime"
onFocus="WdatePicker({dateFmt:'yyyyMMdd HH:mm:ss',minDate:'1900-01-01'})"/>
                    到
                <input   class="Wdate" name="endTime"
id="endBeginTime"
onFocus="WdatePicker({dateFmt:'yyyyMMdd HH:mm:ss',minDate:'1900-01-01'})"/></td>
                </tr>
```

```
                <tr>
                    <td>
                        <input type="submit" value="提交">
                        <input type="reset" value="重置">
                    </td>
                </tr>
            </table>
        </form>
        <br/>
        <br/>
        <table border="1" width="100%" cellpadding="2" cellspacing="1">
            <tr>
                <td>选择</td>
                <td>姓名</td>
                <td>年龄</td>
                <td>性别</td>
                <td>生日</td>
                <td>备注</td>
                <td>操作</td>
            </tr>
            <%
                //遍历结果集 ResultSet
                while(rs.next()){
                    int id = rs.getInt("id");                      //获取 ID
                    String name2  = rs.getString("name");   //获取姓名
                    int age2 = rs.getInt("age");                 //获取年龄
                    String sex2  = rs.getString("sex");      //获取性别
                    Date birthday = rs.getDate("birthday"); //获取出生日期
                    String description2 = rs.getString("description");
                    //获取备注
                    out.println("<tr>");
                    out.println("<td><input type=\"checkbox\"
                   name=\"checkPerson\" value=\""+id+"\"></td>");
                    out.println("<td >"+name2+"</td>");
                    out.println("<td >"+age2+"</td>");
                    out.println("<td >"+sex2+"</td>");
                    out.println("<td >"+birthday+"</td>");
                    out.println("<td >"+description2+"</td>");
                    out.println("<td><a href=\"modify.jsp?id="+id+"\">
                    修改</a>  "+
                    "<a href=\"delete.jsp?id="+id+"\"
                    onclick=\"return confirm('确定删除该记录？')\"
                    >删除</a></td>");
```

```
140                     out.println("</tr>");
141                 }
142             %>
143         </table>
144         <table>
145             <tr>
146                 <td>
147                 全选 <input  type="checkbox" onclick="selectperson(this);"/>
148 
149                 </td>
150             </tr>
151         </table>
152         <%
153             } catch (SQLException e) {
154 
155                 e.printStackTrace();
156             }finally{
157                 //记住关闭连接
158                 try {
159                     rs.close();
160                     st.close();
161                     con.close();
162                 } catch (SQLException e) {
163 
164                     e.printStackTrace();
165                 }
166             }
167         %>
168     </BODY>
169     <script type="text/javascript">
170         function selectPerson(obj){
171             $('input[name="checkPerson"]').attr("checked",obj.checked);
172         }
173     </script>
174 </HTML>
```

上述代码中相比 listPerson.jsp，增加了查询条件的 form 跟从页面获取参数的代码，其中代码第 35~41 行是获取提交的参数值，第 46~63 行是组合查询的 SQL 条件语句，其中日期条件可以直接使用“>=”“<=”符号，因为这是建立在 MySQL 数据库查询上，如果是 Oracle 数据库就要进行转换，所以当遇到某些特殊类型查询的时候要考虑到数据库是否支持。代码第 120~127 行遍历查询的结果集，其运行结果如图 11.11 所示。

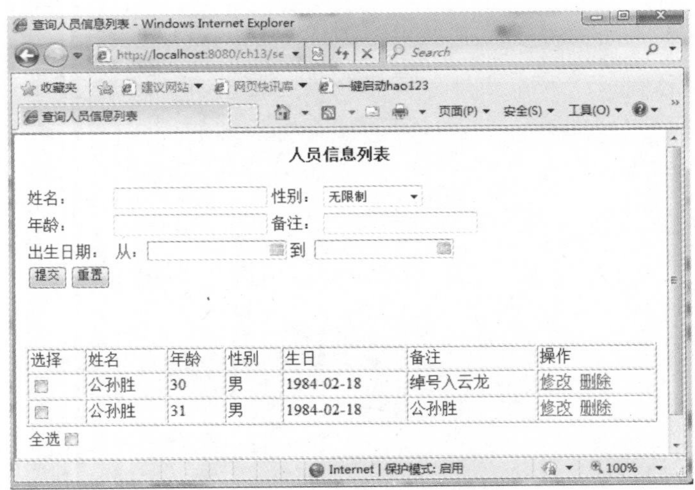

图 11.11　searchPerson.jsp 运行结果

在进行条件查询时，注意特殊类型的转换，例如日期函数、clob、blob 类型的数据等。

11.4.4　ResultSetMetaData 元数据

通过 ResultSetMetaData 可以获得元数据内容，通过它可以直接得知列名，可以动态的显示查询各列内容。下面就来了解下元数据里面的内容。

【例 11.7】根据 ResultSetMetaData 元数据输出列值

```
-----------------------getMetaData.jsp--------------------------
<%@ page language="java" contentType="text/html; charset=UTF-8"%>
<%@ page import="java.sql.*" %>
<!DOCTYPE HTML PUBLIC "-//W3C//DTD HTML 4.01 Transitional//EN">
<HTML>
    <HEAD>
        <TITLE>人员信息列表</TITLE>
    </HEAD>
    <BODY>
        <center>
            <h4>人员信息列表</h4>
        </center>
<%
            Connection con = null;
            Statement st = null;
            ResultSet rs = null;
            try {
             Class.forName("com.mysql.jdbc.Driver");     //注册数据库
                 con = DriverManager.getConnection(      //获取数据库连接
                 "jdbc:mysql://localhost:3306/testWeb","root","admin");
```

```
20                  st = con.createStatement();               //获取 Statement
21                  rs = st.executeQuery("select * from person");
                    //执行查询，返回结果集;
22                ResultSetMetaData rsmd = rs.getMetaData();
23                int columnCount= rsmd.getColumnCount();     //获取列数
24                String[] columnNames = new String[columnCount];
25                for(int i=0;i<columnCount;i++){             //获取列对应的列名
26                 columnNames[i] = rsmd.getColumnName(i+1);
27                }
28                out.println("<table border=\"1\" width=\"100%\"
29              cellpadding=\"2\" cellspacing=\"1\"><tr>");
30                for(int i=0;i<columnCount;i++){             //输出列名
31                   out.println("<td>"+columnNames[i]+"</td>");
32                }
33                out.println("</tr>");
34                //遍历结果集 ResultSet
35                 while(rs.next()){
36                     out.println("</tr>");
37                     //根据列名取得对应列的值
38                     for(int i=0;i<columnCount;i++){
39                     out.println("<td>"+rs.getString(columnNames[i])+
                       "</td>");
40                     }
41                   out.println("</tr>");
42                 }
43                 out.println("</table>");
44             } catch (SQLException e) {
45
46                 e.printStackTrace();
47             }finally{
48                 //记住关闭连接
49                 try {
50                     rs.close();
51                     st.close();
52                     con.close();
53                 } catch (SQLException e) {
54
55                     e.printStackTrace();
56                 }
57             }
58         %>
```

上述代码中第 22 行就是获得元数据，第 23~27 行分析元数据的列数以及将列数对应的列名存放在数组中，第 28~48 行显示数据结果集，其运行结果如图 11.12 所示。

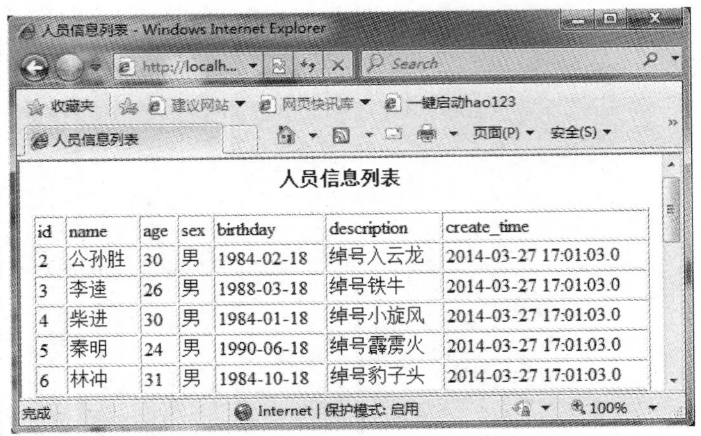

id	name	age	sex	birthday	description	create_time
2	公孙胜	30	男	1984-02-18	绰号入云龙	2014-03-27 17:01:03.0
3	李逵	26	男	1988-03-18	绰号铁牛	2014-03-27 17:01:03.0
4	柴进	30	男	1984-01-18	绰号小旋风	2014-03-27 17:01:03.0
5	秦明	24	男	1990-06-18	绰号霹雳火	2014-03-27 17:01:03.0
6	林冲	31	男	1984-10-18	绰号豹子头	2014-03-27 17:01:03.0

图 11.12　getMetaData.jsp 页面效果图

ResultSetMetaData 元数据中不仅包含列名，还包含列类型、长度等其他信息。

11.4.5　直接显示中文列名

查询 SQL 语句时，可以声明别名。例如：

```
select id as 主键,name as 姓名,age as 年龄,sex as 年龄 from person;
```

使用元数据，可以直接显示中文列名。如果取得的列名是乱码这个时候需要进行转码。例如：

```
String name = rs.getMetaData().getColumnName(2);          //取得列名
name= new String(name.getBytes("latin1"),"UTF-8");        //进行转码
```

11.5　上机实践

1. 建立数据库连接的步骤是什么？请尝试连接数据库，并取出一条数据。
2. 自定义学生成绩表并编写 JSP 页面来查询学生成绩，查询方式可以是多样的。

第 12 章

XML文件格式

在现在的开发系统中，总是会有很多 XML 文件，比如 struts.xml、spring.xml、web.xml、server.xml 以及自定义的 XML 文件，可以说 XML 文件时无处不在，那 XML 是什么？应该怎样去编写？如何应用？本章将解答这些问题。

本章首先向读者介绍什么是 XML、XML 的用途以及技术框架，然后详细介绍 XML 的基本语法包括元素命名规则、元素的定义规范等。在了解这些 XML 的基础后，延伸介绍 3 种 XML 的解析方法：DOM、SAX 和 DOM4j。

12.1 初识 XML

XML 是一种可扩展的标记语言，它被设计用来传输和存储数据，是有万维网协会推出的一套数据交换标准。它可以用于定义 Web 网页上的文档元素以及复杂数据的表述和传输。在 W3C 的官网(http://www.w3.org)上有其更多的描述，也可以去 W3C 在线学校网站去学习 XML 更多的内容。

12.1.1 什么是 XML

XML 是指可扩展标记语言（Extensible Markup Language），是一种标记语言与 HTML 类似；设计的宗旨是传输数据；它没有规定的标签体，需要自定义标签；是一种自我描述的语言，可以储存数据和共享数据。

XML 与 HTML 的主要差异在于：HTML 用来显示数据，XML 是用来传输和存储数据；HTML 用来显示信息，XML 用来传输信息。

XML 最大的特点是它的自我描述和任意扩展，当用其描述数据时，用户可以根据需要，组织符合 XML 规范形式的任意内容，并且标签的名称也可以由用户指定。下面以例子来说明 XML 的定义格式，例子如下：

```
<?xml version="1.0" encoding="UTF-8"?>
<user>
   <name>张三</name>
   <english_name>zhangsan</english_name>
```

```
    <age>20</age>
    <sex>男</sex>
    <address>广东省广州市</address>
    <description>他是一个工程师</description>
</user>
```

它定义的是一用户的基本信息包括用户的姓名、英文名称、性别、年龄、住址、描述等信息。同样是上述的内容，可以用另外的自定义形式进行描述，比如：

```
<?xml version="1.0" encoding="UTF-8"?>
<user>
 <property name="name" value="张三"/>
 <property name="english_name" value="zhangsan"/>
 <property name="age" value="20"/>
 <property name="sex" value="男"/>
 <property name="address" value="广东省广州市"/>
 <property name="description" value="他是一个工程师"/>
</user>
```

不论用哪种结构格式，它都能清楚的描述用户基本信息目的，这就体现了 XML 可扩展和自定义标签的特点。

12.1.2　XML 的用途

XML 设计的宗旨是用来传输和存储数据，其用途也包括了这些，它不仅具有一般纯文本文件的用途还具有自身的特点。下面介绍其在开发系统过程中常见的用途：

1. 传输数据

通过 XML 可以在不同的系统之间传输数据，在开发过程中难免会遇到多个系统之间相互通信，且各系统的存储的数据又是多种多样的，对于开发者而言，这些工作量是巨大的，通过转换为 XML 格式来传输数据可以减少传输数据时的复杂性，并且还可以具备通用性。

比如说，目前流行的 SOA 协议、Web Service 服务、json、Ajax 等，其实都是利用 XML 数据格式，在不同的系统之间交互数据。

2. 存储数据

用 XML 来存储数据是其最基本的用途，因为它可以作为数据文件，所以当需要持久化保存数据时，可以利用 XML 数据格式进行存储，例如，web.xml、struts.xml、spring.xml 等。下面就是经常见到的 web.xml 文件内容：

```
<?xml version="1.0" encoding="UTF-8"?>
<web-app version="2.5"
 xmlns="http://java.sun.com/xml/ns/javaee"
 xmlns:xsi="http://www.w3.org/2001/XMLSchema-instance"
```

```
xsi:schemaLocation="http://java.sun.com/xml/ns/javaee
http://java.sun.com/xml/ns/javaee/web-app_2_5.xsd">
  <welcome-file-list>
    <welcome-file>index.jsp</welcome-file>
  </welcome-file-list>
</web-app>
```

12.1.3　XML 的技术架构

XML 的技术架构如图 12.1 所示，它也展示了 XML 中用到的技术和术语。

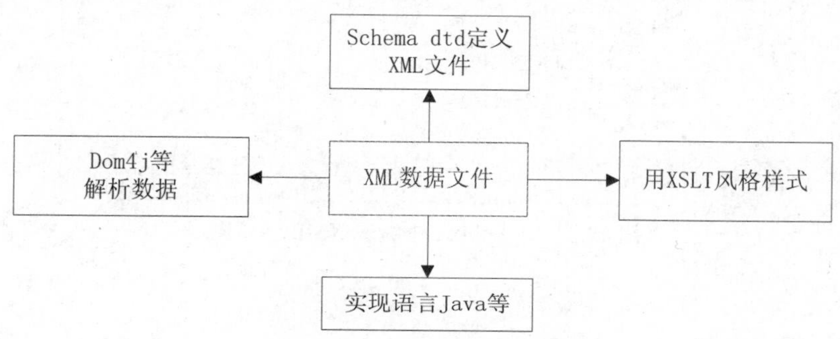

图 12.1　XML 的技术架构

- 数据定义 Schema、DTD：XML 数据文件也是要按照一定的协议进行定义的，它有两种可遵循的定义规则 DTD 和 Schema。DTD 是早期的语言，Schema 是后期发展的语言也是现在用得最多的定义 XML 语言。
- 数据风格样式 XSLT：XSLT 是可扩展样式转换（extensible Stylesheet Language Transformation）使用 XSLT 可以将 XML 中存放的内容按照指定的样式转换为 HTML 页面。
- 解析 XML 文件工具：目前比较盛行的工具是使用 dom、dom4j、SAX，它们各有特点，后续的章节将详细介绍。
- 操作 XML 数据：目前将 XML 数据作具体操作，实现其功能的，都是由额外的程序实现，一般采用 Java 比较多，也可以使用 JavaScript。

12.1.4　XML 开发工具

XML 其实就是一个文本文件，其开发工具有很多，例如：普通的文本编辑器、Editplus、UEStudio、MyEclipse 的 XML 编辑器、XMLSpy 等。在 MyEclipse 中编辑 XML 的情况如图 12.2 所示。

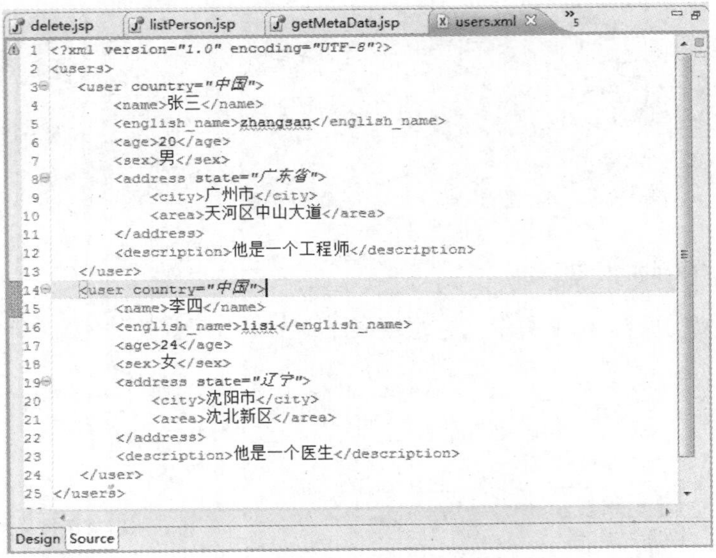

图 12.2　MyEclipse 中编辑 XML 示意图

12.2 XML 基本语法

上一节介绍了 XML 的一些用途、技术架构和开发工具，让读者对 XML 有些基本概念，本章将着重介绍 XML 的基本语法，开发者必须熟悉这些语法规范，才可以正确使用 XML。

12.2.1 XML 文档的基本结构

XML 文档的基本结构如下所示：

```
01  <?xml version="1.0" encoding="UTF-8"?>
02  <users>
03  <user>
04          <name>张三</name>
05          <english_name>zhangsan</english_name>
06           <age>20</age>
07           <sex>男</sex>
08           <address>广东省广州市</address>
09           <description>他是一个工程师</description>
10  </user>
11  </users>
```

如上 XML 文档，首先必须要有 XML 文档的声明，如第 1 行的声明：

```
<?xml version="1.0" encoding="UTF-8"?>
```

其中 version 是指该文档遵循的 XML 标准的版本，encoding 指明文档使用的字符编码格式。

12.2.2 标记必须闭合

在 XML 文档中，除 XML 声明外，所有元素都必须有其结束标识，如果 XML 元素没有文本节点时，采用自闭合的方式关闭节点，例如：<note/>这样的形式进行自闭合。

正常的标记闭合形式如下：

```
<age>20</age>
<sex>男</sex>
<address>广东省广州市</address>
```

age、sex、address 都有相应的结束标记。

12.2.3 必须合理地嵌套

在 XML 文档中，元素的嵌套必须合理，例如如下的嵌套就不合理：

```
<age>30<sex>
</age>女</sex>
```

这样会导致 XML 错误，且描述不清。

正确的描述如下：

```
<age>30</age>
<sex>女</sex>
```

12.2.4 XML 元素

XML 元素是指成对标签出现的内容，且每个元素之间有层级关系。例如：<age>元素指的是<age>20</age>。

<user>元素指的是：

```
<user>
    <name>张三</name>
    <english_name>zhangsan</english_name>
     <age>20</age>
     <sex>男</sex>
     <address>广东省广州市</address>
     <description>他是一个工程师</description>
</user>
```

其中，<age>元素是<user>元素的子元素，<user>元素时<age>元素的父元素；两个<user>元素时并列关系。

元素的命名规则如下：

- 可以包含字母、数字、和其他字符；
- 不能以 xml 开头，包括其大小写例如：XML、xMl 等；
- 不能以数字或者标点符号开头，不能包含空格；
- XML 文档除了 XML 以外，没有其他的保留字，任何的名字都可以使用，但是应该尽量使元素名字具有可读性；
- 尽量避免使用“-”和“.”，可能会引起混乱，可以使用下划线；
- 在 XML 元素命名中不要使用“:”，因为 XML 命名空间需要用到这个特殊字符。
- 例如：下面这些命名是不合法的。

```
<2title>
<xmlTtle>
<titel name>
<.age>
```

正确的命名如下：

```
<title2>
<title_name>
```

12.2.5 XML 属性

XML 元素可以在开始标签中包含属性，类似 HTML，属性 (Attribute) 提供关于元素的额外（附加）信息，它被定义在 XML 元素的标签中，且自身有对应的值，例如<user>元素的属性名和属性值（字体加粗部分）如下：

```
<user  language="java">
   <name>张三</name>
   <english_name>zhangsan</english_name>
    <age>20</age>
    <sex>男</sex>
    <address>广东省广州市</address>
    <description>他是一个工程师</description>
</user>
```

属性的命名规则跟元素的命名规则一样。

属性值必须使用引号，不过单引号和双引号均可使用，如果属性值本身包含双引号，那么有必要使用单引号包围它或者可以使用实体（"）引用。

12.2.6 只有一个根元素

所有的 XML 文档有且只有一个根元素来定义整个文档，例如，在 web.xml 代码中，可以看到< web-app>就是它的根元素。例如：web.xml 下面的定义方式是错误的。

```
<?xml version="1.0" encoding="UTF-8"?>
```

```
<web-app version="2.5"
 xmlns="http://java.sun.com/xml/ns/javaee"
 xmlns:xsi="http://www.w3.org/2001/XMLSchema-instance"
 xsi:schemaLocation="http://java.sun.com/xml/ns/javaee
 http://java.sun.com/xml/ns/javaee/web-app_2_5.xsd">
……
</web-app>
<web-app>
……
</web-app>
```

12.2.7　大小写敏感

XML 文档是大小写敏感的，包括其标签名称、属性名和属性值都是大小写的影响。例如，<age>与<Age>是不同的。

一般初学者刚写 XML 文档时常犯的错就是开始标记与结束标记的大小写不一致而导致 XML 错误。例如：

```
<name>Jonh</Name>
```

<name>与</Name>不能相互匹配，而导致<name>没有正确的被关闭。

12.2.8　空白被保留

空白被保留是指在 XML 文档中，空白部分并不会被解析器删除，而是被当作数据一样完整的保留。例如：

```
<description>好好学习      天天向上</ description>
```

“好好学习　　天天向上”中的空白会被当作数据被保留。

12.2.9　注释的写法

XML 注释形式如下：

```
<!-- 注释单行 -->
<!--
    注释多行
-->
```

12.2.10　转义字符的使用

XML 中有些特殊字符需要转义，比如“>”、“<”、“&”、单引号、双引号等等，其转义字符和 HTML 中的转义字符是一样的。

12.2.11 CDATA 的使用

CDATA 是用于需要原文保留的内容，尤其是在解析 xml 过程中产生歧义的部分，当某个节点的数据有大量需转义的字符，那么 CDATA 就可以发挥其作用，其用法如下：

```
<![CDATA[
    内容
   ]]>
```

例如：

```
<![CDATA[
If(m>n){
  alert(m 大于 n);
}else if(m<n&&m!=0){
   alert(m 小于 n)
}
   ]]>
```

CDATA 是不能嵌套的。

12.3 JDK 中的 XML API

JDK 中涉及 XML 的 API 有两个，它们分别是：

- The Java API For XML Processing：负责解析 XML；
- Java Architecture for XML Binding：负责将 XML 映射为 Java 对象。

它们所涉及到的类包有：javax.xml.*、org.w3c.dom.*、org.xml.sax.*和 javax.xml.bind.*。其中经常用到的类参见表 12.1。

表 12.1 常用的 JDK XML API 类

类	说明
javax.xml.parsers.DocumentBuilder	从 XML 文档获取 DOM 文档实例
javax.xml.parsers.DocumentBuilderFactory	XML 文档获取生成 DOM 对象树的解析器
javax.xml.parsers.SAXParser	获取基于 SAX 的解析器 XML 文档实例
javax.xml.parsers. SAXParserFactory	获取基于 SAX 的解析器以解析 XML 文档
org.w3c.dom.Document	整个 XML 文档
org.w3c.dom.Element	XML 文档中的一个元素
org.w3c.dom.Node	Node 接口是整个文档对象模型的主要数据类型
org.xml.sax. XMLReader	是 XML 解析器的 SAX2 驱动程序必须实现的接口

12.4　最常见的 XML 解析模型

上一节介绍了 XML 的基本语法，了解其命名规则和语法规范，使得读者能对 XML 的文档格式有了基本了解。本节将在此基础上，叙述如何对 XML 进行解析。由于 XML 结构基本上是一种树型结构，因此处理 XML 的步骤都差不多，Java 已经将它们封装成了现成的类库。目前流行解析的方法有 DOM、SAS 和 DOM4j 这 3 种。下面逐一介绍。

12.4.1　DOM 解析

DOM（Document Object Model，即文档对象模型）是 W3C 组织推荐的处理 XML 的一种方式。它是一种基于对象的 API，把 XML 内容加载到内存中，生成一个 XML 文档相对应的对象模型，根据对象模型，以树节点的方式对文档进行操作。下面以例子说明解析步骤：

【例 12.1】DOM 解析 XML 文件

假设 XML 文件如下：

```
------------------------users.xml--------------------------
<?xml version="1.0" encoding="UTF-8"?>
<users>
 <user country="中国">
     <name>张三</name>
     <english_name>zhangsan</english_name>
     <age>25</age>
     <sex>男</sex>
     <address state="广东省">
         <city>广州市</city>
         <area>天河区中山大道</area>
     </address>
     <description>他是一个工程师</description>
 </user>
 <user country="中国">
     <name>李四</name>
     <english_name>lisi</english_name>
     <age>30</age>
     <sex>女</sex>
     <address state="辽宁省">
         <city>沈阳市</city>
         <area>沈北新区</area>
     </address>
     <description>他是一个医生</description>
 </user>
</users>
```

编写解析类 JAXBDomDemo 的代码如下：

```
-----------------------JAXBDomDemo.java---------------------------
public class JAXBDomDemo {

     /**
      * 用 dom 解析 XML 文件
      */
     public static void main(String[] args) {
          //创建待解析的 XML 文件，并指定目录
        File file = new File("F:\\users.xml");
        //用单例模式创建 DocumentBuilderFactory 对象
        DocumentBuilderFactory factory = DocumentBuilderFactory.newInstance();
        //声明一个 DocumentBuilder 对象
        DocumentBuilder documentBuilder =null;
        try {
                //通过 DocumentBuilderFactory 构建 DocumentBuilder 对象
                documentBuilder = factory.newDocumentBuilder();
                //DocumentBuilder 解析 xml 文件
                Document document = documentBuilder.parse(file);
                //获得 xml 文档中的根元素
                Element root = document.getDocumentElement();
                //输出根元素的名称
                System.out.println("根元素："+root.getNodeName());
                //获得根元素下的子节点
                NodeList childNodes = root.getChildNodes();
                //遍历根元素下的子节点
                for(int i=0;i<childNodes.getLength();i++){
                     //获得根元素下的子节点
                     Node node = childNodes.item(i);
                     System.out.println("节点的名称为"+node.getNodeName());
                     //获得子节点的 country 属性值
                     String attributeV = node.getAttributes().
                     getNamedItem("country").getNodeValue();
                     System.out.println(node.getNodeName()+"节点的 country 属性值为
                     "+attributeV);
                     //获得 node 子节点下集合
                     NodeList nodeChilds = node.getChildNodes();
                     //遍历 node 子节点下集合
                     for(int j=0;j<nodeChilds.getLength();j++){
                          Node details = nodeChilds.item(j);
                          String name = details.getNodeName();
                          //判断如果是 address 元素，获取其子节点
                          if("address".equals(name)){
                               NodeList addressNodes = details.getChildNodes();
                              //遍历 address 元素的子节点
                               for(int k=0;k<addressNodes.getLength();k++){
```

```
45                          Node addressDetail = addressNodes.item(k);
46                          System.out.println(node.getNodeName()+
47                  "节点的子节点"+name+"节点的子节点"+
48                          addressDetail.getNodeName()+
49                  " 节点内容为: "+addressDetail.getTextContent());
50                      }
51                      String addressAtt =
52                  details.getAttributes().getNamedItem("state").
                    getNodeValue();
53                  System.out.println(name+"节点的 state
                    属性值为"+addressAtt);
54                  }
55                  System.out.println(node.getNodeName()+
56                  "节点的子节点"+details.getNodeName()+
57                          " 节点内容为: "+details.getTextContent());
58              }
59          }
60      } catch (ParserConfigurationException e) {
61          e.printStackTrace();
62      }catch (IOException e) {
63          e.printStackTrace();
64      }catch (SAXException e) {
65          e.printStackTrace();
66      }
67  }
68 }
```

上述代码中，详细的描述了解析步骤，其中代码第 09~15 行分别创建了解析工程类 DocumentBuilderFactory、DocumentBuilder 对象，第 17 行传入解析的 XML 文件，第 19 行开始就逐个遍历整个 Document 树，得到 XML 数据。代码运行结果如图 12.3 所示。

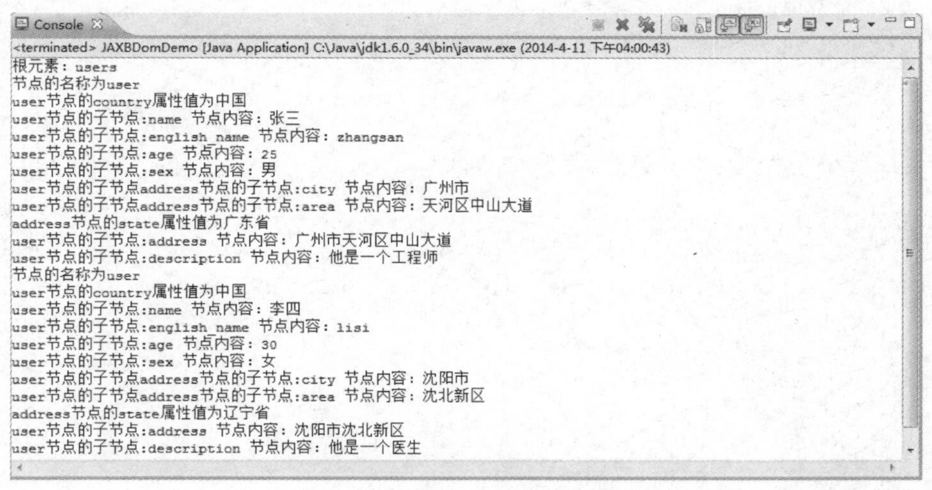

图 12.3　DOM 解析 XML 文件运行结果

通过上述代码，不难发现 DOM 解析 XML 时，主要是以下几步：

（1） 创建 DocumentBuilderFactory 对象

```
//用单例模式创建 DocumentBuilderFactory 对象
 DocumentBuilderFactory factory = DocumentBuilderFactory.newInstance();
```

（2）通过 DocumentBuilderFactory 构建 DocumentBuilder 对象

```
DocumentBuilder documentBuilder =factory.newDocumentBuilder();
```

（3）DocumentBuilder 解析 xml 文件变为 Document 对象

```
Document document = documentBuilder.parse(file);
```

取得 Document 对象之后就可以用 Document 中的方法获取 XML 数据，有关 Document 的详细方法，参见 JAXB API。

12.4.2 SAX 解析

SAX（Simple API for XML）是另外一种解析 XML 文件方法，它虽然不是官方标准，但它是 XML 社区上的事实标准，大部分 XML 解析器都支持它。SAX 与 DOM 所不同的是，它不是一次性将 XML 加载到内存中，而是从 XML 文件开始位置进行解析，根据已经定义好的事件处理器，来决定当前解析的部分是否有必要存储。下面以例子说明 SAX 解析 XML 过程。

【例 12.2】SAX 解析 XML 文件

还是以例 12.1 中的 XML 文件作为源文件。编写解析类 JAXBSAXDemo 代码如下：

```
-----------------------JAXBSAXDemo.java--------------------------
import org.xml.sax.Attributes;
import org.xml.sax.SAXException;
import org.xml.sax.XMLReader;
import org.xml.sax.helpers.DefaultHandler;
import org.xml.sax.helpers.XMLReaderFactory;

public class JAXBSAXDemo extends DefaultHandler{

     private String preTag;

     //接收文档开始的通知
     @Override
    public void startDocument() throws SAXException {
        preTag = null;

    }
     //接收元素开始的通知
     @Override
    public void startElement(String uri, String localName, String qName,
            Attributes attributes) throws SAXException {
        if("user".equals(qName)) {
            System.out.println(qName+"节点的 country 属性值为:
```

```
"+attributes.getValue("country"));
        }
        if("address".equals(qName)){
         System.out.println(qName+"节点的 state 属性值为："+attributes.
         getValue("state"));
        }
        preTag = qName;
    }

    //接收元素结束的通知
    @Override
    public void endElement(String uri, String localName, String qName)
            throws SAXException {
        preTag = null;
    }

    //接收元素中数据的通知，在执行完 startElement 和 endElement 方法之后执行
    public void characters(char ch[], int start, int length)throws
    SAXException {
        String value = new String(ch, start, length);
        if("name".equals(preTag)) {
            System.out.println("name 节点的值为："+value);
        } else if("english_name".equals(preTag)) {
         System.out.println("english_name 节点的值为："+value);
        }else if("age".equals(preTag)){
         System.out.println("age 节点的值为："+value);
        }else if("sex".equals(preTag)){
         System.out.println("sex 节点的值为："+value);
        }else if("description".equals(preTag)){
         System.out.println("description 节点的值为："+value);
        }
        if("city".equals(preTag)) {
         System.out.println("city 节点的值为："+value);
        } else if("area".equals(preTag)) {
         System.out.println("area 节点的值为："+value);
        }
    }

    public static void main(String[] args) throws Exception {
        //由 XMLReaderFactory 类 创建 XMLReader 实例
        XMLReader xmlReader = XMLReaderFactory.createXMLReader();
        //创建一事件监听类
        JAXBSAXDemo handler = new JAXBSAXDemo();
        //XMLReader 解析类设定事件处理类
        xmlReader.setContentHandler(handler);
        //XMLReader 解析类解析 XML 文件
        xmlReader.parse("F:\\users.xml");
    }
}
```

上述代码中介绍了，用 SAX 如何解析 XML 文件的步骤，其中代码第 61 行由 XMLReaderFactory 工厂类创建解析器 XMLReader，第 63 行创建事件处理类，第 67 行开始解析 XML 文件，从第 11~36 行分别重写 DefaultHandler 处理类中的文档开始通知、元素开始通知、元素结束通知等方法。上述代码运行结果如图 12.4 所示。

通过上述代码可以看出，使用 SAX 解析 XML 时，主要是以下几步：

（1）由 XMLReaderFactory 类 创建 XMLReader 实例

```
XMLReader xmlReader = XMLReaderFactory.createXMLReader();
```

（2）创建一事件监听类

```
JAXBSAXDemo handler = new JAXBSAXDemo();
```

（3）为解析类设定事件处理类

```
xmlReader.setContentHandler(handler);
```

（4）解析 XML 文件

```
xmlReader.parse("F:\\users.xml");
```

想要了解更多的 SAX 内容，请查询 org.xml.sax 的 API。

```
Console
<terminated> JAXBSAXDemo [Java Application] C:\Java\jdk1.6.0_34\bin\javaw.exe (2014-4-12 上午12:22:53)
user节点的country属性值为：中国
name节点的值为：张三
english_name节点的值为：zhangsan
age节点的值为：25
sex节点的值为：男
address节点的state属性值为：广东省
city节点的值为：广州市
area节点的值为：天河区中山大道
description节点的值为：他是一个工程师
user节点的country属性值为：中国
name节点的值为：李四
english_name节点的值为：lisi
age节点的值为：30
sex节点的值为：女
address节点的state属性值为：辽宁省
city节点的值为：沈阳市
area节点的值为：沈北新区
description节点的值为：他是一个医生
```

图 12.4　SAX 解析 XML 文件效果图

例 12.2 中应用的是 XMLReader 而不是 SAXParser，是因为在 SAX2 中实现解析的接口名称重命名为 XMLReader。在使用 SAX 解析 XML 资源文件时，默认使用 SAXParser 实现类，它继承自 AbstractSAXParser。同样的，工厂类也是使用 XMLReaderFactory 而不是 SASParserFactory 来创建解析类。在 SAX2 中它们已经过时了。

12.4.3 DOM4j 解析

DOM 在解析的时候把整个 XML 文件映射到 Document 的树型结构中，XML 中的元素、属性、文本都能在 Document 中看清，但是它消耗内存，查询速度慢。SAX 是基于事件的解析，解析器在读取 XML 时根据读取的数据产生相应的事件，由应用程序实现相应的事件处理，所以它解析速度快，内存占用少。但是它需要应用程序自身处理解析器的状态，实现起来比较麻烦，而且它只支持对 XML 文件的读取，不支持写入。

Dom4j 是一个简单、灵活的开源库。前身是 JDOM，与它所不同的是，dom4j 使用接口和抽象类基本方法，使用了大量的 Collections 类，提供些替代方法以允许更好的性能或更直接的编码方法。它不仅可以读取 XML 文件而且还可以写入 XML 文件。目前越来越多的 Java 软件都在使用 Dom4j 来读写 XML，例如 Hibernate，包括 Sun（被 Oracle）公司自己的 JAXM 也用了 Dom4j。可以通过下载地址 http://sourceforge.net/projects/dom4j 下载最新的 Dom4j 包。下面以例子来说明其使用方法。

【例 12.3】Dom4j 解析 XML 文件

同样 XML 文件的内容还是上述中的内容，编写解析类 Dom4jDemo，其代码如下：

```
----------------------Dom4jDemo.java--------------------------
import java.io.File;
import java.util.List;

import org.dom4j.Document;
import org.dom4j.DocumentException;
import org.dom4j.Element;
import org.dom4j.io.SAXReader;

public class Dom4jDemo {

    /**
     * Dom4j 解析 XML 文件
     */
    @SuppressWarnings("unchecked")
    public static void main(String[] args) {
        //创建待解析的 XML 文件，并指定目录
        File file = new File("F:\\users.xml");
        //指定 XML 解析器 SAXReader
        SAXReader saxReader = new SAXReader();
        try {
            //SAXReader 解析 XML 文件
            Document document = saxReader.read(file);
            //指定要解析的节点
            List<Element> list = document.selectNodes("/users/user" );
            for(Element element:list){
                //获得节点 country 属性值
                System.out.println("country----"+element.
```

```
                  attributeValue("country"));;
28                //获得节点的子节点
29                List<Element> childList = element.elements();
30                //遍历节点的子节点
31                for(Element childelement:childList){
32                    //如果是 address 子节点，遍历 address 的子元素
33                    if("address".equals(childelement.getName())){
34                        //获得节点 state 属性值
35               System.out.println("state----"+childelement.
                 attributeValue("state"));
36                        //遍历 address 元素的子元素
37                        List<Element> addressElements =  childelement.
                          elements();
38                        for(Element e:addressElements){
39                            System.out.println(e.getName()+"----"
                              +e.getText());
40                        }
41                    }
42             System.out.println(childelement.getName()+"----"
               +childelement.getText());
43                }
44             }
45         } catch (DocumentException e) {
46             e.printStackTrace();
47         }
48     }
49 }
```

上述代码，讲解了用 dom4j 来解析 XML 资源文件，其中代码第 19 行新建 SAXReader 解析器，第 22 行解析指定文件，从第 24~42 行利用获得到的 Document 不断的循环遍历出节点的属性和内容。上述代码的运行结果如图 12.5 所示。

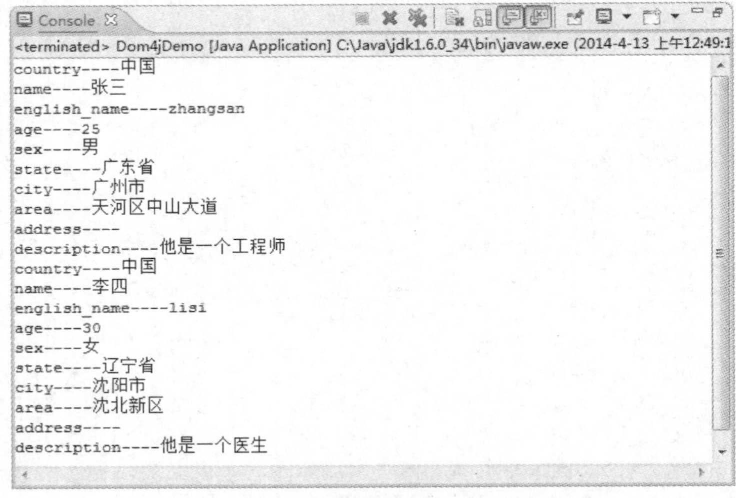

图 12.5 Dom4j 解析 XML 文件效果图

通过上述代码，使用 Dom4j 解析 XML 时，主要是以下几步：

（1）创建 SAXReader 实例

```
SAXReader saxReader = new SAXReader();
```

（2）用 SAXReader 获取 xml 的 Document

```
Document document = saxReader.read(file);
```

从上述步骤可以看出，用 dom4j 解析 XML 文件十分的便捷且极易使用。想要了解更多的 dom4j 内容，请查询 dom4j 的 API 和相关例子。

上述例子中，采用的解析器是 SAXReader，并通过它来获取 XML 文件的 Document，在 dom4j 中还可以用 DocumentHelper.parseText(text);来获取 XML 的 Document。

12.5 XML 与 Java 类映射 JAXB

上一章节叙述了 XML 的解析方法，分别使用 dom、dom4j 和 sax3 种方法解析 XML，从而可以从 XML 数据文件中获得到想要的数据，这非常有用但同时发现获取数据需要编写大量的代码，工作量巨大。有没有更加简单的方法获得 XML 数据和生成 XML 呢？答案是肯定的，用 JAXB 的 API 就可以解决这个问题。

12.5.1 什么是 XML 与 Java 类映射

所谓的映射即一一对应关系。例如：有一 XML 数据文件如下：

```
<?xml version="1.0" encoding="UTF-8"?>
<user>
   <name>张三</name>
   <english_name>zhangsan</english_name>
    <age>20</age>
    <sex>男</sex>
    <address>广东省广州市</address>
    <description>他是一个工程师</description>
</user>
```

另有一个 User 的 Java 类如下：

```
---------------------- User.java--------------------------
public class User {
   String name;           //姓名
   String english_name;       //英文名
   String age;            //年龄
```

```
    String sex;            //性别
    String address;        //地址
    String description;            //描述

  public User() {
      super();
      // TODO Auto-generated constructor stub
  }
  public User(String name, String englishName, String age, String sex,
          String address, String description) {
      super();
      this.name = name;
      english_name = englishName;
      this.age = age;
      this.sex = sex;
      this.address = address;
      this.description = description;
  }

    //省略 get、set 方法
}
```

在开发过程中，要将 XML 中的 name 元素与 User 类中的 name 属性对应、english_name 元素与 User 类中的 english_name 属性对应、age 元素与 User 类中的 age 属性对应等等，那么 XML 数据与 Java 类的对应关系就是一种映射。

12.5.2 JAXB 的工作原理

JAXB 映射主要有 4 个部分构成：Schema、JAXB 映射类、XML 文档、Java 对象，其工作原理的示意图如图 12.6 所示。

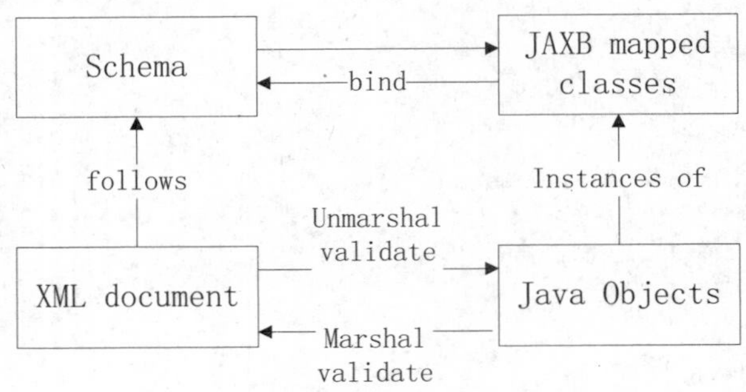

图 12.6　JAXB 的工作原理

Schema 可以理解为表结构、XML Document 是数据来源、JAXB 提供类映射方法、Object 是 Java 中对应的类。

12.5.3 Java 对象转化成 XML

将 Java 对象转化为 XML 的过程为 marshal，可以通过 annotation 注入的方式将 Java 类映射成 XML 文件。

【例 12.4】Java 对象转化为 XML

首先，先将 User 类更改如下：

```
import javax.xml.bind.annotation.XmlRootElement;
@XmlRootElement
public class User {
    String name;          //姓名
    String english_name;      //英文名
    String age;          //年龄
    String sex;          //性别
    String address;         //地址
    String description;        //描述

   //属性的 get 和 set 方法省略
}
```

如上代码：用“@XmlRootElement”方式注入 XML 的根元素，表示这个是 XML 文件的根元素，那么类中的属性就是 XML 文档中的元素。

其次，编写转化类 JAXBMarshalDemo 如下：

```
---------------------- JAXBMarshalDemo.java--------------------------
import java.io.File;

import javax.xml.bind.JAXBContext;
import javax.xml.bind.JAXBException;
import javax.xml.bind.Marshaller;

import com.eshore.pojo.User;

public class JAXBMarshalDemo {

    public static void main(String[] args) {
        //创建 XML 对象，将它保存在 F 盘下
        File file = new File("F:\\user.xml");
        //声明一个 JAXBContext 对象
        JAXBContext jaxbContext;
        try {
         //指定映射的类创建 JAXBContext 对象的上下文
            jaxbContext = JAXBContext.newInstance(User.class);
            //创建转化对象 Marshaller
            Marshaller m = jaxbContext.createMarshaller();
```

```
21                //创建 XML 文件中的数据
22                User user = new User("张三", "zhangsan", "30", "男",
23                "广州市天河区", "一名出生的工程师");
24                //将 Java 类 User 对象转化到 XML
25                m.marshal(user, file);
26            } catch (JAXBException e) {
27                e.printStackTrace();
28            }
29        }
30    }
```

从上述代码中可以得出转化 XML 数据的一般步骤：

（1）创建 JAXBContext 上下文对象，参数为映射的类，如代码第 18 行所示；

（2）通过 JAXBContext 对象的 createMarshaller()方法，创建 Marshaller 转换对象，如代码第 20 行所示；

（3）为转化的类设置内容，如代码第 22 行所示；

（4）通过方法 marshal 将 Java 对象输出到指定位置，如代码第 25 行所示，参数是映射的类和输出文件。

运行上述代码，将在 F 盘下生成 user.xml 文件，其内容如下：

```
<?xml version="1.0" encoding="UTF-8" standalone="yes" ?>
<user>
  <address>广州市天河区</address>
  <age>30</age>
  <description>一名出生的工程师</description>
  <english_name>zhangsan</english_name>
  <name>张三</name>
  <sex>男</sex>
 </user>
```

12.5.4 XML 转化为 Java 对象

上述介绍将 Java 对象转化为 XML 数据，本小节将介绍如何使 XML 对象转化为 Java 对象。转化的过程跟上述过程正好是相反的。

【例 12.5】XML 数据转化为 Java 对象

User 类的代码不变，变的是 JAXBMarshalDemo 中的内容。编写类 JAXBUnmarshalDemo 类如下：

```
-----------------------JAXBMarshalDemo.java---------------------------
import java.io.File;

import javax.xml.bind.JAXBContext;
```

```
04  import javax.xml.bind.JAXBException;
05  import javax.xml.bind.Unmarshaller;
06
07  import com.eshore.pojo.User;
08
09  public class JAXBUnmarshalDemo {
10
11      public static void main(String[] args) {
12           //创建 XML 对象，将它保存在 F 盘下
13          File file = new File("F:\\user.xml");
14          //声明一个 JAXBContext 对象
15          JAXBContext jaxbContext;
16          try {
17           //指定映射的类创建 JAXBContext 对象的上下文
18              jaxbContext = JAXBContext.newInstance(User.class);
19              //创建转化对象 Unmarshaller
20              Unmarshaller u = jaxbContext.createUnmarshaller();
21              //转化指定 XML 文档为 Java 对象
22              User user = (User)u.unmarshal(file);
23              //输出对象中的内容
24              System.out.println("姓名----"+user.getName());
25              System.out.println("英文名字----"+user.getEnglish_name());
26              System.out.println("年龄----"+user.getAge());
27              System.out.println("性别----"+user.getSex());
28              System.out.println("地址----"+user.getAddress());
29              System.out.println("描述----"+user.getDescription());
30           } catch (JAXBException e) {
31
32              e.printStackTrace();
33           }
34      }
35  }
```

从上述代码中，可以发现代码变化不大，其转化的步骤分解如下：

（1）创建 JAXBContext 上下文对象，参数为映射的类，如代码第 18 行所示；

（2）通过 JAXBContext 对象的 createUnmarshaller ()方法，创建 Unmarshaller 转换对象，如代码第 20 行所示；

（3）用方法 unmarshal 将指定的 XML 文件转化为 Java 对象，转化时需强制转换为映射类对象。

运行上述代码，在 Java 控制台输出如下：

```
姓名----张三
英文名字----zhangsan
年龄----30
```

```
性别----男
地址----广州市天河区
描述------一名出生的工程师
```

12.5.5 更为复杂的映射

例 12.1 中的 xml 文件就是一个复杂的 XML 文件，现在要将这复杂的 XML 数据转化为 Java 对象，从数据上发现 user 元素不仅有元素而且还有属性，子元素 address 中包含有子元素 city 和 area，那么原有的类结构已经不适用了，要进行改造，下面就逐一介绍。

【例 12.6】复杂 XML 数据转化为 Java 对象

首先，将 User 类进行改造，使其包含有属性 country，代码如下：

```
-----------------------User.java---------------------------
import javax.xml.bind.annotation.XmlAccessType;
import javax.xml.bind.annotation.XmlAccessorType;
import javax.xml.bind.annotation.XmlAttribute;
import javax.xml.bind.annotation.XmlElement;

@XmlAccessorType(XmlAccessType.FIELD)
public class User {
    @XmlAttribute
    private String country;          //国家
    @XmlElement
    private String name;             //姓名
    @XmlElement
    private String english_name;     //英文名
    @XmlElement
    private String age;              //年龄
    @XmlElement
    private String sex;              //性别
    @XmlElement
    private Address address;       //地址
    @XmlElement
    private String description;      //描述
    public User() {
        super();

    }
    public User(String name, String englishName, String age, String sex,
            Address address, String description,String country) {
        super();
        this.name = name;
        english_name = englishName;
        this.age = age;
```

```
32          this.sex = sex;
33          this.address = address;
34          this.description = description;
35          this.country = country;
36      }
37      @Override
38      public String toString() {
39          String str=name+" 来自于 "+country+" 英文名: "+english_name+"
            性别:"+sex+
40          " 年龄: "+age+" 现住"+address.toString()+","+description;
41          return str;
42      }
43  }
```

上述代码中用“@XmlAttribute”代表这是元素的属性，“@XmlElement”代表元素的子元素，代码第 38~41 行是重写类的 toString 方法。

“@XmlAccessorType(XmlAccessType.FIELD)”代表 User 类中每个非静态字段会字段绑定到 XML 中，即将类中的属性定义为 private，因为 JAXB 默认的情况下是每个公共字段和每个 get、set 方法会自动绑定到 XML 中

Address 类的代码如下：

```
------------------------Address.java--------------------------
import javax.xml.bind.annotation.XmlAccessType;
import javax.xml.bind.annotation.XmlAccessorType;
import javax.xml.bind.annotation.XmlAttribute;
import javax.xml.bind.annotation.XmlElement;

@XmlAccessorType(XmlAccessType.FIELD)
public class Address {
    @XmlAttribute
    private String state;                //国家
    @XmlElement
    private String city;                 //城市
    @XmlElement
    private String area;                 //地区
    public Address() {
        super();
    }
    public Address(String state, String city, String area) {
        super();
        this.state = state;
        this.city = city;
        this.area = area;
```

```
22          }
23          @Override
24          public String toString() {
25              String str=state+" "+city+" "+area;
26              return str;
27          }
28      }
```

上述代码中第 8~9 行定义 Address 元素的 state 属性，第 10~13 行定义 city 和 area 元素，第 24~27 行重写 Address 类的 toString()方法。

因为 XML 文件中有多个 user 元素，所有需要定义一个容器和根元素的类 Users，代码如下：

```
-----------------------Users.java---------------------------
01  @XmlRootElement(name="users")
02  @XmlAccessorType(XmlAccessType.FIELD)
03  public class Users {
04
05      @XmlElement(name="user")
06      private List<User> list = new ArrayList<User>();
07
08      public Users() {
09          super();
10
11      }
12
13      public Users(List<User> list) {
14          super();
15          this.list = list;
16      }
17
18      public void setList(List<User> list) {
19          this.list = list;
20      }
21
22      public List<User> getList() {
23          return list;
24      }
25
26  }
```

上述代码中第 1 行指定元素节点名称为“users”，第 5~6 行定义 list 容器并指定元素名称为“user”，代码第 13~24 行给出 get 和 set 方法。

其次，建立转化类 JAXBComplexUnmarshalDemo，代码如下：

```
-----------------------
```

```
JAXBComplexUnmarshalDemo.java---------------------------
01  public class JAXBComplexUnmarshalDemo {

02      public static void main(String[] args) {
03          //创建 XML 对象，将它保存在 F 盘下
04          File file = new File("F:\\users.xml");
05          //声明一个 JAXBContext 对象
06          JAXBContext jaxbContext;
07          try {
08              //指定映射的类创建 JAXBContext 对象的上下文
09              jaxbContext = JAXBContext.newInstance(Users.class);
10              //创建转化对象 Unmarshaller
11              Unmarshaller u = jaxbContext.createUnmarshaller();
12              //转化指定 XML 文档为 Java 对象
13              Users users = (Users)u.unmarshal(file);
14              List<User> list = users.getList();
15              for(User user:list){
16                  //输出对象中的内容
17                  System.out.println("输出----"+user.toString());
18              }
19
20          } catch (JAXBException e) {
21
22              e.printStackTrace();
23          }
24      }
25  }
```

上述代码中，代码第 15~18 行输出对象数据，其余代码步骤跟类 JAXBUnmarshalDemo 一样，运行上述代码，在 Java 控制台输出结果为：

```
输出----张三 来自于 中国 英文名：zhangsan 性别:男 年龄：25 现住广东省 广州市 天河区中山大道,他是一个工程师
输出----李四 来自于 中国 英文名：lisi 性别:女 年龄：30 现住辽宁省 沈阳市 沈北新区,他是一个医生
```

12.6 上机实践

1. 判断下面 XML 元素定义中哪些是错误的？

```
a. <user_name>  b. <user name>  c. <1address>  d. <&and>  e. <and-or>
f. <xmlTel>  g. <XMLTel>  h. <.phone>  i. <phone>
```

2. 有一 XML 文件如下，用 dom、sax 和 dom4j 分别解析。

```
<?xml version="1.0" encoding="UTF-8"?>
<user>
 <property name="name" value="张三"/>
 <property name="english_name" value="zhangsan"/>
 <property name="age" value="20"/>
 <property name="sex" value="男"/>
 <property name="address" value="广东省广州市"/>
 <property name="description" value="他是一个工程师"/>
</user>
```

3. 将上述的 XML 文件，映射为 Java 的 User 类。

第 13 章

◄ 资源国际化 ►

通常一个 Web 程序是应用在互联网中，从理论上讲它可以被全球所有的网络在线用户所访问。但是不同国家地区的访问者都有自己的语言，Web 应用就需要根据访问者的语言和习惯来自动调整页面的显示内容，那么这就需要用到资源国际化编程。本章将介绍资源国际化编程，使读者学习完本章后可以进行简单的国际化编程和本地化编程，开发出适应性更强的网站。

13.1 资源国际化简介

如上所述，资源国际化就是要解决不同国家与地区之间的文化差异问题，包括其语言的差异、生活习惯的差异等等。在 Java 语言中，提供了相关的方法来支持资源国际化，例如资源绑定类 ResourceBundle、地区 Locale 类等。

资源国际化一般有两种编程，国际化编程（I18N）和本地化编程（L10N）。

13.1.1 国际化编程 I18N

I18N 是 Internationlization 的缩写，其含义是指让软件产品随着国家和地区中语言的不同能够自动显示其相适应的内容，而不是用代码将这些不同的信息写在程序中。通常需要进行国际化编程的信息包括数字、货币信息以及日期与时间等。

13.1.2 本地化编程 L10N

L10N 为资源本地化，全称为 Localization。资源本地化编程就是要向使用不同语言和处于不同地区的访问者提供适合的页面。

一个 Web 应用程序只有当需要实现多种不同语言版本时才有必要进行本地化编程，当然某些应用程序当前只需要一种语言，但考虑到以后可能需要多语言的需求，那么也有必要进行本地化编程的方式来实现，方便将来进行程序的拓展。

13.2 资源国际化编程

Java 语言从它诞生开始，就为资源国际化做了准备，在 Java 中提供了一些类，用于对程序进行国际化，使得实现国际化变得容易；又因为它是基于 Unicode 编码设计的编程语言，因此 Java 程序可以支持目前世界上所有的语言。看个简单国际化例子。

13.2.1 资源国际化示例

先看一个原始的例子，即没有用国际化的方式：

```
01  <%@ page language="java" contentType="text/html; charset=UTF-8"%>
02  <!DOCTYPE HTML PUBLIC "-//W3C//DTD HTML 4.01 Transitional//EN">
03  <html>
04    <head>
05        <title>资源国际化编程示例</title>
06    </head>
07      <body>
08            <p>您好，资源国际化编程示例</p>
09      </body>
10  </html>
```

上述代码中第 8 行代码字符串“您好，资源国际化编程示例”是写在程序中，当某用户访问该页面时，它只会显示中文，要是在没有中文字库的系统中则会显示乱码。现在把字符串用 key-value 方式写入到 properties 资源文件中：

```
   messages_zh_CN.properties:
   helloInfo=
\u60A8\u597D\uFF0C\u8D44\u6E90\u56FD\u9645\u5316\u7F16\u7A0B\u793A\u4F8B
   messages.properties:
   helloInfo = Hello, Internationalization Example
```

一般我们用*_zh_CH.properties 表示为中文资源文件，*.properties 表示默认的资源文件，* 表示的内容即资源文件名称。这两个文件经编译都位于 Web 项目的 classpath 中，在 IDE 中一般是位于 src 包下面。同时把程序变更为：

```
<%@ page language="java" contentType="text/html; charset=UTF-8"%>
<%@taglib prefix="fmt" uri="http://java.sun.com/jsp/jstl/fmt" %>
<!DOCTYPE HTML PUBLIC "-//W3C//DTD HTML 4.01 Transitional//EN">
<html>
  <head>
    <title>资源国际化编程示例</title>
  </head >
    <body>
       <fmt:bundle basename="messages">
```

```
            <fmt:message key=" helloInfo"/>
        </fmt:bundle>
    </body>
</html>
```

上述代码中引用了 JSTL 中的 fmt 标签库<fmt:bundle>标签，其中指定绑定的资源名为 messages。在中文系统中用户访问时会显示“您好，资源国际化编程示例”效果见图 13.1 所示，非中文系统则显示“Hello, Internationalization Example”效果图 13.2 所示。这样就实现了资源国际化。

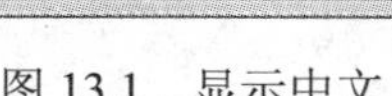

图 13.1　显示中文

图 13.2　显示英文

<fmt:bundle>标签的具体用法参加第 9.3.3 节 。在调试时，可以通过设定浏览器中显示语言选项卡进行调试。

13.2.2　资源文件编码

上一小节介绍了资源国际化的简单示例，让读者了解资源国际化是什么样的，本节介绍有关资源文件的编码内容。

一般而言我们是采用 properties 文件来保存资源文件，properties 文件是以键-值的形式来保存文件。messages_zh_CN.properties 中保存的是经过 UTF-8 编码之后的 ASCII 字符，Unicode 字符中是不允许出现中文、日文等其他字符的文字。经过编码之后的文件就可以直接在程序中被引用。

Unicode 是为了解决传统的字符编码不统一而产生的，它为每种语言中的每个字符设定了统一并且唯一的二进制编码。在字节数上，Unicode 字符占有两个字节，而 ASCII 码字符只占 1 个字节。

JDK 自带的 native2asii.exe 工具可以将 Unicode 码转为 ASCII 码。它位于 JDK 安装目录的 bin 文件夹下。双击运行，输入转换的中文，按回车后效果如图 13.3 所示。

图 13.3　native2asii.exe 运行结果

该工具可以将任意字符转化为 ASCII 字符(英文除外)。图中显示的是将中文转化为 ASCII 的效果。Unicode 字符都被转化为以“\u”开头的一串字符。

一个个语句的转换比较麻烦而且容易遗漏，native2ascii 支持将整个 properties 文件进行转换。语法格式如下：

```
native2ascii  -options  inputfile  outputfile
```

上述语法中-options 包括-encoding encoding_name 与- reverse。-encoding encoding_name 是将文件按照 encoding_name 指定的编码进行转化。-reverse 是将编码后的文件还原。

inputfile 是要转换的源文件路径，outputfile 是转换后的文件输出路径。例如，将源文件 message.properties 按 UTF-8 编码转化到 message_zh_CN.propertes（源文件是 UTF-8 编码的）：

```
C:\>native2ascii -encoding UTF-8 c:\message.properties
c:\message_zh_CN.properties
```

使用 native2ascii 命令，需要将 JDK 的 bin 目录加到系统环境变量 path 中。

13.2.3　显示所有 Locale 代码

在 Java 中，与国际化编程关系比较密切的类就是 Locale，Locale 的一个实例代表了一个特定的语言编码。它提供了语言参数构造方法，例如 public Locale(String language)。

通过以下构造方法可以产生一个美国英语的 Locale 对象：

```
Locale loc = new Locale("en","US");
```

在运行 Java 程序时，Locale 类由 Java 虚拟机提供；运行 Web 程序时，Locale 由浏览器提供。Locale 里面记录着客户的地区与语言信息。Locale 代码格式形如 zh_CN，其中小写的 zh 为语种，表示简体中文，大写的 CN 为国家（或地区），表示中国。

下面的程序显示了 Java 支持的所有的 Locale。

【例 13.1】输出所有的 Locale 代码

showAllLocale.jsp 是显示所有的 Locale，源代码如下：

```
----------------------- showAllLocale.jsp---------------------------
<%@ page language="java" contentType="text/html; charset=UTF-8"%>
<%@page import="java.util.Locale" %>
```

```
03  <%@taglib prefix="c" uri="http://java.sun.com/jsp/jstl/core" %>
04  <!DOCTYPE HTML PUBLIC "-//W3C//DTD HTML 4.01 Transitional//EN">
05  <html>
06    <head>
07      <title>资源国际化显示所有的 Locale 代码</title>
08    </head>
09    <%
10       Locale[] availableLocales = Locale.getAvailableLocales();
11       request.setAttribute("availableLocales",availableLocales);
12    %>
13      <body>
14           <table border="1" width="100%" cellpadding="2" cellspacing="1">
15               <tr>
16                     <td>名称</td>
17                     <td>国家</td>
18                     <td>国家名称</td>
19                     <td>语言</td>
20                     <td>语言名称</td>
21                     <td>别名</td>
22                </tr>
23                <c:forEach items="${availableLocales}" var="locale">
24                    <tr>
25                       <td>${locale.displayName}</td>
26                       <td>${locale.country}</td>
27                       <td>${locale.displayCountry}</td>
28                       <td>${locale.language}</td>
29                       <td>${locale.displayLanguage}</td>
30                       <td>${locale.variant}</td>
31                    </tr>
32                </c:forEach>
33           </table>
34      </body>
35  </html>
```

上述代码中第 9~12 行代码，获得所有本地的 Locale 类，并将传入页面中，代码第 23~32 行利用 forEach 标签进行遍历输出 Locale 类中所包含的名称、国家、国家名称、语言、语言名称以及别名等信息。其运行结果如图 13.4 所示。

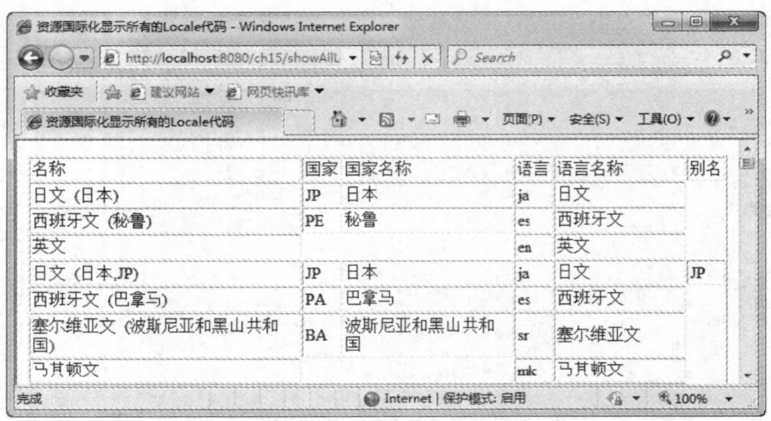

名称	国家	国家名称	语言	语言名称	别名
日文 (日本)	JP	日本	ja	日文	
西班牙文 (秘鲁)	PE	秘鲁	es	西班牙文	
英文			en	英文	
日文 (日本,JP)	JP	日本	ja	日文	JP
西班牙文 (巴拿马)	PA	巴拿马	es	西班牙文	
塞尔维亚文 (波斯尼亚和黑山共和国)	BA	波斯尼亚和黑山共和国	sr	塞尔维亚文	
马其顿文			mk	马其顿文	

图 13.4　showAllLocale.jsp 运行结果图

将语言代码与国家代码用下划线连接，就是该 Locale 的代码。例如中国大陆是 zh_CN、美国是 en_US、日本是 ja_JP。一般默认的资源文件名为*.properties，例如 message.properties，而国家或者地区的资源文件名为*_Locale.properties，例如中国大陆资源文件名为 message_zh_CN.properties，美国资源文件名为 message_en_US.properties。

若一个系统支持多个 Locale 功能，则它一般有多个资源文件。例如，若某 Locale 对应的 properties 文件存在，则会优先显示该 properties 文件里的对应内容。若 properties 文件不存在或者该文件存在但是对应的 key-value 属性对不存在，则会获取默认资源文件里的内容。若默认资源文件不存在指定的内容，则抛出异常。

13.2.4　带参数的资源

带参数的资源，其实就是资源内容时动态的，部分内容时可以变化的，变化的部分用参数指定。资源的参数通过{0}、{1}等指定。在第 9.3.3 节中就有介绍。

例如：params.properties

```
message= Hello, your IP address is: {0}, your Locale: {1}, your language:{2} ,Today is:{3}
```

【例 13.2】输出带参数的资源

showParam.jsp 输出带参数资源的页面，其页面源代码如下：

```
----------------------- showParam.jsp--------------------------
<%@ page language="java" contentType="text/html; charset=UTF-8"%>
<%@page import="java.util.Date" %>
<%@taglib prefix="fmt" uri="http://java.sun.com/jsp/jstl/fmt" %>
<%@taglib prefix="c" uri="http://java.sun.com/jsp/jstl/core" %>
<!DOCTYPE HTML PUBLIC "-//W3C//DTD HTML 4.01 Transitional//EN">
<html>
  <head>
```

```
08      <title>显示带参数的资源</title>
09    </head>
10
11      <body>
12          带参数的资源示例<br/>
13          <fmt:bundle basename="param">
14          <c:set var="todayT" value="<%=new Date()%>"/>
15           <fmt:message key="message">
16               <!-- 输出地址 -->
17                <fmt:param value="${pageContext.request.remoteAddr }" />
18                <!-- 输出 Locale -->
19                <fmt:param value="${pageContext.request.locale }" />
20                <!-- 输出浏览器显示语言 -->
21                <fmt:param value="${pageContext.request.locale.
                  displayLanguage }" />
22                <!-- 输出日期 -->
23                <fmt:param value="${todayT}" />
24           </fmt:message>
25          </fmt:bundle>
26      </body>
27  </html>
```

上述代码中第 15~23 行输出参数中的信息，有请求的 IP 地址，浏览器语言，日期。其运行结果如图 13.5 所示。

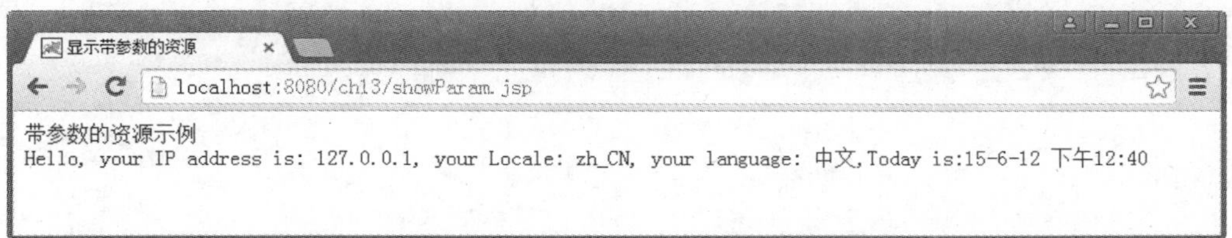

图 13.5　showParam.jsp 运行结果图

13.2.5　ResourceBundle 类

本地化编程需要用到名称为 ResourceBundel 类，它位于 java.util 之下，这个类实际上只是一个抽象的父类，真正操作的是其两个子类 ListResourceBundle 和 PropertyResourceBundle。ResourceBundle 类中最重要的方法是 getBundle，通过指定参数调用 getBundle 方法获取当前 Locale 对应的本地化资源。ResouceBundle 提供了 6 个 getBundle 方法，其中最为常用的是包含 baseName 和 locale 参数的方法。

```
publc static final ResouceBundel getBundle(String baseName,Locale locale)
```

参数 baseName 用于指定资源的路径；参数 locale 用于指定当前的 Locale。

【例 13.3】Java 实现获取资源内容

```
-----------------------JavaGetResouceBundle.java---------------------------
package com.eshore;

import java.text.MessageFormat;
import java.util.Date;
import java.util.MissingResourceException;
import java.util.ResourceBundle;

public class JavaGetResouceBundle {
    // 资源名称
    private static final String BUNDLE_NAME = "message";
    // 资源绑定
private static final ResourceBundle RESOURCE_BUNDLE =
ResourceBundle.getBundle(BUNDLE_NAME);
    // 返回不带参数的资源
    public static String getMessage(String key) {

        try {
            return RESOURCE_BUNDLE.getString(key);
        } catch (MissingResourceException e) {
            return key;
        }
    }
    // 返回带任意个参数的资源
    public static String getMessage(String key, Object... params) {

        try {
            String value = RESOURCE_BUNDLE.getString(key);
            return MessageFormat.format(value, params);
        } catch (MissingResourceException e) {
            return key;
        }
    }

    public static void main(String args[]) {
     System.out.println(getMessage("helloInfo"));
     System.out.println(getMessage("message", "127.0.0.1", "en_US",
     "英文",new Date()));
    }
}
```

代码中 ResourceBundle 类会绑定 message 的资源，然后根据用户不同的 Locale 选择显示不同的 properties 文件内容。

13.2.6 Servlet 的资源国际化

Servlet 的资源国际化就是在 Servlet 类中运用 ResourceBundle 类绑定资源文件，并输出到页面中，例如将例 13.3 进行改造，将 JavaGetResouceBundle 类改造成 Servlet 类并输出到页面就行，具体代码如下。

【例 13.4】Servlet 实现资源国际化

```
----------------------GetResourceBundelServlet.java---------------------
package com.eshore;
import java.io.IOException;
import java.io.PrintWriter;
import java.util.Locale;
import java.util.ResourceBundle;
import javax.servlet.ServletException;
import javax.servlet.http.HttpServlet;
import javax.servlet.http.HttpServletRequest;
import javax.servlet.http.HttpServletResponse;
public class GetResourceBundelServlet extends HttpServlet {
    private static final long serialVersionUID = 1L;
    public GetResourceBundelServlet() {
        super();
    }
    public void destroy() {
        super.destroy(); // 父类销毁
    }
    public void doGet(HttpServletRequest request, HttpServletResponse
    response)
            throws ServletException, IOException {
         doPost(request,response);
    }
    public void doPost(HttpServletRequest request, HttpServletResponse
    response)
            throws ServletException, IOException {
        //设定页面请求的 Locale
        Locale loc = request.getLocale();
        //绑定 welcome 资源文件
```

```
36          ResourceBundle rb = ResourceBundle.getBundle("welcome", loc);
37          //获取文件上 welcomeinfo 内容
38          String welcomeinfo = rb.getString("welcomeinfo");
39          //获取文件上 message 内容
40          String message = rb.getString("message");
41          response.setContentType("text/html;charset=utf-8");
42          PrintWriter out = response.getWriter();
43    out.println("<!DOCTYPE HTML PUBLIC \"-//W3C//DTD HTML 4.01 Transitional//EN\">");
44          out.println("<HTML>");
45          out.println("  <HEAD><TITLE>welcomeinfo</TITLE></HEAD>");
46          out.println("  <BODY>");
47          out.println(" <h2>"+welcomeinfo+"</h2>");
48          out.println("  http://blog.sina.com.cn/u/1268307652</br>");
49          out.println(" <h4>"+message+"</h4>");
50          out.println("  </BODY>");
51          out.println("</HTML>");
52          out.flush();
53          out.close();
54      }
55
56      public void init() throws ServletException {
57
58      }
59
60  }
```

上述代码中第 34 行取得页面请求的 Locale，第 35 行用指定的 Locale 获取绑定的资源文件。代码第 44~52 行将文件内容输出到页面中。完成上述代码并编译与部署后，在浏览器中访问该 Servlet，可以发现当浏览器发送不同的 Locale 时，它的页面将分别显示中文和英文文字。本例中文运行效果如图 13.6 所示。

图 13.6　Servlet 资源国际化中文显示效果

13.2.7 显示所有 Locale 的数字格式

一般 Web 程序中需要格式化的数据包括数字、日期、时间、百分数、货币等。不同地区或国家的显示方法会稍有不同，也有可能截然不同。例如“2,000”，美国人认为是 2 千，但是欧洲人认为是 2。如果显示不对，差别就很大。因此 JSP 中尽量使用 JSTL 标签，它支持资源国际化，能根据用户的 Locale 自动选择合适的数据格式。在第 9.3.5 节中也有介绍显示各国的数字格式，读者也可以参阅其说明。

【例 13.5】显示所有 Locale 的数字格式

showNumber.jsp 页面显示所有地区的 Locale 数字格式，其源代码如下：

```
------------------------showNumber.jsp---------------------------
<%@ page language="java" contentType="text/html; charset=UTF-8"%>
<%@ taglib uri="http://java.sun.com/jsp/jstl/core" prefix="c"%>
<%@ taglib uri="http://java.sun.com/jsp/jstl/fmt" prefix="fmt"%>
<%@page import="java.util.*"%>
<%
   request.setAttribute("availableLocales",
    Locale.getAvailableLocales());
%>
<!DOCTYPE HTML PUBLIC "-//W3C//DTD HTML 4.01 Transitional//EN">
<html>
  <head>
    <title>显示所有的日期格式</title>
  </head>
    <body>
        <table border="1" width="100%" cellpadding="2" cellspacing="1">
            <tr>
                <td>Locale 码</td>
                <td>语言</td>
                <td>日期时间</td>
                <td>数字</td>
                <td>货币</td>
                <td>百分比</td>
            </tr>
            <c:set var="date" value="<%=new Date()%>"/>
            <c:forEach var="locale" items="${availableLocales}">
                <fmt:setLocale value="${ locale }" />
                <tr>
                    <td align="left">
                        ${ locale.displayName }
                    </td>
                    <td align="left">
                        ${ locale.displayLanguage }
                    </td>
                    <td>
                        <fmt:formatDate value="${date}" type="both" />
                    </td>
```

```
36                          <td>
37                              <fmt:formatNumber value="100000.5" />
38                          </td>
39                          <td>
40                              <fmt:formatNumber value="100000.5" type="currency" />
41                          </td>
42                          <td>
43                              <fmt:formatNumber value="100000.5" type="percent" />
44                          </td>
45                      </tr>
46                  </c:forEach>
47              </table>
48          </body>
49      </html>
```

上述代码中第 5~7 行是向页面存放本地的 Locale 集合，JSP 运用<c:forEach>标签循环遍历出各地的日期、货币、百分比格式。其运行结果如图 13.7 所示。

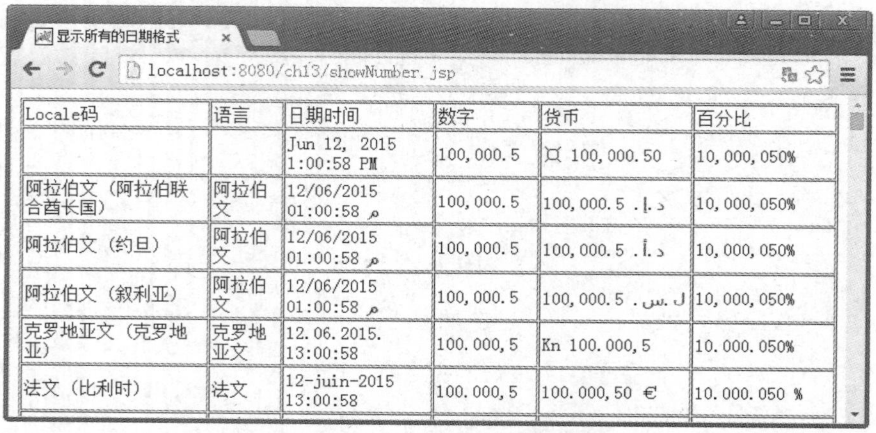

显示所有的日期格式

localhost:8080/ch13/showNumber.jsp

Locale码	语言	日期时间	数字	货币	百分比
		Jun 12, 2015 1:00:58 PM	100,000.5	¤ 100,000.50	10,000,050%
阿拉伯文（阿拉伯联合酋长国）	阿拉伯文	12/06/2015 01:00:58 م	100,000.5	100,000.5 د.إ.	10,000,050%
阿拉伯文（约旦）	阿拉伯文	12/06/2015 01:00:58 م	100,000.5	100,000.5 د.أ.	10,000,050%
阿拉伯文（叙利亚）	阿拉伯文	12/06/2015 01:00:58 م	100,000.5	100,000.5 ل.س.	10,000,050%
克罗地亚文（克罗地亚）	克罗地亚文	12.06.2015. 13:00:58	100.000,5	Kn 100.000,5	10.000.050%
法文（比利时）	法文	12-juin-2015 13:00:58	100.000,5	100.000,50 €	10.000.050 %

图 13.7　showNumber.jsp 页面运行结果

13.2.8　显示全球时间

在第 9.3.9 节中，已经介绍了如何用<fmt:timeZone>标签显示全球时间，详细请查看第 9.3.9 节中<fmt:timeZone>标签的内容。

TimeZone 类代表时区，用来表示不同地区间的时间差异。地球上共有 24 个时区，每两个相邻的时区间相差 1 个小时。看下面的程序：

【例 13.6】显示全球时间

```
-----------------------showTimeZone.jsp---------------------------
<%@ page language="java" contentType="text/html; charset=UTF-8"%>
<%@page import="java.util.*"%>
<%@taglib prefix="c" uri="http://java.sun.com/jsp/jstl/core"%>
<%@taglib prefix="fmt" uri="http://java.sun.com/jsp/jstl/fmt"%>
<!DOCTYPE HTML PUBLIC "-//W3C//DTD HTML 4.01 Transitional//EN">
```

```
06 <html>
07     <head>
08         <title>显示全球时间</title>
09     </head>
10     <%
11         Map<String, TimeZone> hashMap = new HashMap<String, TimeZone>();
12         for (String id : TimeZone.getAvailableIDs()) { // 所有可用的 TimeZone
13             hashMap.put(id, TimeZone.getTimeZone(id));
14         }
15         request.setAttribute("timeZoneIds", TimeZone.
           getAvailableIDs());
16         request.setAttribute("timeZone", hashMap);
17         request.setAttribute("date",new Date());//当前时间
18     %>
19     <body>
20         <fmt:setLocale value="zh_CN" />
21         现在时刻：<%=TimeZone.getDefault().getDisplayName()%>
22         <fmt:formatDate value="${date}" type="both" />
23         <br />
24         <table border="1">
25             <tr>
26                 <td>时区 ID</td>
27                 <td>时区</td>
28                 <td>现在时间</td>
29                 <td>时差间隔</td>
30             </tr>
31             <c:forEach var="id" items="${ timeZoneIds }"
               varStatus="status">
32                 <tr>
33                     <td>${id}</td>
34                     <td>${timeZone[id].displayName}</td>
35                     <td>
36                         <!-- 用 fmt 标签格式化日期输出 -->
37                     <fmt:timeZone value="${id}">
38                 <fmt:formatDate value="${date}" type="both"
                   timeZone="${id}" />
39                     </fmt:timeZone>
40                     </td>
41                     <td>
42                         ${ timeZone[id].rawOffset / 60 / 60 / 1000 }
43                     </td>
44                 </tr>
45             </c:forEach>
46         </table>
47     </body>
48 </html>
```

上述代码中第 11~17 行向页面中传递 TimeZone 类中所包含的全球时间内容，代码第 31~45 行遍历输出 TimeZone 类中的内容。其运行结果如图 13.8 所示。

像上述情况就不能直接输出日期时间，而是根据不同的 Locale 显示不同的日期时间。JSP 中可以使用<fmt:timeZone />标签输出日期，它会根据客户所在的时区自动输出当地时间。

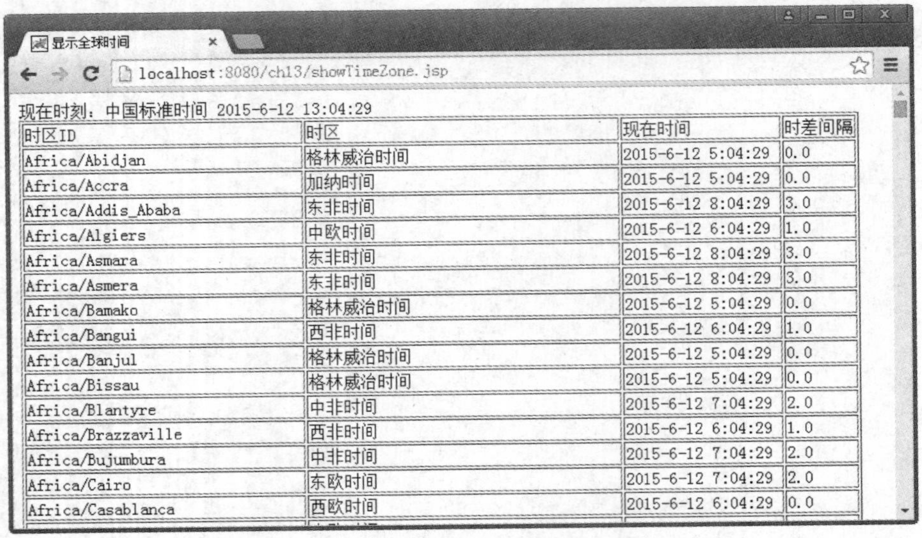

显示全球时间
localhost:8080/ch13/showTimeZone.jsp

现在时刻：中国标准时间 2015-6-12 13:04:29

时区ID	时区	现在时间	时差间隔
Africa/Abidjan	格林威治时间	2015-6-12 5:04:29	0.0
Africa/Accra	加纳时间	2015-6-12 5:04:29	0.0
Africa/Addis_Ababa	东非时间	2015-6-12 8:04:29	3.0
Africa/Algiers	中欧时间	2015-6-12 6:04:29	1.0
Africa/Asmara	东非时间	2015-6-12 8:04:29	3.0
Africa/Asmera	东非时间	2015-6-12 8:04:29	3.0
Africa/Bamako	格林威治时间	2015-6-12 5:04:29	0.0
Africa/Bangui	西非时间	2015-6-12 6:04:29	1.0
Africa/Banjul	格林威治时间	2015-6-12 5:04:29	0.0
Africa/Bissau	格林威治时间	2015-6-12 5:04:29	0.0
Africa/Blantyre	中非时间	2015-6-12 7:04:29	2.0
Africa/Brazzaville	西非时间	2015-6-12 6:04:29	1.0
Africa/Bujumbura	中非时间	2015-6-12 7:04:29	2.0
Africa/Cairo	东欧时间	2015-6-12 7:04:29	2.0
Africa/Casablanca	西欧时间	2015-6-12 6:04:29	0.0

图 13.8　显示全球时间效果图

13.3 上机实践

1. 编码获取带参数的资源文件内容。

2. 编写一个网页，用中文浏览器进来显示中文“欢迎”，用英文浏览器进来显示英文“Welcome”。

第 14 章

简易的网上购物系统

当下，形形色色的网上购物网站遍地都是，人们也开始习惯网上购物带来的便利。因此网站的建设要求也越来越复杂。本章将和读者一起完成一个简易的网上购物网站。本系统采用JSP+Servlet+JavaBean技术完成，JSP页面负责展示数据，业务逻辑则在Servlet中实现，JavaBean则负责数据的处理。这也是JSP的小型项目所常用的分层思想，也是现在三大框架（Struts、Spring、Hibernate）常用的技术，希望读者能够熟练掌握这种分层技术，对以后学习大型项目的开发将有事半功倍的作用。

14.1 系统需求分析

某电商为加快人们对自己生产的商品了解，想开发一个在线网上购物系统，让用户可以在线直接购买商品。有了在线网上购物系统，用户可以直接在网上查询商品的相关信息，并对自己满意的商品进行下单购买。

系统分成4大模块，即用户登录模块、用户管理模块、购物车模块、支付模块，如图14.1所示。

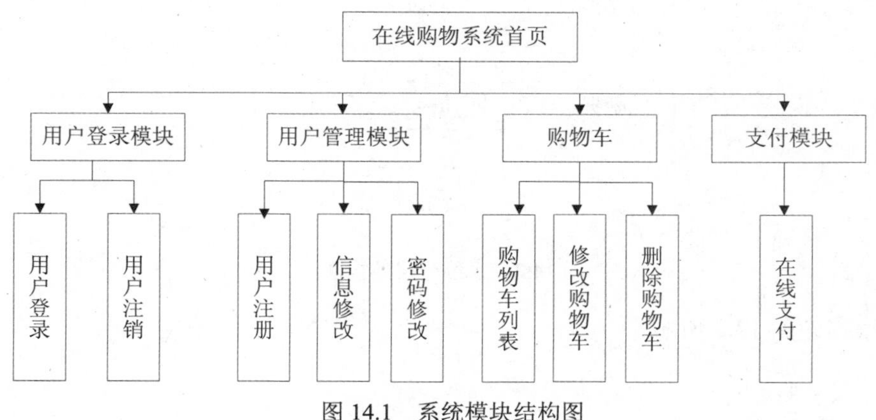

图 14.1　系统模块结构图

（1）用户登录模块

用户登录模块，主要作用是判断用户是否登录与用户退出系统的操作。只有当用户登录系统才可以进行购买商品和支付操作。

（2）用户管理模块

用户管理模块负责对注册的用户进行管理，其功能有：用户的注册、用户信息修改、密码修改等操作。

（3）购物车模块

购物车模块负责对已登录的用户购买商品的管理，其功能有：查看购物车列表、修改购物车列表、删除购物车列表等操作。

（4）支付模块

支付模块负责对购买商品的支付操作。

本系统只对购物车商品进行状态处理，与银行的支付接口没有实现。

14.2 系统总体架构

根据用户的需求，目前只是做个简单的网上购物系统，包括 4 个模块的功能，业务逻辑比较简单，容易理解，采用目前比较流行的 3 层架构就能解决问题。

系统的层次结构如图 14.2 所示。

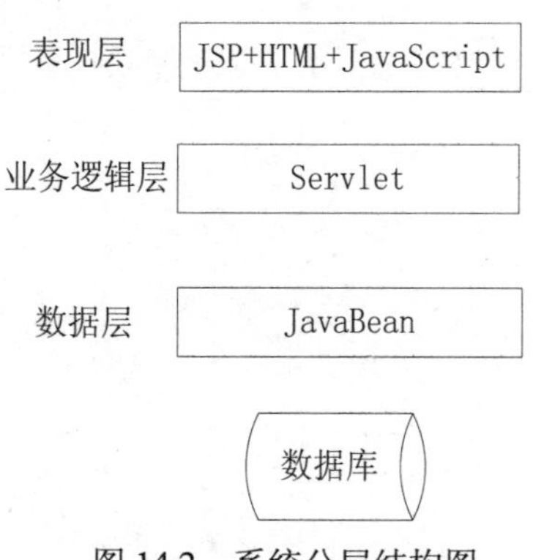

图 14.2　系统分层结构图

系统的流程如图 14.3 所示。

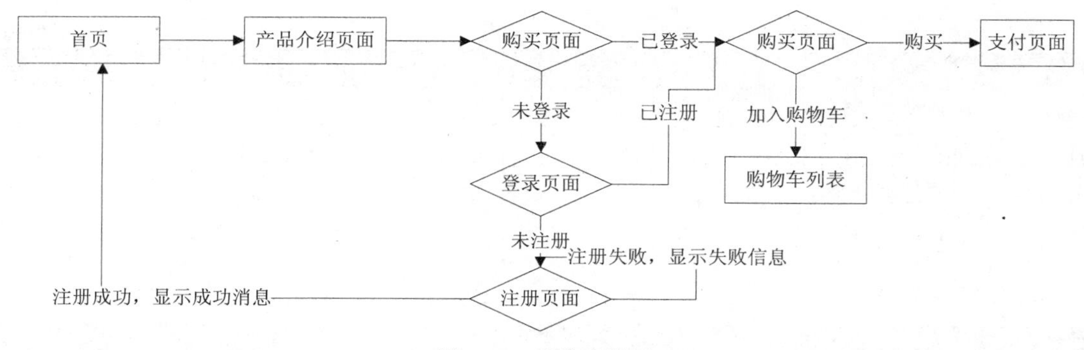

图 14.3　系统流程图

14.3 数据库设计

数据库设计时，一般先构建 E-R 图，再根据 E-R 图创建数据库表、视图等。当然可以借助一些 OOA 工具（例如 PowerDesiger、Rose 等），进行数据库的设计。

14.3.1　E-R 图

这边介绍的是简易在线购物，所以涉及到的表结构很简单，就 3 张表：用户、购物车、商品。其 E-R 关系图如图 14.4 所示。

图 14.4　系统 E-R 图

一个用户只能有一个购物车，购物车支付商品之后就清空这个用户的购物车数据，购物车里可以拥有多个商品 ID，同样一个商品 ID 可以存在多个购物车中。所以购物车与商品的关系是多对多的关系。

14.3.2　数据物理模型

根据图 14.4 的 E-R 图，可以设计出系统的数据物理模型。

下面对关系图中表的设计作个简要分析：

（1）用户表（见表 14.1）

表 14.1　users表

字段名称	含义	数据类型	是否主键	是否为空	是否外键	其他约束
uid	用户 ID	int	Yes	No	No	AUTO_INCREMENT
uname	用户姓名	varchar(20)	No	No	No	

（续表）

字段名称	含义	数据类型	是否主键	是否为空	是否外键	其他约束
passwd	用户登录密码	varchar(50)	No	No	No	
email	用户 Email	varchar(50)	No	No	No	
lastlogin	最后登录时间	datetime	No	No	No	

（2）商品表（见表 14.2）

表 14.2 goods 表

字段名称	含义	数据类型	是否主键	是否为空	是否外键	其他约束
gid	物品 ID	int	Yes	No	No	AUTO_INCREMENT
kinds	商品类别	varchar(50)	No	No	No	
gname	商品名称	varchar(100)	No	No	No	
gphoto	商品照片	varchar(100)	No	No	No	
types	商品型号	varchar(100)	No	No	No	
producer	生产商	varchar(50)	No	No	No	
price	商品价格	float(10,2)	No	No	No	
carriage	商品运费	float(10,2)	No	No	No	
pdate	生产日期	datetime	No	No	No	
paddress	生产地址	varchar(100)	No	No	No	
described	商品描述	varchar(200)	No	No	No	

（3）购物车表（见表 14.3）

表 14.3 shoppingcart 表

字段名称	含义	数据类型	是否主键	是否为空	是否外键	其他约束
id	购物车 ID	int	Yes	No	No	AUTO_INCREMENT
uid	用户 ID	int	No	No	Yes	
gid	物品 ID	int	No	No	Yes	
status	购物车状态 1	int	No	No	No	0：未支付商品 1：已支付商品
number	物品数量	int	No	No	No	

14.4 系统详细设计

下面详细列出本系统的所有源代码和所有文件。

14.4.1 系统包的介绍

网上在线购物系统，都是采取三层架构的模式设计系统，所以读者们可以照搬第 9 章中的目录结构来建立该系统。系统开发的 JavaBean 和 Servlet 类包如图 14.5 所示。

其中，action 包中存放对系统所有操作的相关类；dao 包中存放的是对系统数据操作的相关类；db 包中存放的是对系统数据库链接的相关类；factory 包中存放的是对数据实现的相关类；filters 包中存放的是系统的 Servlet 过滤器相关类；pojo 包中存放的是系统的 JavaBean 相关类；service 包中存放的是对数据真实操作的相关类；tag 包中存放的是系统的自定义标签相关类；utils 包中存放的是系统的工具类。

有关页面中的相关代码存放在 WebRoot 目录下，其中 common 包中存放的页面共有的页面；css 包中存放的是页面样式；js 存放的是页面用到的 JavaScript；其他目录是按照各功能模块存放相应的页面。

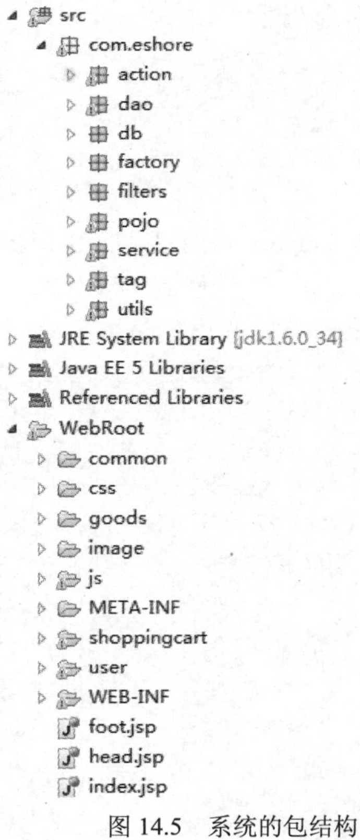

图 14.5 系统的包结构

14.4.2 系统的关键技术

1. 数据库连接

数据库连接是引用第 11 章中的数据库连接代码，代码的说明详见第 11 章。

2. 系统分页技术

在本系统中运用自定义标签的方式进行分页。首先，定义标签基本类，为的是以后可以扩展它。其原代码如下：

```
----------------------- BaseTagSupport.java---------------------------
01  import javax.servlet.ServletRequest;
02  import javax.servlet.jsp.tagext.TagSupport;
03
04  /**
05   *
06   *基本的标签类
07   */
08  public class BaseTagSupport extends TagSupport{
09
10      private static final long serialVersionUID = 1L;
11
12      protected ServletRequest getRequest(){
13          return pageContext.getRequest();
14      }
15  }
```

上述代码中，代码第 8 行继承 TagSupport 类，代码第 13 行获取 ServletRequest 请求对象，以后类标签类只要继承 BaseTagSupport 就可以获取到请求对象。

其次，建议分页标签类 PageTag.java 类，其原代码如下：

```
-----------------------PageTag.java---------------------------
……

public class PageTag extends BaseTagSupport {

    private static final Logger log = Logger.getLogger(PageTag.class);

    private PageObject object;      //分页对象

    private String link;                  //分页链接

    private String script;          //页面 javaScript 方法名
    //参数的 get 和 set 方法
     ……

    public int doStartTag() throws JspException {
        int[] iparams={0,0,0};
        String[] sparams={"",""};
        if(object!=null && object.getData()!=null){
            iparams[0]=object.getDataCount();
```

```
18                iparams[1]=object.getPageCount();
19                iparams[2]=object.getCurPage();
20                if(link!=null && link!=""){
21                    sparams[0]=link;
22                }
23                if(script!=null && script!=""){
24                    sparams[1]=script;
25                }
26
27            }
28            getRequest().setAttribute("iPageObjectTag", iparams);
29            getRequest().setAttribute("sPageObjectTag", sparams);
30            return EVAL_BODY_INCLUDE;
31        }
32
33        public int doEndTag() throws JspException {
34            getRequest().removeAttribute("iPageObjectTag");
35            getRequest().removeAttribute("sPageObjectTag");
36            return EVAL_PAGE;
37        }
38    }
```

上述代码中，代码第 05~09 行定义标签中用到的属性包括 PageObject、link、script，代码第 14~27 行用 int 类型数组分别存放分页中的数据、分页的页数、当前页，String 类型数组分别存放跳转链接和页面的方法。分页对象 PageObject 源代码如下：

```
-----------------------PageObject.java--------------------------
……
public class PageObject {

    private final int DEFAULT_PAGE_SIZE = 10;  // 默认显示记录数
    private final int DEFAULT_CUR_SIZE = 1;    // 默认当前页
    private List data;                           //数据列表
    private int dataCount;                       //数据总数
    private int pageSize;                            //显示记录数
    private int pageCount;                       //总页数
    private int curPage;                             //当前页

    public int getCurPage() {                        //获得当前页
        if (curPage < DEFAULT_CUR_SIZE) {
            curPage = DEFAULT_CUR_SIZE;
        }
        return curPage;
    }

    //参数的 get 和 set 方法省略
```

```
......
    public int getPageCount() {                //获得页数
        if (dataCount > 0) {
            pageCount = dataCount % pageSize == 0 ?
(dataCount / pageSize) : (dataCount / pageSize + 1);
        }
        return pageCount;
    }
    public int getPageSize() {                          //获得每页的显示数量
        if (pageSize < 1) {
            pageSize = DEFAULT_PAGE_SIZE;
        }
        return pageSize;
    }

    public void reqProperty(HttpServletRequest request) {
        String curPage = null, pageSize = null, dataCount = null;

        curPage = request.getParameter("curPage"); //设定当前页数
        if (curPage != null && curPage != "") {
            try {
                this.curPage = Integer.valueOf(curPage).intValue();
            } catch (NumberFormatException ex) {
            }
        }

        pageSize = request.getParameter("pageSize"); //设定每页的显示数量
        if (pageSize != null && pageSize != "") {
            try {
                this.pageSize = Integer.valueOf(pageSize).intValue();
            } catch (NumberFormatException ex) {
                ex.printStackTrace();
            }
        }

        dataCount = request.getParameter("dataCount");//设定总数量
        if (dataCount != null && dataCount != "") {
            try {
                this.dataCount = Integer.valueOf(dataCount).intValue();
            } catch (NumberFormatException ex) {
                ex.printStackTrace();
            }
        }
    }

```

```
64     public int getBeginPoint() {                    //获取开始的数据点
65         return (getCurPage() - 1) * getPageSize();
66     }
67     //获得 PageObject 对象
68     public static PageObject getInstance(HttpServletRequest request) {
69         PageObject pageObject = new PageObject();
70         pageObject.reqProperty(request);
71         return pageObject;
72     }
73 }
```

上述代码中，设定分页中需要用到的属性例如：数据列表、数据总数、总页数、当前页等如代码第 5~9 行所示；代码第 11~16 行获得当前页；代码第 20~26 行获得页数；代码第 45~62 行设定分页中数据总数、总页数、显示数量等值；代码第 67~71 行获取分页对象。

最后，定义一个标签文件 lms.tld 存放在 WEB-INF 目录下。其文件的内容如下：

```
----------------------lms. tld--------------------------
<?xml version="1.0" encoding="UTF-8"?>
<!DOCTYPE taglib PUBLIC "-//Sun Microsystems, Inc.//DTD JSP Tag Library 1.2//EN"
      "http://java.sun.com/dtd/web-jsptaglibrary_1_2.dtd">
<taglib>
    <tlib-version>1.0</tlib-version>
    <jsp-version>1.2</jsp-version>
    <short-name>lms</short-name>
    <uri>/lms-tags</uri>
    <tag>
        <name>page</name>
        <tag-class>com.eshore.tag.PageTag</tag-class>
        <body-content>JSP</body-content>
        <description>分页</description>
        <attribute>
            <name>object</name>
            <required>true</required>
            <type>com.eshore.tag.PageObject</type>
        </attribute>
        <attribute>
            <name>link</name>
            ……
        </attribute>
        <attribute>
            <name>script</name>
            ……
        </attribute>
    </tag>
```

```
</taglib>
```

如上代码，标签定义了 3 个属性 object、link、script，并设置它们的属性类型。具体说明可以参考第 10 章内容介绍。分页页面 page.jsp，其源代码如下：

```
-----------------------page.jsp---------------------------
<%@ page language="java" pageEncoding="UTF-8"%>
<%@taglib prefix="c" uri="http://java.sun.com/jsp/jstl/core" %>
<style type="text/css">
.page_bg td{ background:#d9ecf2; height:24px; border-bottom:1px solid #abc4de;}
</style>
<table width="100%" align="center" cellpadding="0" cellspacing="0">
            <tr class="page_bg">
                <td style="width:10%; " height="35" nowrap="nowrap"
>  共有 <strong><span id="count">
<c:out value="${iPageObjectTag[0]}"/></span></strong>条记录 </td>
                <td style="width:12%;" nowrap="nowrap"> 当前第 <strong>
<c:out value="${iPageObjectTag[2]}"/></strong> 页/共 <strong>
<c:out value="${iPageObjectTag[1]}"/></strong> 页</td>
                <td style="width:63%; line-height:35px; vertical-align:middle;
padding-top:5px; padding-right:5px;" valign="middle" align="right" nowrap="nowrap">
                <img title="第一页"
src="${pageContext.request.contextPath}/common/images/dg_btn_lt_end.gif"
border="0" onclick="toPage(1)" style="cursor:hand"/> 
                <img title="上一页"
 src="${pageContext.request.contextPath}/common/images/dg_btn_lt.gif"
border="0" onclick="toPage(${iPageObjectTag[2]-1})" style="cursor:hand"/> 
                <img title="下一页"
src="${pageContext.request.contextPath}/common/images/dg_btn_rt.gif"
border="0" onclick="toPage(${iPageObjectTag[2]+1})" style="cursor:hand"/> 
                <img title="最后一页"
src="${pageContext.request.contextPath}/common/images/dg_btn_rt_end.gif"
border="0" onclick="toPage(${iPageObjectTag[1]})" style="cursor:hand"/></td>
                <td style="width:9%;" nowrap="nowrap">到第
<input id="tagCurPage" size="2" maxlength="3"
onkeypress="return myKeyPress(event);" style="cursor:hand; position:relative;
```

```
top:2px; height:16px; padding:0px; margin:0px;"/> 页</td>
            <td style="width:5%;" colspan="6" nowrap="nowrap">
 <img
src="${pageContext.request.contextPath}/common/images/turnpage_go.gif"
 width="24" height="20" border="0" title="GO" onclick="toPage(-1)"
style="cursor:hand; position:relative; top:2px;"/></td>
         </tr>
</table>
```

上述代码就是分页页面的通用代码，在运用的时候就可以直接引入到要分页的页面中。

3. 自定义版权标签

自定义版权标签和标签文件 copryright..tld 可以参见第 10 章中自定义版权标签的介绍，这里就不给出它的源码。

4. 业务操作类

在开发系统中，尽量采用解耦方式开发程序，这样可以提高系统的可读性和可维护性。同样在本系统开发中个，在业务操作这部分中，尽可能分清数据的操作和数据库的操作，因此先定义一个业务操作类 DAOFactory 类，其作用是获取各个业务的数据操作方法，源代码如下：

```
----------------------- DAOFactory.java--------------------------
01  ……
02  public class DAOFactory {
03      //取得 Good 业务操作类
04      public static GoodDao getGoodDAOInstance()throws Exception {
05          return new GoodService();
06      }
07      //取得 shoppingcart 业务操作类
08      public static ShoppingCartDao getShoppingCartDAOInstance()throws
        Exception {
09          return new ShoppingCartService();
10      }
11      //取得用户业务操作类
12      public static UserDao getUserDAOInstance()throws Exception {
13          return new UsersService();
14      }
15  }
```

上述代码中，第 4~13 行分别获得商品操作类 GoodService、购物车操作类 ShoppingCartService、用户操作类 UsersService。在随后的章节中，会进一步展示它们的作用和代码。

14.4.3 过滤器

1. 字符过滤器

在第 5 章中，介绍过 Servlet 的过滤器使用，在本系统中也运用到字符过滤器和登录过滤器。其中，字符过滤器代码跟第 5 章中的过滤代码一样，具体说明参加第 5 章介绍。

2. 登录过滤器

登录过滤器的作用是为了防止用户没有登录系统，就对商品进行支付或者查看购物车中的商品，而设置的系统过滤。过滤的中判断 session 中是否有用户如果没有用户，则说明用户没有登录，让系统返回到登录页面。LoginFilter 源代码如下：

```
-----------------------LoginFilter.java---------------------------
……
@WebFilter(
        description = "登录过滤",
        filterName = "loginFilter",
        urlPatterns = { "/user/*","/shoppingcart/*" }
    )
public class LoginFilter implements Filter {

private static Logger log = Logger.getLogger("LoginFilter");
private String filterName="";//过滤器名称
public void destroy() {
    log.debug("请求销毁");
}
public void doFilter(ServletRequest req, ServletResponse res,
        FilterChain chain) throws IOException, ServletException {
    HttpServletRequest request = (HttpServletRequest) req;
    HttpServletResponse response = (HttpServletResponse) res;
    log.debug("请求被"+filterName+"过滤");
    String uname =(String) request.getSession().getAttribute("uname");
        //请求过滤，如果用户为空，返回登录页面
        if (uname == null) {
            request.setAttribute("status", "请先登录");
            request.getRequestDispatcher("/login.jsp")
            .forward(request, response);
        } else {
            chain.doFilter(req, res);
        }
    }

    public void init(FilterConfig filterConfig) throws ServletException {
        ……
```

```
    }
}
```

14.5 系统首页与公共页面

系统首页 index.jsp，是展示商品的页面，其源代码如下：

```
-----------------------index.jsp--------------------------
<%@ page language="java" import="java.util.*" pageEncoding="UTF-8"%>
<!DOCTYPE HTML PUBLIC "-//W3C//DTD HTML 4.01 Transitional//EN">
<html>
    <head>
        <title>淘淘网—开心淘！</title>
        <jsp:include page="common/common.jsp"/>
        <script type="text/javascript" src="js/common/index.js"></script>
    </head>

    <body>
        <div align="center">
            <div id="top">
                <jsp:include page="head.jsp"></jsp:include>
            </div>
            <p>
            <div id="logoselect">
                <jsp:include page="logo_select.jsp"></jsp:include>
            </div>
            <input id="status" type="hidden" name="status" value="${status}">
            <div id="main">
                <div>
                    <br>
                    <table border="1" id="list">
                        <tr class="goodlist">
                            <td>
                                <br/>
                                数
                                <br/>
                                <br/>
                                码
                                <br/>
                            <td>
                            <td>
                                <a
href="goods?keyWord=cellphone&keyClass=2&action=index-select">品牌手机</a>
                                <br>
                                <a
href="goods?keyWord=nokia&keyClass=4&action=index-select">诺基亚</a>|
```

```
39                                  <a
40  href="goods?keyWord=iphone&keyClass=4&action=index-select">iphone</a>|
41                              </td>
                              ……//代码跟<td>中类似
42                          </tr>
43                      </table>
44                  </div>
45              </div>
46              <div id="foot">
47                  <jsp:include page="foot.jsp"></jsp:include>
48              </div>
49          </div>
50      </body>
51  </html>
```

如上代码，首页有 3 个部分构成，代码第 12~14 行是首页的顶部内容；16~18 行是首页的搜索框内容；第 20~69 行是首页的正文内容，用超链接来显示各个商品；第 70~72 行是首页的底部内容，显示版权信息；代码第 6 行是引入公共的样式页面，其代码如下：

```
01  <link rel="stylesheet" type="text/css" href="css/styles.css">
02  <script type="text/javascript" src="js/jquery.js"></script>
03  <script type="text/javascript" src="js/jquery.validate.js"></script>
04  <script type="text/javascript" src="js/messages_cn.js"></script>
```

上述代码第 1 行，导入系统用到的样式文件；第 2 行引入 JQuery.js。head.jsp 页面中包含有判断用户是否已经登录的判断，以及显示超链接等信息。

底部 foot.jsp 页面的源代码如下：

```
-----------------------foot.jsp--------------------------
01  <%@ taglib prefix="linl" uri="/copyright-tags" %>
02  <!DOCTYPE HTML PUBLIC "-//W3C//DTD HTML 4.01 Transitional//EN">
03  <html>
04      <head>
05          <title>淘淘网—开心淘！</title>
06      </head>
07      <body>
08          <div align="center">
09          <hr>
10              <font size="2" color="black">
11              <linl:copyright startY="2014" user="linl"/>
12                          <a
13                  href="swarding99@163.com">联系我们</a> </font>
14          </div>
15      </body>
16  </html>
```

上述代码中，代码第 1 行用自定义版权标签；代码第 11~13 行用版权标签显示内容。首页的效果如图 14.6 所示。

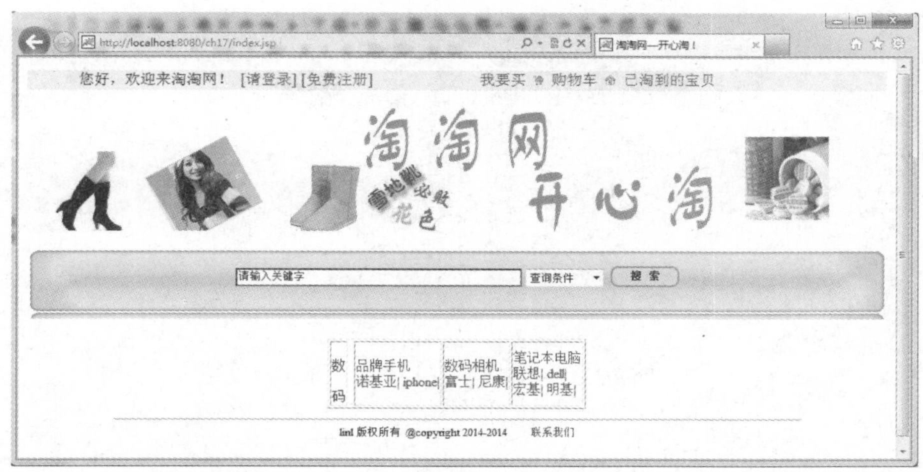

图 14.6　在线购物系统首页

14.6 用户登录模块

用户登录界面 login.jsp，其页面提供用户名跟密码输入框，输入用户名跟密码，提交验证成功则跳转到系统首页，否则提示相应的错误信息。前面章节有类似的代码，这里由于篇幅原因就不再叙述。

form 表单中跳转的 Servlet 类 LoginServlet.java，其源代码如下：

```
-----------------------LoginServlet.java--------------------------
package com.eshore.action;

@WebServlet(
    urlPatterns = { "/login" },
    name = "loginServlet"
)
public class LoginServlet extends HttpServlet {
    public void doPost(HttpServletRequest request, HttpServletResponse response)
            throws ServletException, IOException {
        String uname = request.getParameter("uname");  //获取用户名
        String passwd = request.getParameter("passwd");    //获取用户密码
        String action = request.getParameter("action"); //获取 action 类型
        String path = null;
        try{
            if (action.equals("login")) {          //如果是登录
                Users user = DAOFactory.getUserDAOInstance().
                    queryByName(uname);            //根据用户名查询用户
                if (passwd.equals(user.getPasswd())) {
                //输入的密码与数据库中的一致
```

```
19                      request.getSession().setAttribute("uname", uname);
20                      request.getSession().setAttribute("uid",
                        user.getUid());
21                      path = "index.jsp";
22                  } else {
23                      request.setAttribute("status", "用户名或密码错误！");
24                      path = "login.jsp";
25                  }
26              } else if (action.equals("logout")) {
                //用户退出，注销 session 中的用户
27                  request.getSession().removeAttribute("uname");
28                  request.getSession().removeAttribute("uid");
29                  path = "login.jsp";
30              }
31          }catch(Exception e){
32              e.printStackTrace();
33          }
34          request.getRequestDispatcher(path).forward(request, response);
35      }
36  }
```

上述代码中第 3~6 行用注入的方式声明了 Servlet；代码第 15~30 行，判断页面中的 action 参数是登录或是退出，如果是登录操作则验证密码是否输入正确，并保存到 session 中，如果是注销操作则注销 session 中的用户。登录界面如图 14.7 所示。

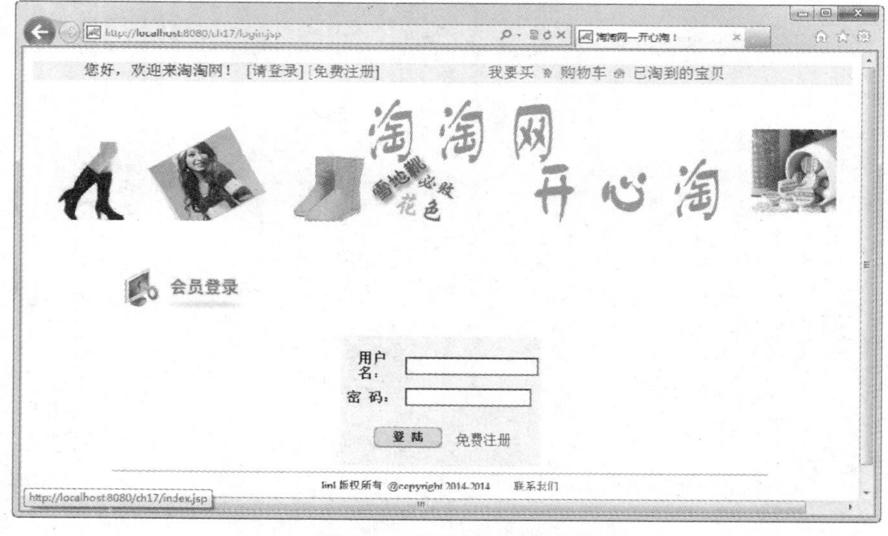

图 14.7　登录界面效果图

14.7 用户管理模块

用户管理模块内容包括用户注册、用户信息修改、用户密码修改等操作。下面逐一进行介绍。

14.7.1 用户注册

用户注册页面 register.jsp，该页面提供给用户注册系统用户，需要用户提供用户名、密码、邮箱等信息，其中用户名和邮箱必须是在本系统中未被注册过的。用 form 表单提供用户名、密码、邮箱输入框且都是必输项，action 路径是 register，提交方法时 POST，页面的底部公共页即显示版权页面。由于篇幅问题，页面源代码可以参考前面章节中的部分代码，这里省略源代码。注册提交的 RegisterServlet 类源代码如下：

```
---------------------- RegisterServlet.java--------------------------
@WebServlet(
    urlPatterns = { "/register" },
    name = "registerServlet"
)
public class RegisterServlet extends HttpServlet {

    public void doPost(HttpServletRequest request, HttpServletResponse response)
            throws ServletException, IOException {
        //获取页面参数，包括用户名密码邮箱
        String uname = request.getParameter("uname");
        String passwd = request.getParameter("passwd");
        String email = request.getParameter("email");
        String path = null;
        //为用户设定属性值
        Users user = new Users(uname,passwd,email);
        if (DAOFactory.getUserDAOInstance().
                queryByName(uname).getUid() == 0) {          //用户名可用
            if (DAOFactory.getUserDAOInstance().
                    queryByEmail(email).getUid() == 0) {         //邮箱可用
                if (DAOFactory.getUserDAOInstance().addUser(user) == 1) {
                    request.getSession().setAttribute("uname", uname);
                    request.getSession().setAttribute("uid",
            DAOFactory.getUserDAOInstance().queryByName(uname).getUid());
                    path = "index.jsp";
                    request.setAttribute("status", "恭喜您，注册成功！");
                } else {
                    path = "register.jsp";
```

```
28                      request.setAttribute("status", "注册失败，请重试……");
29                  }
30              } else {
31                  path = "register.jsp";
32                  request.setAttribute("status", "电子邮箱已被注册");
33              }
34          }else{
35              path = "register.jsp";
36              request.setAttribute("status", "用户名已被注册");
37          }
38          request.getRequestDispatcher(path).forward(request, response);
39      }
40  }
```

上述代码中，代码第 01~04 行声明方式声明 Servlet，并配置 url 为 register；代码第 10~13 行接收页面的传递的用户名、密码、邮箱等参数；代码第 17~36 行，判断用户名和邮箱是否被注册过，如果被注册过提示错误消息，如果都未被注册过则提示“注册成功”并跳转到首页。DAOFactory 获得用户操作类 UsersService，其源代码如下：

```
-----------------------UsersService.java---------------------------
01  public class UsersService implements UserDao {
02
03      private DBConnection dbconn = null;               //定义数据库连接类
04      private UserDao dao = null;                   //声明 DAO 对象
05      // 在构造方法中实例化数据库连接，同时实例化 dao 对象
06      public UsersService() throws Exception {
07          this.dbconn = new DBConnection();
08          this.dao = new UserDaoImpl(this.dbconn.getConnection());
09        // 实例化 GoodDao 的实现类
10      }
11      public int addUser(Users user) throws Exception {
12          ……
13      }
14  ……
15  }
```

上述代码中，代码第 3~4 行，声明一个数据库连接和数据操作 UserDao 对象；代码第 6~7 行在构造方法中初始化数据库连接和 UserDao 对象实例化。UsersService 类的主要作用是启动数据库连接和关闭数据库连接，有关数据的具体操作在 UserDaoImpl 中进行。

UserDao 接口源代码如下：

```
----------------------- UserDao.java---------------------------
public interface UserDao {
    //添加用户
    public int addUser(Users user) throws Exception;
```

```
    //修改用户信息
    public int editInf(int uid,String uname,String email) throws Exception;
    //修改用户密码
    public int editPasswd(int uid,String passwd) throws Exception;
    //根据用户 id，删除用户
    public int deleteUser(int uid) throws Exception;
    //根据用户名查询用户
    public Users queryByName(String uname) throws Exception;
    //根据用户 Email 查询用户
    public Users queryByEmail(String email) throws Exception;
}
```

上述接口代码中，是定义用户注册中需要用到的方法。有关具体对数据的操作在实现类 UserDaoImpl 中，部分源代码如下：

```
----------------------- UserDaoImpl.java---------------------------
01 public class UserDaoImpl implements UserDao {
02 
03     private Connection conn = null;              //数据库连接对象
04     private PreparedStatement pstmt = null;     // PreparedStatement 对象
05     ResultSet rs = null;
06 
07     // 通过构造方法取得数据库连接
08     public UserDaoImpl(Connection conn) {
09         this.conn = conn;
10     }
11     public int addUser(Users user) throws Exception{
12         String sql = "insert into users(uname,passwd,email,lastlogin)
13                 values(?,?,?,sysdate())";
14         int result = 0;
15         pstmt = this.conn.prepareStatement(sql);     //获取 PreparedStatement 对象
16         pstmt.setString(1, user.getUname());    //设定用户用户名
17         pstmt.setString(2, user.getPasswd());   //设定用户密码
18         pstmt.setString(3, user.getEmail());    //设定用户 Email
19         result = pstmt.executeUpdate();         //执行数据库操作
20         pstmt.close();
21         return result;
22     }
23 ……
24 }
```

上述代码中，代码第 3~4 中声明数据库连接对象、PreparedStatement 对象；代码第 11~22 行实现新增用户的方法，代码第 12~13 行写出 sql 语句，第 15 行获得 PreparedStatement 对象，第 15~18 行向 sql 语句中填入参数值，第 19 行执行数据库操作方法并返回结果值。用户实体

类 User.java 的源代码如下：

```
-----------------------User.java---------------------------
01  public class Users {
02
03      private int uid;                               //用户 id
04      private String uname;                          //用户名
05      private String passwd;                         //用户密码
06      private String email;                          //用户的 Email
07      private Date lastLogin;                        //最后的登录时间
08       //省略 get 和 set 方法
09  }
```

上述代码中，代码第 3~7 行列出 User 对象所用到的 JavaBean 属性，并给出属性的 get 和 set 方法。

注册页面效果图见图 14.8 所示。

图 14.8　注册页面效果图

14.7.2　修改用户信息

修改用户信息页面 editinfo.jsp，其内容相对简单，form 表单提供用户名、邮箱等输入框，action 则跳转到 UserServlet 类中。其页面的内容可以参考用户 register.jsp 注册页面，这里省略其源代码。

因为用户管理模块中将会有很多个操作方法例如修改用户信息、修改用户名密码、查看用户信息等动作，为了降低耦合度并提高代码的复用性和可读性，本系统建立一个 UserServlet，以页面传递的 action 为标识，来判断具体执行的是哪个操作。UserServlet.java 的源代码如下：

```
---------------------- UserServlet.java--------------------------
01  @WebServlet(
02          urlPatterns = { "/user" },
03          name = "userServlet"
04      )
05  public class UserServlet extends HttpServlet {
06  
07      public void doPost(HttpServletRequest request, HttpServletResponse
        response)
08              throws ServletException, IOException {
09  
10          String action = request.getParameter("action");
11          Action targetAction =null;
12          String path = null;
13          if (action.equals("show")) {                    //查看用户列表
14              targetAction = new ShowUserAction();
15              path=targetAction.execute(request, response);
16          } else if (action.equals("editinf")) {          //修改用户信息
17              targetAction = new EditinfUserAction();
18              path=targetAction.execute(request, response);
19          }……
20          request.getRequestDispatcher(path).forward(request, response);
21      }
22  }
```

上述代码的逻辑是十分清晰的，代码 1~3 行声明一个 Servlet；代码第 10 行获得页面传送的 action 参数值；代码第 13~19 行根据 action 值判断执行的是哪步操作，显然代码执行的是第 16 行所示的修改用户信息操作，所以程序跳转到 EditinfUserAction 方法中并返回页面的跳转信息。EditinfUserAction 类的是具体执行修改用户信息的方法，其源代码如下：

```
---------------------- EditinfUserAction.java--------------------------
public class EditinfUserAction implements Action{
    public String execute(HttpServletRequest request,
        HttpServletResponse response) throws ServletException,
        IOException {
        //获取用户的 id 值
        int uid = Integer.parseInt(String.valueOf(
        request.getSession().getAttribute("uid")));
        //获取用户的用户名
        String uname = request.getParameter("uname");
        //获取用户 email
        String email = request.getParameter("email");
```

```
11            //根据用户名查询用户
12            Users user=DAOFactory.getUserDAOInstance().queryByName(
13                    String.valueOf(request.getSession().getAttribute("uname")));
14            if(user.getUname().equals(uname)||
15                    DAOFactory.getUserDAOInstance().
16                    queryByName(uname).getUid()==0){          //用户名未注册
17                if(user.getEmail().equals(email)||
18                        DAOFactory.getUserDAOInstance().
19                        queryByEmail(email).getUid()==0){   //邮箱未被注册
20                    if(DAOFactory.getUserDAOInstance().
21                            editInf(uid, uname, email)==1){//用户信息修改成功
22                        request.getSession().setAttribute("uname", uname);
23                        request.setAttribute("status", "信息修改成功！");
24                    }else{//用户信息修改失败
25                        request.setAttribute("status","修改操作失败,请重试!");
26                    }
27                }else{//邮箱已经被注册
28                request.setAttribute("status", "电子邮箱账号已被注册,请换一个！");
29                }
30            }else{//判断用户名已经存在
31                request.setAttribute("status", "用户名已存在，请换一个！");
32            }
33            return "shoppingcart?action=lookbus";
34        }
35    }
```

上述代码中，第 5~6 行，先从 session 中取得登录用户的 id 值，代码第 8~16 行取得页面输入的用户名和邮箱；然后利用 UserService 中的方法对用户进行判断，如果用户名已经存在，提示用户名已存在例如代码第 31 行；如果邮箱已经被注册，提示邮箱账号已经被注册例如代码第 28 行所示；如果用户修改信息成功，则提示信息修改成功如代码第 24~30 行所示。

接口 Action 中的方法很简单，就是提供一个 String 的返回值，传入 HttpRequest、HttpResponse。其源代码如下：

```
---------------------- Action.java--------------------------
/**
 * 业务操作的接口类
 */
public interface Action {
    public String execute(HttpServletRequest request,
            HttpServletResponse response)
            throws ServletException, IOException;
}
```

跳转至修改页面的 action 方法 EditAction.java 的源代码如下：

```
---------------------- EditUserAction.java--------------------------
```

```
09  public class EditUserAction implements Action {
10
11      public String execute(HttpServletRequest request,
12              HttpServletResponse response) throws ServletException,
                IOException {
13          //根据用户名查询用户
14          Users user = DAOFactory.getUserDAOInstance().queryByName(
15          String.valueOf(request.getSession().getAttribute("uname")));
16          System.out.println(user.getEmail());
17          request.setAttribute("email", user.getEmail());
18          return "user/editinf.jsp";
19      }
20  }
```

上述代码中，第 15 行从 session 中查询已经登录的用户，通过 UserService 查询出该用户，获得该用户的邮箱并输出到页面中。

修改用户信息的页面效果如图 14.9 所示。

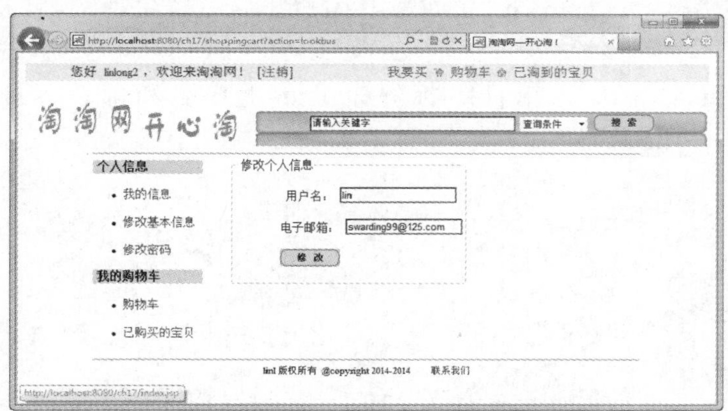

图 14.9　修改用户信息页面效果图

14.7.3　查看用户信息

查看用户信息的结果页面 myinf.jsp，其主要功能是显示用户信息，这里为了简单就显示用户名和邮箱等信息，其源代码如下：

```
-----------------------myinf.jsp--------------------------
<%@ page language="java" import="java.util.*" pageEncoding="UTF-8"%>
<!DOCTYPE HTML PUBLIC "-//W3C//DTD HTML 4.01 Transitional//EN">
<html>
    <head>
        <title>查看用户</title>
    </head>

    <body>
```

```
09          <div align="center" style="width: 60%; padding-left: 10%">
10              <fieldset>
11                  <legend>
12                      个人信息
13                  </legend>
14                  <div align="left" style="padding-left: 20%">
15                      <p>
16                          <label>
17                                用户名:
18                          </label>${uname }<br/>
19                          <label>
20                              电子邮箱:
21                          </label>${email }
22                      <p>
23                  </div>
24              </fieldset>
25          </div>
26      </body>
27  </html>
```

上述代码，第 18~21 行用 EL 标签来显示用户名和邮箱。业务中的方法类 ShowUserAction.java 的源代码如下：

```
-----------------------ShowUserAction.java---------------------------
01  public class ShowUserAction implements Action{
02  
03      public String execute(HttpServletRequest request,
04              HttpServletResponse response) throws ServletException,
                IOException {
05              //根据用户名查询用户
06              Users user = DAOFactory.getUserDAOInstance().queryByName(
07              tring.valueOf(request.getSession().getAttribute("uname")));
08              request.setAttribute("email", user.getEmail());
09          return "user/myinf.jsp";
10      }
11  }
```

上述代码中，第 6 行根据 session 中已经登录的用户，通过 UserService 方法查询出用户，并将邮箱输出到页面中。

14.7.4 修改用户密码

修改用户密码的页面 editpasswd.jsp 页面，提供 form 表单，表单中有密码输入框，根据比对密码的合法性，判断是否修改成功，页面跳转到 EditPasswdAction 类方法中，源码可参见 register.jsp 页面，页面效果见图 14.10 所示。

EditPasswdAction 类的源代码如下：

```
-----------------------EditPasswdAction.java--------------------------
01  public class EditPasswdAction implements Action{
02      public String execute(HttpServletRequest request,
03              HttpServletResponse response) throws ServletException,
                IOException {
04          //获取用户的 id 值
05          int uid = Integer.parseInt(String.valueOf(
06                  request.getSession().getAttribute("uid")));
07          //获取旧密码
08          String oldPasswd = request.getParameter("oldPasswd");
09          //获取新密码
10          String passwd = request.getParameter("passwd1");
11          String confirdPasswd = request.getParameter("passwd2");
12              //根据用户名查询用户
13              Users user =DAOFactory.getUserDAOInstance().
14                  queryByName(String.valueOf(
15                          request.getSession().getAttribute("uname")));
16              //判断输入的旧密码跟原来的旧密码是否一致,
17              //如果一致进行修改
18              if(user.getPasswd().equals(oldPasswd)){
19                  if(isValidPassword(passwd,confirdPasswd)){      //验证密码
20                      request.setAttribute("status", "密码为空或者密码不一致！");
21                  }
22                  if(DAOFactory.getUserDAOInstance().
23                          editPasswd(uid, passwd)==1){    //密码修改成功
24                      request.setAttribute("status", "密码修改成功！");
25                  }else{                                  //密码修改失败
26                      request.setAttribute("status","密码修改操作失败，请重试！");
27                  }
28              }else{                                      //输入密码错误
29                  request.setAttribute("status", "原密码错误，你不能修改密码！");
30              }
31          return "shoppingcart?action=lookbus";
32      }
33      //验证密码，如果密码为空且长度小于6并且跟确认密码不统一
34      //返回 true
35      public boolean isValidPassword(String passwd,String confirdPasswd){
36          return passwd==null||confirdPasswd==null
37          ||passwd.length()<6||confirdPasswd.length()<6
38          ||!passwd.equals(confirdPasswd);
39      }
40  }
```

上述代码中，代码第 5 行获得 session 中已登录的用户 id 值；代码第 8~11 行获取页面中

获得的输入密码值；代码第 22~30 行通过 UserService 方法验证密码的合法性，验证通过则修改成功，否则提示相应的错误信息。

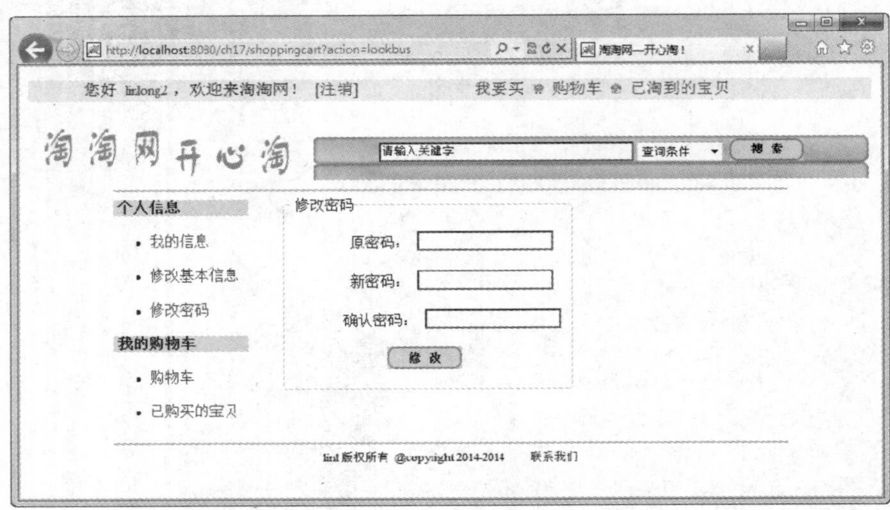

图 14.10　editpasswd.jsp 页面效果图

14.8 购物车模块

购物车模块内容包括添加购物车、删除购物车、查看购物车等功能。

14.8.1　添加购物车

同样的，为了代码的可维护性和可读性，只建立一个 ShoppingCartServlet，用 action 来标识跳转到哪个方法中。ShoppingCartServlet.java 的源代码如下：

```
----------------------- ShoppingCartServlet.java--------------------------
@WebServlet(
    urlPatterns = { "/shoppingcart" },
    name = "shoppingCartServlet"
)
public class ShoppingCartServlet extends HttpServlet {

    public void doPost(HttpServletRequest request, HttpServletResponse response)
            throws ServletException, IOException {
        String uids = String.valueOf(request.getSession().getAttribute("uid"));
        String action = String.valueOf(request.getParameter("action"));
        Action targetAction =null;
```

```
12          String path = null;
13          try{
14              if (uids == null || uids.equals("null")) {
15                  path = "login.jsp";
16              } else {
17                  if (action.equals("deletebus")) {       //删除购物车
18                      targetAction = new DeletShoppingCartAction();
19                      path=targetAction.execute(request, response);
20              } else if (action.equals("intobus")) { //点击加入购物车时处理
21                      targetAction = new InsertShoppingCartAction();
22                      path=targetAction.execute(request, response);
23                  }………
24              }
25          }catch(Exception e){
26              e.printStackTrace();
27          }
28          request.getRequestDispatcher(path).forward(request, response);
29      }
30  }
```

上述代码中，第 10 行获得页面传入的 action 标识，根据 action 值跳转到相应的方法中。InsertShoppingCartAction.java 代表添加到购物车中，其源代码如下：

```
----------------- InsertShoppingCartAction.java--------------------------
public class InsertShoppingCartAction implements Action {

    public String execute(HttpServletRequest request,
            HttpServletResponse response) throws ServletException,
            IOException {
                                    //获取商品的 ID
        int gid = Integer.parseInt(String.valueOf(request
                .getParameter("gid")));
                                    //获取商品的数量
        int number = Integer.parseInt(String.valueOf(request
                .getParameter("number")));
                                    //获取登录用户的 ID
        String uids = String.valueOf(request.getSession().getAttribute
        ("uid"));
        int uid = Integer.parseInt(uids);
        ShoppingCart bus = DAOFactory.getShoppingCartDAOInstance().
                getGoodsId(uid, gid, 0);
        if (bus.getId() == 0) {               // 如果购物车中不存在则加入购物车
            DAOFactory.getShoppingCartDAOInstance().addBus(gid, uid,
            number);

        } else {                              // 否则修改未付款的商品数量
```

```
20            DAOFactory.getShoppingCartDAOInstance().updatebus(bus.getId(),
21                    bus.getNumber() + number, 0);
22            }
23            request.setAttribute("status", "已将该宝贝添加到您的购物车");
24            return "goods?sid=" + gid
25                    + "&action=goodslist-select";
26        }
27    }
```

如上代码中，第 6~10 行分别获取商品的 id、商品的数量、登录用户的 id；代码第 16~22 行判断该商品是否存在于购物车中，如果不存在则直接添加，否则修改购物车中该商品的数量；最后返回商品的列表页面。DAOFactory 获得 ShoppingCartService 类，其作用就是对购物车进行连接数据库和关闭数据库的作用，源代码如下：

```
---------------------- ShoppingCartService.java--------------------------
01  public class ShoppingCartService implements ShoppingCartDao {
02
03      private DBConnection dbconn = null;          // 定义数据库连接类
04      private ShoppingCartDao dao = null;          // 声明 DAO 对象
05      // 在构造方法中实例化数据库连接，同时实例化 dao 对象
06      public ShoppingCartService() throws Exception {
07          this.dbconn = new DBConnection();
08          this.dao = new ShoppingCartDaoImpl(this.dbconn.getConnection());
09          // 实例化 GoodDao 的实现类
10      }
11      ………
12      public PageObject getPageObject(String curPage, PageObject pageObject,
13              List<Object> listObject) {
14          pageObject = this.dao.getPageObject(curPage, pageObject, listObject);
15          return pageObject;
16      }
17  }
```

上述代码中，第 7~8 行新建数据库连接，而后在实现具体的数据操作。具体的数据执行动作在 ShoppingCartDaoImpl 类中，其源代码如下：

```
-----------------------ShoppingCartDaoImpl.java--------------------------
public class ShoppingCartDaoImpl implements ShoppingCartDao {

    private Connection conn = null;                  //数据库连接对象
    private PreparedStatement pstmt = null;              //数据库操作对象

    ResultSet rs = null;
    Vector<ShoppingCart> busVector = new Vector<ShoppingCart>();

    // 通过构造方法取得数据库连接
```

```
    public ShoppingCartDaoImpl(Connection conn) {
        this.conn = conn;
    }
    //删除指定的购物车信息
    public int deleteGoods(int gid, int uid,int status) throws Exception{
        String sql = "delete from shoppingcart where uid=? and gid=? and status=?";
        int result = 0;
        this.pstmt=this.conn.prepareStatement(sql);
        //获取 PreparedStatement 对象
        this.pstmt.setInt(1, uid);
        this.pstmt.setInt(2, gid);
        this.pstmt.setInt(3, status);
        result = pstmt.executeUpdate();      //执行数据库操作
        this.pstmt.close();                  //关闭 PreparedStatement 操作

        return result;
    }
    ……
}
```

上述代码中就是对购物车中的数据进行具体的数据操作。接口 ShoppingCartDao 的源代码如下：

```
---------------------- ShoppingCartDao.java--------------------------
public interface ShoppingCartDao {

    ///根据购物车状态，用户 id 查询购物车
    public Vector<ShoppingCart> getAppointedGoods(int uid, int
    status)throws Exception;
    //根据用户 id 获取所有的商品
    public Vector<ShoppingCart> getAllGoods(int uid)throws Exception;
    //根据购物车状态，商品 id，用户 id 查询购物车
    public ShoppingCart getGoodsId(int uid, int gid, int status)throws
    Exception;
    //根据购物车状态，商品 id，用户 id 删除购物车
    public int deleteGoods(int gid, int uid, int status)throws Exception;
    //根据用户 id，购物车状态删除购物车
    public int deleteAll(int uid, int status)throws Exception;
    //添加购物车
    public int addBus(int gid, int uid, int number)throws Exception;
    //修改购物车信息
    public int updatebus(int id, int number, int status)throws Exception;
    //更新购物车信息
    public int updateShopcarts(String ids,int status) throws Exception;
    //购物车的分页对象
    public PageObject getPageObject(String curPage,PageObject
```

```
        pageObject,List<Object> listObject);
}
```

上述代码中，定义接口所用到的方法。其他要使用购物车中的方法可以直接调用该接口的方法。

14.8.2 删除购物车

DeletShoppingCartAction 类是删除购物车的方法以及跳转的页面标识，首先获取商品的 ID 值和登录用户的 ID，从库中查询出指定的商品，然后再删除，源代码如下：

```
---------------------- DeletShoppingCartAction.java------------------------
01  public class DeletShoppingCartAction implements Action {
02
03      public String execute(HttpServletRequest request,
04              HttpServletResponse response) throws ServletException,
                IOException {
05          //获取商品的 ID
06          int gid = Integer.parseInt(String.valueOf(request
07                  .getParameter("gid")));
08          //获取登录用户的 ID
09          String uids = String.valueOf(request.getSession().getAttribute("uid"));
10          int uid = Integer.parseInt(uids);
11              if (DAOFactory.getShoppingCartDAOInstance().
12                      deleteGoods(gid, uid, 0) == 1) {     //删除购物车中指定的商品
13                  request.setAttribute("status", "已从购物车中删除商品");
14              } else {                                    //删除失败
15                  request.setAttribute("status", "删除商品操作失败，请重试");
16              }
17          return "shoppingcart?action=lookbus";
18      }
19  }
```

上述代码中，第 6 行获得商品的 ID 值，代码第 9 行获得登录用户的 ID 值，代码第 11~14 行根据 ShoppingCartService 删除指定商品。

14.8.3 查看购物车

ShowShoppingcartAction 类用于查看购物车列表类，根据登录用户的 ID 获取购物车列表，然后进行遍历获取相关的信息，源代码如下：

```
----------------------ShowShoppingcartAction.java------------------------
public class ShowShoppingcartAction implements Action {

    public String execute(HttpServletRequest request,
            HttpServletResponse response) throws ServletException,
```

```
            IOException {
05      //新建 TempGoods 对象
06      Vector<TempGoods> tempVector = new Vector<TempGoods>();
07      //获取登录用户的 ID
08      String uids = String.valueOf(request.getSession().getAttribute
        ("uid"));
09      int uid = Integer.parseInt(uids);
10      float countPrice = 0.0f;
11          //获取用户的所有未支付的购物车列表
12          Vector<ShoppingCart> busVector =
13          DAOFactory.getShoppingCartDAOInstance().
14                  getAppointedGoods(uid, 0);
15          for (int i = 0; i < busVector.size(); i++) {
16              ShoppingCart cart = new ShoppingCart();
17              cart = (ShoppingCart) busVector.get(i);     //获取购物车
18              Goods good=new Goods();
19              TempGoods tempGoods = new TempGoods();
20              Vector<Goods> gVector=DAOFactory.getGoodDAOInstance().
21                  queryGoodBySid(cart.getGid());      //获取指定商品
22              if(gVector.size()>0&&gVector!=null)
23                  good =(Goods)gVector.get(0);
24              //组合 TempGoods 对象
25              tempGoods.setGood(good);
26              tempGoods.setNumber(cart.getNumber());
27              tempVector.add(tempGoods);
28              countPrice+=cart.getNumber()*good.getPrice();//计算价格
29          }
30      request.setAttribute("goods", tempVector);
31      request.setAttribute("countPrice",countPrice);
32      return "shoppingcart/bus.jsp";
33    }
34  }
```

上述代码中，第 17 行获得购物车列表集合；代码第 19~32 行遍历购物车列表集合，并获得指定商品；代码第 29~32 行组合 TempGoods 对象。在该方法中，还使用到了两个类 TempGoods 和 ShoppingCart 的 JavaBean，源代码分别如下：

```
----------------------- ShoppingCart.java---------------------------
public class ShoppingCart {

    private int id;                                  //购物车 id
    private int gid;                                 //商品 id
    private int uid;                                 //用户 id
    private int number;                              //物品的数量
    private int status;                              //1：已付款；0：未付款
    //省略 get 和 set 方法
```

```
09  }
```

代码第 1~9 行列出 ShoppingCart 对象的属性并给出它们的 get 和 set 方法。

临时购物车对象 TempGoods 源代码如下：

```
01  public class TempGoods {
02      private Goods good;                          //商品对象
03      private int number;                          //购买的数量
04       //省略 get 和 set 方法
05  }
```

代码第 1~5 行列出 TempGoods 对象的属性并给出它们的 get 和 set 方法。

物品对象 Goods 的源代码如下：

```
01  public class Goods {
02
03      private int gid;
04      private String kinds;        // 类型
05      private String gname;        // 名字
06      private String gphoto;       // 实物图片
07      private String types;        // 型号
08      private String producer;     // 生产商
09      private String paddress;     // 出产地
10      private String described;    // 描述
11      private Date pdate;          // 生产日期
12      private float price;         // 单价
13      private float carriage;      // 运费
14      private int keyclass;        // 小类别
15      private int big_keyclass;    // 大类别
16      private String keyword;      // 类别名称
17     //省略 get 和 set 方法
18  }
```

代码第 1~18 行列出 Goods 对象的属性并给出它们的 get 和 set 方法。

14.8.4 修改购物车

EditShoppingCartAction 类是修改购物车类，获取商品的 ID 值和商品的数量，再根据登录用户的 ID 获取购物车列表，取得购物车列表信息，源代码如下：

```
public class EditShoppingCartAction implements Action {

    public String execute(HttpServletRequest request,
            HttpServletResponse response) throws ServletException,
            IOException {
        //获取商品的 ID
        int gid = Integer.parseInt(String.valueOf(request
```

```
                    .getParameter("gid")));
        //获取商品的数量
        int number = Integer.parseInt(String.valueOf(request
                    .getParameter("number")));
        //获取登录用户的 ID
        String uids = String.valueOf(request.getSession().getAttribute("uid"));
        int uid = Integer.parseInt(uids);
            //获得指定的购物车列表
            ShoppingCart bus = DAOFactory.getShoppingCartDAOInstance().
                    getGoodsId(uid, gid, 0);
            DAOFactory.getShoppingCartDAOInstance().  //更新购物车中商品数量
                    updatebus(bus.getId(), number, 0);
        return "shoppingcart?action=lookbus";
    }
}
```

上述代码中，分别获取页面中商品的 ID 和商品的数量，然后对指定的购物车进行修改，修改成功后跳转到查看购物车方法中。

14.8.5　删除购物车所有商品

DeleteallAction 类是删除购物车中所有商品，根据登录用户的 ID 值获取其购物车对象进行删除操作，源代码如下：

```
01  public class DeleteallAction implements Action {

02      public String execute(HttpServletRequest request,
03              HttpServletResponse response) throws ServletException,
                IOException {
04          //获取登录用户的 ID
05          String uids = String.valueOf(request.getSession().getAttribute("uid"));
06          int uid = Integer.parseInt(uids);
07          if (DAOFactory.getShoppingCartDAOInstance().
08                  deleteAll(uid, 0) > 0) {                    //删除购物车商品
09              request.setAttribute("status", "您的购物车中没有商品。");
10          } else {                                        //删除失败
11              request.setAttribute("status", "删除商品操作失败，请重试。");
12          }
13          return "shoppingcart?action=lookbus";
14      }
15  }
```

上述代码中，代码第 7~8 行删除购物车中的商品。

14.8.6 购物车中的页面

查看购物车页面 bus.jsp，主要是显示购物车中商品信息，包括商品的价格，购买的数量等，源代码如下：

```
------------------------bus.jsp---------------------------
<!DOCTYPE HTML PUBLIC "-//W3C//DTD HTML 4.01 Transitional//EN">
<html>
    <head>
        <title>淘淘网—开心淘！</title>
        <jsp:include page="../common/common.jsp"/>
        <script type="text/javascript" src="js/shopcart/bus.js"></script>
    </head>

    <body>
        <div id="top">
            <jsp:include page="../head.jsp"/>
        </div>
        <p>
        <div>
            <jsp:include page="../logo_select1.jsp"/>
        </div>
        <input id="status" type="hidden" name="status" value="${status }">
        <div align="center">
            <div style="width: 80%; height: 78%;">
                <div id="left" align="left">
                    <div style="padding-top: 2px;">

                        <div id="title">
                        我的购物车
                    </div>
                    <ul>
                        <li>
                            <a href="shoppingcart?action=lookbus">购物车
                                        </a>
                            <p>
                        <li>
                        <a href="shoppingcart?action=paid">已购买的宝贝
                                        </a>
                            <p>
                    </ul>
                </div>
            </div>
            <div id="right" align="left" style="width: 100%;height:100%">
                <div
                   style="padding-right: 3%; padding-left: 5%; width: 92%;
```

```
                    height:
                                                100%;">
                    <div align="center">
                        <div id="title" align="left">
                            <table width="90%">
                                <tr style="text-align: center">
                                    <td width="100px" >图片</td>
                                    <td width="180px">宝贝详细</td>
                                    <td width="90px">单价（元）</td>
                                <td width="150px">数量</td>
                                <td width="100px">总计（元）</td>
                                <td colspan="2" width="150px">操作
                                    </td>
                            </tr>
                        </table>
                    </div>
                    <form action="shoppingcart" method="post" id="bus">
                        <table width="100%" border="0" >
                            <input type="hidden" name="action"
                                    value="editbus">
                                ……
                <div id="foot">
                    <jsp:include page="../foot.jsp"/>
                </div>
            </div>
        </body>
</html>
```

页面效果如图 14.11 所示。

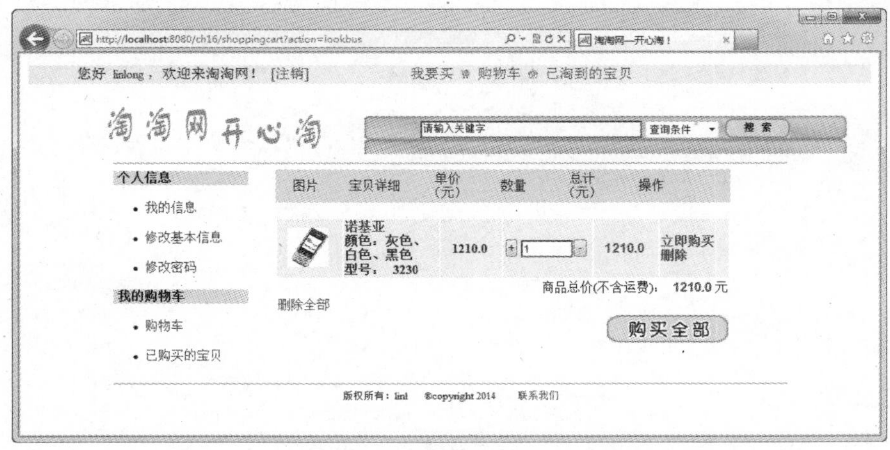

图 14.11　bus.jsp 页面效果图

14.9 商品模块

商品模块内容包括查看商品列表、查找单个商品功能。

14.9.1 查看商品列表

GoodServlet 类是商品的 Servlet 类，同样通过 action 的标识来决定是查看商品列表还是查询单个商品，源代码如下所示：

```
------------------------ GoodServlet.java---------------------------
@WebServlet(
    urlPatterns = { "/goods" },
    name = "goodsServlet"
)
public class GoodServlet extends HttpServlet {
    public void doPost(HttpServletRequest request, HttpServletResponse
    response)
            throws ServletException, IOException {
        //判断 action 类型
        String action=request.getParameter("action");
        String path=null;
        Vector<Goods> gVector=new Vector<Goods>();

        try{
            if(action.equals("index-select")){          //查询商品列表
                String keyWord=request.getParameter("keyWord");
                //获取查询的输入值
                String keyClass=request.getParameter("keyClass");
                //获取查询类别
                gVector=DAOFactory.getGoodDAOInstance().//获得所有商品
                    queryAll(keyWord, keyClass);
                request.setAttribute("goods", gVector);
                path="goods/goodslist.jsp";
            }else if(action.equals("goodslist-select")){
            //指定商品列表
                Goods good=new Goods();
                String sid=request.getParameter("sid");     //获得商品的id
                gVector=DAOFactory.getGoodDAOInstance().//获得指定的商品对象
                    queryGoodBySid(Integer.valueOf(sid));
                if(gVector.size()>0&&gVector!=null)
                    good =(Goods)gVector.get(0);
                request.setAttribute("good", good);
                path="goods/good.jsp";
```

```
30                 }
31             }catch(Exception e){
32                 e.printStackTrace();
33             }
34             request.getRequestDispatcher(path).forward(request, response);
35         }
36     }
```

从上述代码中，第 9 行获取 action 标识，根据 action 值跳转到相应的方法中。index-select 表示查询商品列表，通过 GoodService 中查询所有商品的方法，跳转到 goodslist.jsp 页面中。GoodService 类的源代码如下：

```
-----------------------GoodService.java-------------------------
//省略……
01  public class GoodService implements GoodDao {
02
03      private DBConnection dbconn = null;                 //定义数据库连接类
04      private GoodDao dao = null;                         //声明 DAO 对象
05      // 在构造方法中实例化数据库连接，同时实例化 dao 对象
06      public GoodService() throws Exception {
07          this.dbconn = new DBConnection();
08          this.dao = new GoodDaoImpl(this.dbconn.getConnection());
09                  // 实例化 GoodDao 的实现类
10      }
11      public PageObject getPageObject(String curPage, PageObject pageObject,
12              List<Object> listObject) {
13          pageObject = this.dao.getPageObject(curPage, pageObject, listObject);
14          return pageObject;
15      }
16      public Vector<Goods> queryGoodBySid(int sid) throws Exception {
17          ……
18      }
19  ……
20  }
```

上述代码中个，GoodService 类主要是实现 GoodDao 接口，并在构造方法中实例化 GoodDao 对象例如代码第 8 行，在方法实现中调用 GoodDao 的相应方法。GoodDaoImpl 是具体的商品数据操作类，其源代码如下：

```
---------------------- GoodDaoImpl.java--------------------------
public class GoodDaoImpl implements GoodDao {

        private Connection conn = null;                    //数据库连接对象
        private PreparedStatement pstmt = null;            //数据库操作对象
        ResultSet rs = null;
    // 通过构造方法取得数据库连接
```

```
    public GoodDaoImpl(Connection conn) {
        this.conn = conn;
    }
    //分页显示商品列表
    public PageObject getPageObject(String curPage,PageObject pageObject,
        List<Object> listObject){
        SetPageObject setPageObject = SetPageObject.getInstance();
        //获取分页对象 PageObject
        pageObject = setPageObject.setPageObjectData(curPage, pageObject,
        listObject);
        return pageObject;
    }
    //根据商品 id 查询指定商品
    public Vector<Goods> queryGoodBySid(int sid) throws Exception {
        ……
    }
}
```

上述代码中，主体的思想都是获得 PreparedStatement 对象，然后执行数据库操作方法，得到结果返回值，将返回值设置到容器中。接口 GoodDao 中的方法定义 Goods 中所用到的各个方法，其源代码如下：

```
---------------------- GoodDao.java--------------------------
public interface GoodDao {

    //添加商品
    public int addGood(Goods good) throws Exception;
    //删除指定商品
    public int deleteGood(int gid) throws Exception;
    //更新指定商品
    public int updateGood(Goods good) throws Exception;
    //根据商品的 id 查找商品
    public Vector<Goods> queryGoodBySid(int sid) throws Exception;
    //根据类型，输入关键字查询商品列表
    public Vector<Goods> queryAll(String keyWord, String keyClass) throws
    Exception;
    //分页显示商品列表
    public PageObject getPageObject(String curPage,PageObject
            pageObject,List<Object> listObject);

}
```

从上述代码中，就可以清晰地了解接口中的方法。

goodslist.jsp 页面是显示商品列表的页面，页面用 JSTL 中的 forEach 进行遍历结果集，其源代码如下所示：

```
------------------------ goodslist.jsp---------------------------
<%@ page language="java" import="java.util.*" pageEncoding="UTF-8"%>
<%@taglib prefix="c" uri="http://java.sun.com/jsp/jstl/core" %>
<!DOCTYPE HTML PUBLIC "-//W3C//DTD HTML 4.01 Transitional//EN">
<html>
    <head>
        <title>淘淘网—开心淘！</title>
        <link rel="stylesheet" type="text/css" href="css/styles.css">
    </head>

    <body>
        <div align="center">
            <div id="top">
                <jsp:include page="../head.jsp"/>
            </div>
            <p>
            <div id="logoselect">
                <jsp:include page="../logo_select1.jsp"/>
            </div>
            <div>
                <div style="background-color: #E1F0F0; width: 1000px;
                        height: 35px; font-size: 25px; color: red">
                    <table width="1000px">
                        <tr>
                            <td width="13%">图片</td>
                            <td width="21%">产品</td>
                            <td width="18%">单价</td>
                            <td width="19%">运费</td>
                            <td width="18%">型号</td>
                            <td >出产地</td>
                        </tr>
                    </table>
                </div>
                <div id="main">
                    <table width="1000px" border="0" id="list">
                       <c:choose>
                        <c:when test="${empty goods}">
                            <div align="left">
                                    <span>抱歉，没有找到符合您条件的商品，
                                               请看看别的</span>
                                    <br>
                                    <jsp:include page="../recommend.jsp"/>
                                </div>
                          </c:when>
                          <c:otherwise>
```

```
                                    <c:forEach items="${goods}" var="good">
                                            <tr height="100px">
                                                <td width="13%">
                                                    <a
                            href="goods?sid=${good.gid}&action=goodslist-select">
                                                        <img src="${good.gphoto}"
                                width="100px" height="100px" border="0">
                                                    </a>
                                                </td>
                                                <td width="21%">
                                                    <a
                            href="goods?sid=${good.gid}&action=goodslist-select">
                                                        ${good.gname}</a>
                                                    <br>
                                                    ${good.described}
                                                    <br>
                                                    出厂日期：${good.pdate}
                                                </td>
                                                <td width="18%">${good.price}￥
                                                </td>
                                                <td width="19%">${good.carriage}￥
                                                </td>
                                                <td width="18%">${good.
                                                types}</td>
                                                <td>${good.paddress}</td>
                                            </tr>
                                    </c:forEach>
                                </c:otherwise>
                            </c:choose>
                        </table>
                    </div>
                    <div id="foot">
                        <jsp:include page="../foot.jsp"/>
                    </div>
                </div>
            </div>
        </body>
</html>
```

上述代码中，页面只负责显示商品的列表，而业务中的操作交给后台处理，页面中不出现 JSP 代码。

14.9.2 查看单个商品

查询单个商品的 Servlet 和查询商品列表是一样的，将 goodslist.jsp 页面中循环遍历的代码删除就是 good.jsp 页面的代码。

14.10 支付模块

支付模块内容包括支付商品、支付全部商品、查看已支付商品。

14.10.1 支付商品

PayAction 类是支付单个商品方法，其源代码如下：

```
----------------------PayAction.java---------------------------
public class PayAction implements Action {

    public String execute(HttpServletRequest request,
        HttpServletResponse response) throws ServletException, IOException {
        //获得参数 flag 标识
        String flag = request.getParameter("flag");
        //获取登录用户的 ID
        String uids = String.valueOf(request.getSession().
        getAttribute("uid"));
        int uid = Integer.parseInt(uids);
        if(flag.equals("payall")){                    //支付全部
            String shopcartId = request.getParameter("shopcartId");
            //获得所有购物车的 ID
            try{
                if (DAOFactory.getShoppingCartDAOInstance().
                        updateShopcarts(shopcartId, 1) != 0) {
                        //更新购物车中的状态
                    request.setAttribute("status", "交易成功!您可以继续
                    选购宝贝。");
                }
            }catch(Exception e){
                e.printStackTrace();
            }
        }else{                                        //支付单个商品
            int gid = Integer.parseInt(String.valueOf(request
                    .getParameter("gid")));           //获得商品的 ID
            int number = Integer.parseInt(String.valueOf(request
                    .getParameter("number")));        //获得购买商品的数量
            try{
                ShoppingCart bus = DAOFactory.getShoppingCartDAOInstance().
                        getGoodsId(uid, gid, 0);      //获得指定的购物车
                if (DAOFactory.getShoppingCartDAOInstance().
```

```
32                          updatebus(bus.getId(), number, 1) != 0) {
                            //更新购物车中的状态
33
34                          request.setAttribute("status", "交易成功!
                            您可以继续选购宝贝");
35                      }
36                  }catch(Exception e){
37                      e.printStackTrace();
38                  }
39              }
40              return "index.jsp";
41      }
42
43  }
```

上述代码中，第 9 行用 flag 标识来判断是支付单个商品还是支付所有商品：如果是支付所有商品则获取所有商品的 id 值，调用 ShoppingCartService 支付所有的商品的方法，如果是支付单个商品则获取该商品的 id 值，同样是调用 ShoppingCartService 类支付单个商品的方法。

14.10.2 查看已支付商品

ShowPaidAction 类是查看已支付商品的方法，其源代码如下：

```
---------------------- ShowPaidAction.java---------------------------
public class ShowPaidAction implements Action {

    public String execute(HttpServletRequest request,
            HttpServletResponse response) throws ServletException,
            IOException {

        Vector tempVector = new Vector();
        //获取登录用户的 ID
        String uids = String.valueOf(request.getSession().getAttribute
        ("uid"));
        int uid = Integer.parseInt(uids);
        //获取分页对象 PageObject
        PageObject pageObject = PageObject.getInstance(request);
            //获取该用户已经支付的购物车列表
            Vector<ShoppingCart> busVector = DAOFactory.
                    getShoppingCartDAOInstance().
                getAppointedGoods(uid, 1);
            //遍历购物车列表
            for (int i = 0; i < busVector.size(); i++) {
                ShoppingCart cart = new ShoppingCart();
                cart = (ShoppingCart) busVector.get(i);
                Goods good=new Goods();
```

```
                TempGoods tempGoods = new TempGoods();
                Vector<Goods> gVector=DAOFactory.getGoodDAOInstance().
                    queryGoodBySid(cart.getGid());      //获取指定商品
                if(gVector.size()>0&&gVector!=null)
                    good =(Goods)gVector.get(0);
                //设置 TempGoods 对象值
                tempGoods.setGood(good);
                tempGoods.setNumber(cart.getNumber());
                tempVector.add(tempGoods);
            }
            String curPage = request.getParameter("curPage");//获取当前页
            pageObject = DAOFactory.getGoodDAOInstance().
            //向页面传送分页内容
                getPageObject(curPage, pageObject, tempVector);
        request.setAttribute("pageObject", pageObject);
        return  "shoppingcart/paidbus.jsp";
    }
}
```

上述代码中，遍历已经支付的商品，将商品对象存放到分页对象中，页面用分页的方式显示。

14.10.3　查看已支付商品页面

查看已支付商品列表页面 paidbus.jsp，其源代码和 bus.jsp 类似，只是在 paidbus.jsp 中增加分页标签。

14.10.4　支付中的页面

支付中的页面 pay.jsp，页面显示快递的地址，支付的金额等信息，部分源代码如下：

```
------------------------pay.jsp---------------------------
<html>
    <head>
        <title>淘淘网—开心淘！</title>
    </head>
    <body>
        <div id="top">
            <jsp:include page="../head.jsp"/>
        </div>
        <p>
        <div>
            <jsp:include page="../logo_select1.jsp"/>
        </div>
        <div align="center">
```

```
            <div style="width: 80%; height: 100%;">
                <table width="100%" align="center" border="0">
                    <tr>
                        <td width="30%">
                            <div align="center" style="border: 1px solid
                            #c1eae8;">
                                <div id="title" align="left">
                                    宝贝信息
                                </div>
                                <a id="img-link"

    href="goods?sid=${good.gid}&action=goodslist-select">
                                    <img src="${good.gphoto}"
width="115" height="115" border="0">
                                </a>
                                <div style="padding-left: 5%" align="left">
                                    <p>
                                        宝贝名称：
                                        <a

    href="goods?sid=${good.gid}&action=goodslist-select">

    ${good.gname}<br>${good.described}</a>
                                    <p>
                                        宝贝单价：
                                        <font
id="price" color="blue">${good.price}</font>元
                                    <p>
                                        宝贝运费：
                                        <font
                        id="carriage" color="blue">${good.
                        carriage}</font>元

                                    <p>
                                        出产地：
                                        <font
                        color="blue">${good.producer}</font>
                                    <p>
                                        出厂日期：
                <font color="blue">${good.pdate}</font>
                                </div>
                            </div>
                        </td>
                        <td width="70%">
                            <form action="shoppingcart" method="post"
```

```
                                id="pay" name="pay">
                                    <input type="hidden" name="gid"
                                    value="${good.gid}">
                                    <input type="hidden" name="action" value="pay">
                                   <div style="padding-left: 3%; padding-top: 3px; ">
                                        <div>
                                            <div id="title">
                                                确认收货地址
                                            </div>
                                            <br>
                                    ………

                <input type="submit" value="确认无误,购买"

    style="background-image: url(image/button1.jpg); width: 150px; height: 35px;
                        border-style: none; font-weight: bold;">
                                                        </td>
                                                    </tr>

        </body>
</html>
```

上述代码中，显示购买的地址，显示购买的总金额，然后单击“确认购买”按钮。其页面效果如图 14.12 所示。

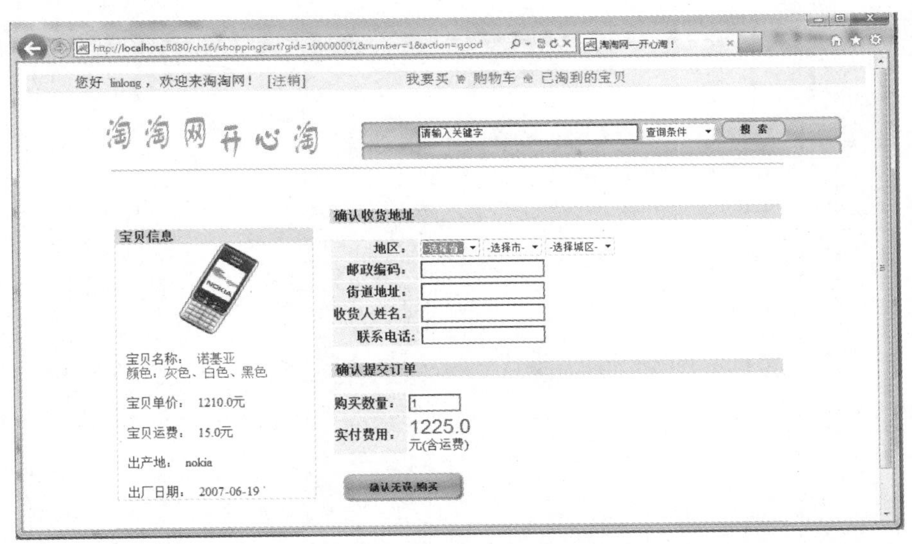

图 14.12　pay.jsp 页面效果图

到目前为止，系统中的代码基本介绍完毕，剩余的只是页面的样式和 JavaScript 代码，这些代码只是为了增加页面的美观性和易操作性，读者可以去掉这部分代码。

14.11 实战总结

本章以“网上购物系统”的开发和实现为主线，从系统需求、系统总体框架、数据库设计、系统详细设计这 4 个方面逐步深入分析，详细地讲解了该系统的实现过程。

读者在本章最需要注意的地方就是，如何将前面学习的内容串联起来，为将来的实战工作打下好的基础。